COASTAL
POLLUTION

Effects on Living Resources and Humans

Marine Science Series

The CRC Marine Science Series is dedicated to providing state-of-the-art coverage of important topics in marine biology, marine chemistry, marine geology, and physical oceanography. The series includes volumes that focus on the synthesis of recent advances in marine science.

CRC MARINE SCIENCE SERIES

SERIES EDITOR

Michael J. Kennish, Ph.D.

PUBLISHED TITLES

Artificial Reef Evaluation with Application to Natural Marine Habitats, William Seaman, Jr.

The Biology of Sea Turtles, Volume I, Peter L. Lutz and John A. Musick

Chemical Oceanography, Second Edition, Frank J. Millero

Coastal Ecosystem Processes, Daniel M. Alongi

Ecology of Estuaries: Anthropogenic Effects, Michael J. Kennish

Ecology of Marine Bivalves: An Ecosystem Approach, Richard F. Dame

Ecology of Marine Invertebrate Larvae, Larry McEdward

Ecology of Seashores, George A. Knox

Environmental Oceanography, Second Edition, Tom Beer

Estuarine Research, Monitoring, and Resource Protection, Michael J. Kennish

Estuary Restoration and Maintenance: The National Estuary Program, Michael J. Kennish

Eutrophication Processes in Coastal Systems: Origin and Succession of Plankton Blooms and Effects on Secondary Production in Gulf Coast Estuaries, Robert J. Livingston

Handbook of Marine Mineral Deposits, David D. Cronan

Handbook for Restoring Tidal Wetlands, Joy B. Zedler

Intertidal Deposits: River Mouths, Tidal Flats, and Coastal Lagoons, Doeke Eisma

Marine Chemical Ecology, James B. McClintock and Bill J. Baker

Ocean Pollution: Effects on Living Resources and Humans, Carl J. Sindermann

Physical Oceanographic Processes of the Great Barrier Reef, Eric Wolanski

The Physiology of Fishes, Second Edition, David H. Evans

Pollution Impacts on Marine Biotic Communities, Michael J. Kennish

Practical Handbook of Estuarine and Marine Pollution, Michael J. Kennish

Practical Handbook of Marine Science, Third Edition, Michael J. Kennish

Seagrasses: Monitoring, Ecology, Physiology, and Management, Stephen A. Bortone

Trophic Organization in Coastal Systems, Robert J. Livingston

COASTAL POLLUTION

Effects on Living Resources and Humans

Carl J. Sindermann

CRC Press
Taylor & Francis Group
Boca Raton London New York

CRC Press is an imprint of the
Taylor & Francis Group, an **informa** business
A TAYLOR & FRANCIS BOOK

CRC Press
Taylor & Francis Group
6000 Broken Sound Parkway NW, Suite 300
Boca Raton, FL 33487-2742

First issued in paperback 2019

ISBN-13: 978-0-8493-9677-9 (hbk)
ISBN-13: 978-0-367-39149-2 (pbk)
Library of Congress Card Number 2005051483

Library of Congress Cataloging-in-Publication Data

Sindermann, Carl J.
    Coastal pollution: effects on living resources and humans / Carl J. Sindermann.
        p. cm. -- (Marine science)
        Rev. and enl. ed. of: Ocean pollution. 1996.
    Includes bibliographical references and index.
    ISBN 0-8493-9677-8
    1. Marine animals--Effect of water pollution on. 2. Seafood--Contamination. I. Sindermann, Carl
J. Ocean pollution. II. Title. III. Marine science series.

QL121.S62 2005
577.7'27--dc22   2005051483

Visit the Taylor & Francis Web site at
http://www.taylorandfrancis.com

and the CRC Press Web site at
http://www.crcpress.com

Dedication

I would like to dedicate this new edition to my dear wife, Joan, who has been and continues to be my severest critic and my most enthusiastic supporter.

Prologue: Menace of the Sludge Monster

Environmental crises are daily events in the New York metropolitan area and its much-abused adjacent waters. During the late 1970s and early 1980s, when human concerns about degradation of the planet were still in their ascendancy, the news media gave unusual attention to problems created by an ocean dumpsite just 12 mi southeast of New York City, where stupendous quantities of sewer sludge, contaminated dredge spoil, toxic industrial wastes, and construction rubble were deposited every day. But it was the sewer sludge — some 5 million tons of it being dumped every year — that particularly fascinated the news people (see Figure P.1).

The dumping had created a zone on the ocean bottom that was deficient in most forms of marine life and was therefore labeled "the dead sea." Bottom samples contained all that is awful about our society's offal but little evidence of life forms, except for a few species of pollution-resistant worms and luxuriant populations of microbes. Furthermore, the sludge was found by scientists to have accumulated to appreciable depths near the dumpsite. Some imaginative reporter with headline possibilities in mind extrapolated the scientific observations to a "sludge monster" lurking just off the coast. To many people the monster was almost real, with a sinister energy derived from the ocean currents. It was out there — huge, black, and menacing — just beyond the surf zone, poised to overwhelm the already marginal beaches of Long Island and New Jersey, ready to make them totally unacceptable for any further human presence.

During the long hot summers of that traumatic period from 1976 to 1984, the state (New York and New Jersey) departments of health and environmental protection and the federal Environmental Protection Agency (EPA) were called upon repeatedly to examine what seemed to be early warning signs of the feared sludge invasion in the form of slimy blobs deposited on the beaches by the tides. These ugly masses (referred to as "tar balls" or "waste balls") were identified consistently by the regulatory and public health agencies as "innocuous material," "decaying mats of algae," or "aggregates of weathered oil," and *not* of human fecal origin — but savvy metropolitan beachgoers knew better. They were not about to be conned by the so-called experts, and many stayed away from those suspect shores. Each year during that time (1976 to 1984), the "sludge monster" frenzy peaked in summer and then dissipated with the onset of cool weather and the withdrawal of people from the beaches, only to reappear in the following spring. But, unaccountably, the major invasion never came. By 1985, there were fewer reports of sludge-like contaminants on the beaches, and talk of the sludge monster began to recede from the morning news.

This relative calm was shattered in the summers of 1987 and especially 1988 by a new coastal crisis: sightings of quantities of medical wastes (including bloody

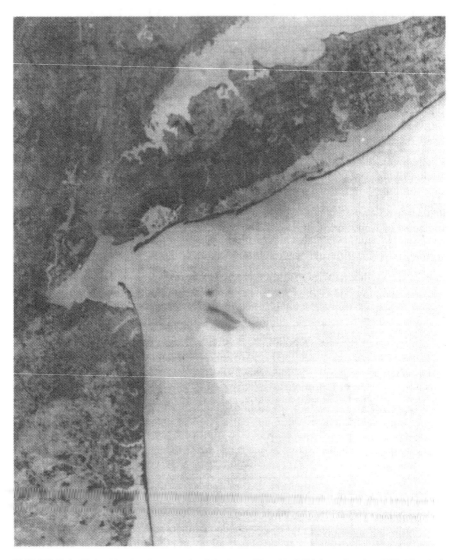

FIGURE P.1 High-altitude photograph of the inner New York Bight, taken in 1977. The dark streaks in the center are surface residues of ocean dumping, after barges have deposited their noisome cargoes.

hospital dressings and used syringes, some containing HIV-positive blood) cast up on a number of bathing beaches of New Jersey and New York, probably as consequences of illegal dumping in coastal waters or equipment failures in municipal sewage treatment facilities. News accounts, including graphic photographs of this revolting new form of shoreline pollution, drove masses of people from the beaches during those dismal summers. The obscene combination of sludge and medical wastes was just too much to tolerate, even for hardened urban sensitivities.

But the medical waste furor also dissipated quickly, leaving only a small residue of heightened vigilance among the few who persisted in visiting those mean shores of the New York Bight. The news media moved nimbly to other crises, helicopter surveillance flights and water sampling surveys by the regulatory agencies were reduced or eliminated, and the coastline slumped back to its usual blighted normalcy. Sludge dumping was, however, banished by EPA from the 12-mi dumpsite to a location 106 mi seaward of New York City, on the edge of the continental shelf, late in 1988, and was officially terminated even there in 1992. Undoubtedly, the sludge monster publicity, regardless of its validity, contributed significantly to attempts by environmental activist groups to stop ocean dumping.

Some day in the distant future, the 12-mi dumpsite will be a rich source of information for cultural anthropologists — a thin black layer of compressed sediments rich in fossilized artifacts that illustrate the nadir of human abuse of the edges of the sea in the 20th century, just offshore of the site where New York City used to stand. Those scientists of the future will never know the excitement and the dread generated by the sludge monster whose essence is captured in those sediments, but the physical evidence will be appalling enough for all time.

From *Field Notes of a Pollution Watcher*
(C.J. Sindermann, 1993)

Preface

Late in the year 1970, a major turning point occurred in my scientific career: I joined the staff of a federal fisheries research center at Sandy Hook, New Jersey. One of the principal programs of that center was to examine the effects of coastal pollution on productive systems of the oceans, especially effects on fish and shellfish resources. The Sandy Hook Laboratory, one of the operating units of the center, was ideally located for such a program, positioned as it was on a sandspit within sight of the smog-dimmed skyline of New York City, at the mouth of the grossly polluted Hudson River. Two important factors added to the logic of doing pollution research there: first, 12 mi seaward of the laboratory was the largest sewer sludge dumpsite on the east coast of the United States, and, second, industrial as well as sewage effluent pipes were (and still are) abundant along the immediate coastline.

One of the most fascinating aspects of this research assignment was that, in the presence of all this degradation from human population pressures and industrial pollution, fish and shellfish stocks existed and were objectives of vigorous sport and commercial fisheries. Several laboratory programs examined the reproduction, life cycles, and abundances of these stressed species, and, when integrated with the ongoing pollution studies, provided a superb opportunity to assess impacts of humans on living resources.

After more than a decade characterized by intense learning experiences about effects of coastal pollution in that unusual research venue, I left Sandy Hook for a briefer assignment in Miami, Florida — also a coastal area troubled by too many people living too close to the ocean. One of the results of those back-to-back research exposures to damaged marine environments and their effects on fish and shellfish was great internal pressure to write a book that would provide its readers with some insights into the history and consequences of human-related modifications of coastal/estuarine waters.

In response to that internal pressure, I published a book in 1996 titled *Our Own Pollution* — a somewhat technical document with a living resource perspective and a persistent emphasis on pathological effects of coastal pollution. The publication you have in hand is an expansion and extensive revision of that earlier book, written with an attempt at greater translucency, while still preserving some of the technical aspects and most of my favorite vignettes about life and death in disturbed marine habitats. After several unsatisfactory earlier drafts, I have settled on what might be described as a semihistorical episodic approach, with a fragile structure based (in Section I) on exploration of eight specific horror stories that have emerged partly as consequences of coastal pollution. Section II considers effects of coastal pollution on resource species and marine mammals, and Section III is concerned with effects of coastal pollution on humans.

Because few people ever read a technical book like this one from cover to cover (and rightly so, for it is, after all, not a novel), I offer seven options:

For the dilettante: Skip lightly through the italicized vignettes in each chapter, and ignore the rest of the text. This approach will give a soupçon — a tiny taste — of the flavor and content of the entire document.

For the casual reader: Read the introductory and concluding chapters, and maybe some of the vignettes; then put the book aside for future reference.

For the selective reader: Look at the table of contents, read only those chapters that seem to be of immediate and compelling interest, and ignore the rest.

For those with wide interests but short attention spans: I recommend a subset of thrillers from Section I, Chapter 1 through Chapter 8.

For resource-oriented people: Focus immediately on Section II, Chapter 9 through Chapter 11.

For those interested in the effects of coastal pollution on humans: Turn to Section III, Chapter 12 through Chapter 14.

For my favorites, the dedicated readers: Read the introduction and follow the chapter sequence in an orderly fashion through to the end. Good luck!

For all readers, I especially urge attention to the more robust and meaty chapters — Chapter 8, "Biological Pollution: Invasions by Alien Species"; Chapter 10, "Effects of Coastal Pollution on Yields from Fish and Shellfish Resources"; and Chapter 12, "Effects of Coastal Pollution on Public Health." From my perspective, these three chapters carry the book, at least in terms of scientific content.

I have resolved, in this revised edition, to include small dabs of history in the anecdotes and the narrative. I do this in part out of conviction that there is too much "now" in today's science and too little "then." I made this profound discovery because of my almost lifelong habit of reading technical journals. At some vague time just before the advent of the new millennium, I began to notice that over 90% of literature citations in the national journals that I read were for papers published *after* 1990 — as if science had appeared by an act of immaculate conception or spontaneous generation during that magic year. Now I recognize that science stumbles along (or maybe races along) at a variable pace in different subdisciplines, but something is wrong here. Science consists of more than today's victories or defeats — it has a long history of successful or failed efforts by countless very good, mediocre, or poor investigators. That history should have some greater recognition by current practitioners, at least in their own journals.

Science practiced without occasional genuflection to its history is too flat and featureless — intense but without depth — stimulating but lacking an important link with the past. We can do better.

I have walked the surface of this planet for enough years now to have discerned phases and trends in the improvement of understanding about coastal pollution. A few that could be mentioned are: the unfolding of knowledge, beginning in the 1950s, about the major role of *Vibrios* in coastal/estuarine waters; the realization, beginning in the 1960s and 1970s, that industrialization and industrial effluents were having significant chemical impact on those same waters (witnessed by such events as Earth Day in 1970 and the great Japanese fish riots of 1973); the more recent realization that nutrient chemicals of human origin (phosphorus and nitrogen in particular) were beginning to unbalance coastal ecosystems; and the findings that persistent toxic chemicals (such as DDT and PCBs) are now global in their distribution, with total effects still not fully understood.

Before plunging ahead, I would like to acknowledge the great benefits of long-term discussions about coastal pollution with Dr. John B. Pearce, formerly with the National Marine Fisheries Service, Woods Hole, Massachusetts, and now Director of the Buzzards Bay Marine Laboratory in Falmouth, Massachusetts.

I also thank the directors, past and present, of the NOAA National Ocean Service's Cooperative Oxford Laboratory (COL) in Oxford, Maryland, for encouraging completion of this long manuscript — recognizing that statements and conclusions in it are my personal responsibility. The manuscript was not reviewed by NOAA, so no official endorsement should be inferred.

I especially thank Mrs. B. Jane Keller, Editorial Assistant, COL, for professional help in the almost endless process of preparing a book manuscript for publication. Her assistance has been critical in bringing us to the present stage.

I also have special thanks for Dr. Aaron Rosenfield, Emeritus Director of the Laboratory, for many useful comments on earlier drafts, and for Mrs. Electa Pace of the University of Miami for advice, comments, and encouragement.

Finally, I would also like to acknowledge the hospitality of the Commonwealth of Massachusetts, for providing facilities for writing and contemplation at South Pond in the Savoy Mountain State Forest high in the northern Berkshires. Without drawing too many gratuitous parallels, South Pond is in many of its characteristics the present-day equivalent of the well-known but now despoiled Walden Pond (located in the eastern part of the Commonwealth) as it was more than a century and a half ago, during Henry David Thoreau's tenancy there.

Carl J. Sindermann
Oxford, Maryland

The Author

Dr. Carl J. Sindermann grew up in the western Massachusetts town of North Adams. During World War II, he served as a medic in an infantry reconnaissance platoon of the 26th (Yankee) Division, with combat experience in France, Luxembourg, Belgium, Germany, and Austria. He was awarded a bronze star medal in action during the Battle of the Bulge.

He received a Bachelor of Science degree with honors in zoology from the University of Massachusetts in 1949 and then an A.M. and Ph.D. in Biology from Harvard University in 1951 and 1953. During the latter part of his graduate program, he was a teaching assistant in parasitology in the Department of Tropical Public Health at Harvard Medical School. Later in his career, he also received an honorary Doctor of Science degree from Monmouth University in recognition of his contributions to marine environmental sciences.

His research specialties have been in the parasitology of marine animals and the effects of coastal pollution on living resources and on humans. He has published more than 150 scientific papers, as well as six technical books and several edited volumes in marine sciences. His principal contribution to the scientific literature was a thousand-page, two-volume book titled *Principal Diseases of Marine Fish and Shellfish*, published by Academic Press in 1990. One of his books (*Principal Diseases of Marine Fish and Shellfish*) received an outstanding scientific publication award from the Wildlife Society of America, and another (*Winning the Games Scientists Play*) was cited by *Library Journal* as one of the best sci-tech books of the publication year.

He has published technical books on such varied topics as coastal pollution, diseases of marine animals, marine aquaculture, drugs and food from the sea, anoxia in coastal environments, and sea herring of the western North Atlantic. Additionally, in another genre, he has published books about scientists at work, with titles such as *Winning the Games Scientists Play, The Joy of Science, Survival Strategies for New Scientists, The Woman Scientist,* and *The Scientist as Consultant.*

During the course of his scientific career, Dr. Sindermann was for several of his early professional years a member of the teaching faculty of Brandeis University in Waltham, Massachusetts, and, later, an adjunct professor at Cornell University, Georgetown University, University of Guelph (Canada), University of Rhode Island, and University of Miami.

Two decades of his scientific career were occupied principally with administration of ocean research laboratories of the federal government — first as director of the Oxford Biological Laboratory, Oxford, Maryland, then as director of the Tropical Atlantic Biological Laboratory, Miami, Florida, and then as center director of the Middle Atlantic Coastal Fisheries Center, Highlands, New Jersey. During his tenure as center director, he received the U.S. Department of Commerce Silver Medal for effective leadership of geographically dispersed research facilities.

Throughout his administrative career, he participated actively in the affairs of several international scientific organizations; he served terms as board member and then president of the World Aquaculture Society; he was for more than a decade chairman of the International Council for the Exploration of the Sea's Working Group on Introductions of Nonindigenous Species; he served as scientific advisor for the United Nations FAO Central West African Fisheries Commission; and he was a long-term member of the U.S.–Japan Joint Panels on Aquaculture.

Also during his research administrative career, Dr. Sindermann served for four years as scientific editor for the National Marine Fisheries Service and editor of the highly respected journal *Fishery Bulletin*. He also served on the editorial boards of other technical journals.

Since his retirement in 1991, he has continued his technical and nonfiction writing, publishing four additional books during that time.

Contents

SECTION III Effects of Coastal Pollution on Humans 211

List of Figures

List of Vignettes

* Journal entry date.

List of Tables

Introduction: Current Health Status of Coastal Waters

The story about the sludge monster told in the prologue to this book illustrates (among other things) the absence of any firmly held, long-term human concern for the well-being of coastal/estuarine waters. In that example, at one very sensitive location — the New York Bight, virtual front yard for millions of quarrelsome, presumably intelligent, and always vocal people — coastal waters were reduced in status, beginning late in the 19th century, to that of a convenient dump for sewer sludge, as well as toxic industrial chemicals, radioactive wastes, construction rubble, and even "cellar dirt" (the descriptor borrowed from an official U.S. Army Corps of Engineers classification). These were the noxious wastes of an uncaring growth-oriented industrial society.

That small snippet about the sludge monster, taken from the anecdotal history of coastal pollution in the United States, contains an obvious horror story — and there are many others ahead of us in this narrative (usually, however, without the strong sensory impact of sewer sludge). Some of them are tales of injury and death as consequences of contamination by humans of the margins of the oceans. Some effects of pollution can be best described and illustrated as damage to individual animals; others are thought to reduce abundance of local populations; and still others may affect entire ecosystems. It is easy to identify specific abnormalities in animals from the most degraded areas: the varieties of deformed fish embryos and larvae that can be seen in plankton collections, the juvenile fish with skin ulcerations and fin erosion, and the adult fish with cancerous or precancerous liver lesions. Effects at a population level are more difficult to detect, except by observing gradual disappearances of less resistant species from areas of heaviest contamination or cataloging the reduced reproductive success of vulnerable resident species. At the ecosystem level, evidence for disturbances may take the form of mass mortalities of fish or shellfish, extensive blooms of toxic algae, or the sudden intrusion and temporary dominance of noxious nonnative organisms.

These and other examples of the consequences of coastal pollution will be encountered at the individual, population, and ecosystem levels again in the chapters that lie immediately ahead. We'll begin by examining some of the impacts of marine pollution on human health and well-being — in keeping with an introductory emphasis on "undersea horrors."

But, regardless of its title, this is not intended to be a book of doom and gloom about the state of the coastal environment. Although it is true that humans have acted (and many still act) irresponsibly in matters concerning the oceans, there are positive things that need to be said — about the gradual emergence of public environmental concerns and pressures to reduce contamination, about the responses of governments to perceived problems in the sea, and about the accumulation of knowledge useful to the management of resource populations and their supporting ecosystems. These topics need to be discussed here, along with some recounting of the consequences of past abuses. We can start with some observations about the present health status of our coastal waters.

Our living space today is frequently invaded by harsh voices crying "environmental crisis." We hear warnings about global warming, ozone depletion, deforestation, atmospheric carbon dioxide increase, outbreaks of cholera, and many other calamities — all pressing for a larger share of public attention and funding for remedial actions. Some of these presumed crises occur in oceans and estuaries; of those, some involve mortalities of marine animals, others imply threats to human health from marine sources, and still others suggest potentially irreversible damage to ocean ecosystems. Events that have been described as environmental crises in the oceans include oil spills, fish kills, red tides, mortalities of marine mammals, contamination of fish and shellfish, refuse on beaches, and oxygen depletion or anoxia. In at least some of these happenings, it is difficult to separate the natural from the human-induced, the major from the minor, the immediate danger from the remote, or the likely consequences from the speculative ones.

Where, then, is "reality" in these ocean environmental events, sometimes labeled as crises, or even catastrophes? How and by whom is this reality determined, and what is the robustness of the database used in making the determination? Answers to these questions have to be given in tentative terms, because available information is inadequate for definite statements. But then, inconclusive answers to technical questions are very common, and they can be frustrating to nonspecialists. In rapidly developing research areas such as the effects of marine pollution, however, uncertainties abound, and this fact shouldn't be obscured by too many speculative opinions and conclusions perched precariously atop fragile databases.

One of the best statements about the health status of the marine waters surrounding the United States was made some time ago by the Office of Technology Assessment (OTA), a prestigious (but now defunct) governmental oversight group that reported directly to Congress.* Its report, published more than a decade ago but still relevant today, reached three general conclusions:

- "Estuaries and coastal waters around the country receive the vast majority of pollutants introduced into marine environments. As a result, many of

---

* "Prestigious" notwithstanding, a Republican-dominated Congress abolished OTA on September 30, 1995, presumably as a cost-cutting measure. At present, a groundswell is developing in Congress to reestablish that important advisory body (Greenberg 2003).

these waters have exhibited a variety of adverse impacts, and their overall health is declining or threatened" (OTA 1987, p. 31).

- "In the absence of additional remedial measures, new or continued degradation will occur in many estuaries and some coastal waters around the country during the next few decades (even in some areas that exhibited improvements in the past)" (p. 31).
- "In contrast, the health of the open ocean generally appears to be better than that of estuaries and coastal waters. Relatively few impacts from waste disposal in the open ocean have been documented, in part because relatively little waste disposal has taken place there and because wastes disposed of there usually are extensively dispersed and diluted" (p. 31).

Statements such as these, sobering though they may be, should be made available to the general public, as a countermeasure against more extreme news media presentations. The conclusions clearly indicated present and future problems, especially in estuaries and coastal waters of the United States that are most vulnerable to degradation by human activities.

One year after that report was published, two of the country's leading news magazines — *Time* and *Newsweek* — both ran cover stories about the deplorable condition of the world's oceans. *Time*'s feature story was titled "The Dirty Seas"; *Newsweek*'s was "Our Ailing Oceans." I think that both articles were precipitated by findings of medical wastes washed up on the shores of the New York Bight earlier that same summer, although a background of previous news reports on sludge dumping, red tides, fish kills, and anoxic waters probably contributed to a sharp increase in public indignation about the degradation by humans of estuarine/coastal waters.

The news media play, and have played, a curiously ambivalent role in ocean pollution matters. On one hand, they do identify the abuses that occur, and they highlight the possible consequences of those abuses. On the other hand, probably because of the felt need to capture and retain even fleeting public attention, they often cast their disclosures in overly dramatic language that may minimize or ignore the very thin factual base for a news story. This is the essence of what can be called "media science." Some of its negative characteristics are:

- Intrusion of subjective views and the assumption of an obvious environmental advocacy position.
- Selection of data that reinforce the storyline being presented.
- Extrapolation of supposed or presumed causes to observed effects.
- Interweaving of fact and opinion (or conjecture) cleverly enough so that the casual reader takes away the impression that the entire story is based on fact.
- Grand disregard for the careful qualifying words and caveats offered to reporters and feature writers by scientists, in their attempts to describe a phenomenon or the findings of their investigations.
- Speed with which a current environmental crisis can be relegated to back pages or can disappear altogether as other more important news is perceived or as public interest wanes.

- Use of an array of deliberately inflammatory words and phrases, such as "spew," "poison," "poisoned sea," "dying sea," "dead sea," "calamity," "laced with pollutants," "awash with poisons," "catastrophe," "resource disaster," and "conservation emergency."
- Slanting a story through careful choice of words, as demonstrated in an excerpt from a news report stating that "chemical pollutants in the North Sea are believed to have been involved in the deaths of 1500 harbor seals." That statement may be technically correct, but viral and other microbial diseases were actually found to be the underlying immediate cause of the mortalities, even though some scientific observers suggested that pollutants might be factors in reducing resistance to disease. Readers of the news story undoubtedly reached the conclusion that pollution killed the seals — which was the reporter's slant but *not* the reality of the event.

So, the blessings and the defects of "media science" must both be considered. The evils of ocean pollution *do* need public scrutiny, but in assessing the extent of damage it is essential to be able to distinguish factually based scientific observations and conclusions from exaggerations and extrapolations that capture public attention. What is required is a calm professional attempt to answer the complex question, "Is there a crisis in our coastal waters as a consequence of pollution, and, if so, what are its proportions and what remedial actions are needed?" Unfortunately, we are confronted with so many claims about catastrophes and crises that we suffer from an "environmental crisis overload." Our senses can be numbed eventually by the persistence and intensity of demands from proponents and opponents of environmental issues.

Scientists can also be guilty of adding to the clamor and confusion. With a constant and seemingly insatiable need for research funding, they often feel constrained to participate in "crisis mongering" in preparing grant proposals and in announcing preliminary findings, as this seems to be the only way to gain any kind of attention from politicians, federal bureaucrats, environmental activists, or the general public. Ocean pollution scientists fare best in the funding quest when they emphasize aspects of their own special crisis that may have direct consequences to human health and well-being; impacts on sea life or ecosystem disruptions may be important, but they seem to excite lesser concern from those who control research funding.

To lend a measure of substance to these generalizations about the interplay of fact and hyperbole in ocean pollution affairs, it might be instructive to take a few retrospective examples from happenings in coastal waters of the western Atlantic in recent decades that were in their time described in the news media as "environmental crises" — large-scale events in which some human involvement was either apparent or inferred. I have selected three episodes:*

- The great IXTOC-1/Gulf of Mexico oil spill of 1979–1980
- "Brown tide" in Long Island (New York) waters in 1985–1987

---

* The three brief episodes that follow — so-called environmental crises — will form integral parts of some of the succeeding chapters of the book, and will be repeated in proper context in those chapters.

- Dolphin mortalities on the Atlantic coast of the United States in 1987–1988

Each of these perceived "crises" can be presented as a brief vignette.

Crisis 1: The Great IXTOC-1 Oil Spill, 1979

With the expansion of offshore petroleum production in the Gulf of Mexico, residents of the southeast coastal zone have become accustomed to the minor spills and leaks that make oil contamination an everyday reality there. Even these pollution-adapted citizens had to be impressed, though, by the magnitude of an oil spill that began on June 3, 1979. IXTOC-1, a Mexican exploratory well located 58 miles off the Gulf of Campeche coast of the Yucatan Peninsula, had blown on that date, and was initially gushing an estimated million gallons of oil per day into the Gulf. Prevailing winds and currents had caused a gigantic oil slick to reach the exquisite white sand beaches of Padre Island off the southern Texas coast by late August of that year.

Totally by chance, arrival of the oil coincided with an annual meeting of the World Aquaculture Society, a technical association with a membership of over 3000 aquatic scientists and entrepreneurs, being held in Corpus Christi, not far from those by then despoiled beaches of Padre Island. The meeting organizing committee quickly laid on field trips to the impacted zone, so that participants could witness at close hand the reality of what is always a concern of aquaculturists — spilled oil polluting productive waters.

The scene on the beaches was awesome. A black rim of oil stretched along the shore in both directions to the farthest horizons; oil fumes and an oil spray overpowered the senses; great floating globs of brown chocolate mousse-like material (called "mousse") sloshed in the surf and coated the sand; front-end loaders charged up and down the beach in a futile effort to scoop up the oil-soaked sand; and NOAA crisis response teams in helicopters and small boats assessed the condition and movement of the oil — all combining to create an unreal science fiction-like stage set. At risk, in many peoples' minds as they looked at the devastation, was the major Gulf shrimp fishery. Predictions of disaster for the shrimp industry from mass mortalities and tainted catches were universal, and the future well-being of these estuarine-dependent resource species in a contaminated environment was thought to be in jeopardy.

The oil flow was reduced by late summer, but the well was not finally capped until March 1980, after releasing an estimated 130 million gallons of oil into the Gulf (some estimates were much higher). The aquaculturists flew home from their meeting in Corpus Christi with many "oh my" photographs of the damaged beaches; a few reports of oil-tainted shrimp catches were highly publicized by the Texas news media; the oil slick dissipated rapidly, and it soon became difficult to locate any bottom deposits of oil, except in the immediate vicinity of the wellhead. The event faded into history almost without a trace. The antici-

pated impacts of the spill on subsequent year classes of shrimp and other resource species of the Gulf were examined by fisheries scientists, with equivocal results, because they were unable to separate effects of the oil from those of many other environmental factors that could influence population abundance (it should be noted, though, that annual Gulf shrimp landings since 1979 have fluctuated around 100,000 tons, with no significant declines).

So, from the perspective of the human intruder, standing on an oil-covered shore, the event seemed catastrophic; but, from that of the resource species and the entire Gulf ecosystem, it probably represented a relatively mild perturbation, quite likely of lesser overall significance than the passage of an average-strength hurricane.

Crisis 2: "Brown Tide" in Long Island Waters

The EPA surveillance helicopter made its final crisis response flight for the month over the abnormally golden brown waters of Great South Bay on the outer coast of Long Island. It was late summer 1986 — the second year that bays on the Island had been discolored and choked by the massive growth of a planktonic microscopic algal species, Aureococcus anophagefferens, an organism not previously known to cause blooms in that area of the coast. The findings from the day's survey were grim: current abundance of the toxic organism in the bay was 1,000,000,000 algal cells per liter, similar to what it had been all summer. This abundance of the microalga reduced light penetration of the water, which in turn caused significant reduction in eel grass abundance and distribution in Long Island bays, and profound disturbances in other components of the shallow-water communities.

One of the animal species most affected by the algal bloom was the bay scallop Argopecten irradians, the base for an important commercial shellfishery. In the summer of 1985, when the bloom began, most of the scallop larvae had died, resulting in a massive recruitment failure. New York scallop landings in that year were only 58% of the average for the preceding four years. Natural replenishment was precluded by recurrence of the bloom in the summer of 1986, and its reappearance in some previously affected areas in 1987.

The problem was not confined to Long Island waters. The same alga, Aureococcus anophagefferens, bloomed from Narragansett Bay, Rhode Island, southward to Barnegat Bay, New Jersey, in 1985. It caused mortalities of mussels Mytilus edulis in excess of 95% in Narragansett Bay, and significant growth suppression in hard clams Mercenaria mercenaria in Long Island bays. Mats of dead eel grass littered the shores in New York and New Jersey.

The problem did not disappear with the passage of time, either. A 1991 "NOAA News Bulletin" reported that an Aureococcus bloom had reoccurred in eastern Long Island Sound in June of that year, with cell densities eight times that known to harm marine animals. Bay scallop larvae were again assumed to have been killed during the early stages of the bloom.

A relationship of recurrent algal blooms such as these to chemical modifications by humans of estuarine/coastal waters is often suggested, and some evidence exists. Nutrient enrichment from agricultural runoff, sewage outfalls, and some industrial effluents may be involved, as may be the transport of toxic algae to new locations in ships' ballast water or with introduced marine animals. Whatever the causation, there seems to be a real increase in the frequency, intensity, and geographic area affected by algal blooms on a worldwide basis, and shellfish populations are among the groups most severely impacted.

Crisis 3: Died at Sea, of Unknown Causes

The early morning mist was disappearing from the beach at Asbury Park, New Jersey, on a late summer day in 1988, and as it dissipated it revealed the bloated bodies of three sea mammals that had floated in with the tide during the night. They were young adult bottlenose dolphins — a species noted for their graceful antics at sea and for crowd-pleasing trained behavior in seaside commercial aquaria. But these specimens were, unaccountably, dead. They were also statistics, because during the summers of 1987 and 1988 more than 700 other members of their species had washed up on Atlantic shores — an unprecedented event.

What killed these intelligent marine mammals? Had they encountered fatal levels of some industrial chemical, or some natural algal toxin? Had a new viral disease reached epizootic levels in the population? Or was some parasitic disease affecting equilibrium or respiration to the point where drowning could occur?

A task force of experts in marine diseases was assembled to examine the problem soon after the deaths began in 1987 — headed by a professor imported temporarily from a Canadian veterinary college. The group established field headquarters in 1987 and 1988 in an oceanfront motel and pursued various hypotheses about causes as specimens were dissected and examined. News media and curious onlookers asked every day for answers, and tourists gathered in subdued clusters around dead and decomposing animals on the beaches. After the various analyses — pathological, microbiological, and chemical — were completed, a tentative (and controversial) diagnosis was made and a final report prepared. According to the report, deaths during the outbreak period seemed associated with high body tissue levels of an algal toxin ("brevetoxin," known as a "red tide" toxin) ingested with food. Affected animals seemed less resistant to the microbial infections that were the immediate cause of death in most cases. Pathological effects may have been exacerbated also by exposure to industrial pollutants such as PCBs, which could create physiological stress and reduce disease resistance.

At any rate, the somewhat inconclusive results of the investigations did not seem to please many observers, and as an interesting but equally baffling sequel to these mortalities on the Atlantic coast, dolphins also died in 1990 in large numbers, estimated to be in excess of 200 — but this time in the Gulf of Mexico.

Most marine mammals — especially the dolphins — have a strange hold on the human conscience. We seem to actually care about their well-being. So it should behoove us, in rare moments of introspection, to examine our possible role in despoiling their habitats — to the degree that they can die in great numbers, as they did in 1987, 1988, and 1990.

Humans, standing uncomprehendingly on the shore in bathing suits while a major ecological event such as an oil spill (or an algal bloom or a mass mortality of fish or marine mammals) is occurring, must have at least some of these thoughts:

"This truly is a marine catastrophe that will be disastrous for sea life."
"How long will this crisis persist, and what will the coast be like after it is over?"
"How will this event affect human health, and what will be its economic impact?"
"Why us here on Padre Island (or Long Island), oh Lord? Go pick on those bums on the New Jersey coast."

Meanwhile, a cluster of oil-covered crabs, trying to adapt to this drastic chemical environmental change, might have quite different perspectives on the happenings. Conversations among them, if any, might include some of these complaints:

The group nerd: "These excess hydrocarbons are difficult to metabolize, and I fear that this environmental level of contamination may be beyond our limits of adaptation."
The old-timer: "Why can't we have a little *stability* around here — I'm sick of adapting to environmental changes."
The know-it-all: "Don't say I didn't warn you. This estuarine existence is full of nasty surprises."
Chorus: "Petroleum, DDT, PCBs, mercury — is there no end to what these upstart humans will dump on us?"

Once the events are over and rational thinking resumes, it should be possible to make at least a few tentative observations about these carefully selected examples of attention-getting disturbances — so-called "crises" or "catastrophes" — in the marine environment. Some immediate thoughts are these:

• "Catastrophe," "disaster," and "crisis" are distinctly human perceptions — not necessarily shared by the rest of the living world, in which environmental disruptions and perturbations are normal aspects of existence.
• Most of the events in coastal waters that are perceived by humans as catastrophes are short-lived, with acute effects that disappear relatively quickly.

- The role of humans in causing certain major coastal disturbances is still subject to conjecture, and some events, such as algal blooms, undoubtedly have a large natural component.
- Awareness of the extent of potential human impacts on the oceans can be increased by sensitive and well-researched news stories about such major disturbances, but, alas, media attention tends to be transient, superficial, and often inflammatory.
- Humans are capable of making noticeable impacts in nearshore waters, but abuses of the marine environment often disclose a remarkable degree of resiliency on the part of the affected species and ecosystems.
- Documentation of quantitative effects of pollution on abundance of resource species is still not very substantial, despite frequent observations of localized mortalities.

Thus far in this introductory chapter, focus has been on the recent history and present status of pollution in coastal/estuarine waters of the United States, but much broader aspects of the health status of the oceans need exploration as well. Examining even a small sample of the recent world literature on marine pollution leads quickly to a very important generalization: *the greatest effects and the greatest risks from pollution are in enclosed or semienclosed seas** (Aral Sea, Black Sea, Mediterranean Sea, North Sea, Baltic Sea, Seto Inland Sea of Japan, and others) *and in estuaries having restricted mouths or extended salinity transition zones with minimal water exchange* (Oslofjord, Puget Sound, Hudson River estuary, and others). Risks are lower in estuaries with short tidal reaches and good water exchange, or in open coastal areas with even modest tidal amplitudes and residual current flow.

The reality of these observations has been emphasized in several publications on the Black Sea (Mee 1992), the Baltic Sea (Hansson & Rudstam 1990, Wulff & Niemi 1992), the Mediterranean Sea (Pearce 1995), and a summary article on seven enclosed and coastal seas (Platt 1995). Significant findings in these and other reports include the following:

- Enclosed or semienclosed seas are more susceptible to pollution damage than open ocean waters because circulation is limited and dilution is reduced (Platt 1995).
- The Black Sea is an extreme example and a microcosm of the worst forms of abuses and exploitation. Mee (1992) listed these as dumping toxic wastes, mineral exploitation, unrestricted shipping activity, and unmanaged fisheries.

---

* Describing the ocean as "the body of salt water covering a large part of the planet" is reasonably satisfying, but a precise definition of "sea" can be elusive. My favorite is "one of the larger bodies of salt water, less than an ocean, on the earth." But other descriptions of a sea take us into murkier depths; try "a body of salt water of second rank, more or less landlocked and generally forming part of, or connecting with, an ocean or a larger sea (as the Mediterranean Sea)." Or try this: "a sea is an inland body of water, especially if large or if salt or brackish (as the Caspian Sea or the Aral Sea); sometimes a small freshwater lake (as the Sea of Galilee)."

- The Black Sea is 90% anoxic — the largest known anoxic water mass on earth, permanently anoxic below 150 to 200 m (Mee 1992, Platt 1995). Recent expansion of the anoxic zone is considered to be largely a consequence of extreme eutrophication resulting from riverine contributions of nutrient chemicals (phosphorous and nitrogen in particular; Green 1993).
- A recently published description of desperate environmental conditions in the Mediterranean Sea (Pearce 1995) points to eutrophication in two main forms — red tides of dinoflagellates and enormous masses of mucus-like foam from secretions (in part) of diatoms — as most serious current problems. This is only the latest in a series of overt consequences of pollution in that body of water; the deaths of several thousand striped dolphins in 1990 to 1992 apparently due to synergistic action of a viral pathogen and physiological effects of industrial contaminants, especially PCBs, was a precursor. Widespread reports of illnesses among bathers and outbreaks of enteric diseases transmitted by contaminated shellfish preceded those events. Sewage, untreated and in overwhelming amounts, is a principal culprit, augmented by industrial pollution and limited water exchange.
- National and international environmental management actions are particularly critical for enclosed or semienclosed seas. Wherever mitigation measures have been instituted (Seto Inland Sea, North Sea), the processes of degradation have been slowed; where such actions are absent (Black Sea, Aral Sea, Caspian Sea), deterioration has accelerated (Platt 1995).
- Studies in the Baltic Sea and the Black Sea indicate that most pollution does not come from estuarine/coastal point sources but from rivers and atmospheric fallout (Mee 1992, Wulff & Niemi 1992).
- International restoration programs proposed for the Mediterranean and Baltic seas have resulted in augmented research but little if any progress in improving environmental conditions. The UN-organized Mediterranean Action Plan, now more than 20 yr old, has produced little action, despite promises made by countries bordering that sea to "take all appropriate measures to prevent, abate, and combat pollution in the Mediterranean Sea area and to protect and improve the marine environment in the area." Baltic Sea restoration has been addressed by an international commission with membership from all states in the Baltic area. Despite decades of research, an action plan to reduce pollution loads has been presented by only one member, and suggested remedial measures are directed principally toward "hot spots" of large industrial or municipal point sources (Cederwall & Elmgren 1990).
- Two semienclosed seas — the Seto Inland Sea and the North Sea — can be described in positive terms, in which actions by humans to correct human abuses have been implemented. Japan began in the 1970s to reduce industrial pollution of the very productive Inland Sea, after increasing frequencies of red tides and reports of chemically contaminated seafood made remedial action a national priority. The North Sea benefited in the 1970s from terms of the Paris Convention, in which Northeast Atlantic nations agreed to reduce dumping of sewage sludge and industrial wastes at sea. A 1993 report to the international commissions with oversight of North Sea envi-

ronmental conditions concluded that "measures already taken to protect the North Sea are effective [but] there is a need for continued and in some cases further action" (Miller 1994). The message from these two examples is that it is possible to reverse the course of environmental degradation, even in semienclosed seas, if a genuine commitment to do so exists.

So much then for this very important distinction between the nature and extent of pollution problems in enclosed or semienclosed seas, as contrasted with those in estuarine/coastal waters where circulation and dilution are adequate. The categories are of course not really discrete; what we have created is an artificial subdivision of a continuum extending from one extreme of an enclosed inland sea with almost intractable problems (the Aral Sea, for example; Brown 1991), to a semienclosed coastal sea (such as the Baltic Sea), to coastal seas partially enclosed by island chains (the Bering Sea and the south China Sea, for example), to estuaries with restricted circulation (such as the Oslofjord), to estuaries with short tidal/salinity reaches and good water exchanges (such as Passamaquoddy Bay on the northeast coast of North America), to open high-energy coasts (such as much of the Pacific coast of the United States), where pollution problems exist but are of smaller magnitude and are more amenable to solution.

Swirling along in center ranks, behind the major problem areas, are lesser bodies of water — the many small bays and estuaries — that have been overwhelmed by massive human interventions of several types:

- Discharge from a major polluted river (the Elbe estuary at Hamburg would be an example)
- Dense human populations crowded on adjacent land masses (the New York metropolitan area is a good example)
- The presence of a large polluting industrial complex (Pamlico Sound, North Carolina, and Minamata Bay, Japan, would be examples)

In all these bodies, the amount of water exchange per unit of time is a most critical element.

We can terminate this too brief introductory discussion of the health status of the world's oceans (and enclosed seas) at the beginning of the 21st century by proposing what could be called "unifying concepts" — principles that may be demonstrated and verified in the future. Every self-respecting reviewer of science literature would have a counterpart list of such concepts in his area of specialization, tucked in an inside pocket, awaiting a request for disclosure. These are mine:

- A combination of human population density and the degree of industrial development of adjacent shoreline and drainage area determines the extent of estuarine/coastal degradation and impacts on resources. Wherever peaks in human populations occur in coastal areas, those are also locations of severe environmental pollution of municipal and industrial origin.
- Pollution problems are most severe in semienclosed coastal seas or in estuaries with restricted circulation. The degree of risk is modified by

water exchange (residence time, flushing rates, tidal amplitude, bottom topography), as well as by riverine contributions of nutrients and toxic chemicals, human population density, and extent of industrialization.

Effect of pollution on world fisheries is a critical consideration related to food supplies, but overfishing of stocks and damage to habitats caused by industrialized fishing are clearly the principal resource problems. In 2001, a definitive study (Watson & Pauly 2001) indicated that world fish catches had peaked in 1988 and had been declining irregularly at a rate of about 0.66 million tonnes/yr since that time, with most traditional wild stocks either fully utilized or overfished, and despite a gradual increase in production from aquaculture.*

It appears now that China had been for years reporting to the United Nations Food and Agriculture Organization (FAO) coastal fish catch statistics that exceeded actual yields by up to 50%. When the new estimates were substituted for the inflated ones from China, it became apparent that the global catch has actually been dropping irregularly since 1988 (Watson & Pauly 2001, p. 536). With this degree of statistical uncertainty (lying) in reporting catches, it is not surprising that coastal pollution has not yet been demonstrated to be a major factor affecting fish production. One observation that emerges from examination of coastal pollution and fisheries is that, reasonable evidence exists for localized effects of pollution on fisheries, there is as yet little specific evidence of widespread damage to major fisheries resource populations resulting exclusively from coastal pollution. There is also evidence that other factors, such as repeated year-class successes or failures, long-term shifts in geographic distribution, and (in particular) overfishing, may cause major changes in fish abundance. (Note that this statement refers to coastal areas and coastal fish populations. The situation is different in enclosed seas, where some species with restricted distributions have disappeared from catches or have declined in abundance so drastically that no viable fishery persists — an example would be sturgeon in the Black Sea.)

Humans refuse — persistently — to accept the reality that they are part of the planetary ecosystem, and, now that they have achieved overwhelming numbers, that their actions can seriously affect all parts of that ecosystem. They have created local and global problems with atmospheric gases, they have destroyed an alarming portion of the earth's forests; they have used major rivers as sewers; and in the last century they began to inflict harm on coastal/estuarine resource populations and their habitats. Furthermore, some of the toxic chemical products of their technology, such as DDT and PCBs, have been disseminated globally and can be found in the inhabitants of all oceans. Rachel Carson, in one of her speeches after publication of *Silent Spring* (Carson 1963), reminded us that "all the life of the planet is interrelated; each species has its own ties to others; and all are related to the earth." Most members

---

* The 2-decade gradual increase in global aquaculture has undoubtedly occurred, but its proportions have been clouded by exaggerated reports from the People's Republic of China, which accounted for an inflated 63% of total world aquaculture production in 1998 (Pauly et al. 2002). These authors state that "it is now known that China not only overreports its marine fisheries catches, but also the production of many other sectors of its economy. Thus, there is no reason to believe that global aquaculture production in the past decades has risen as much as officially reported" (p. 693).

of the human species have, unfortunately, rejected and still reject this thesis, even though evidence is accumulating for its validity.

This, then, is my brief assessment of the health status of the planet's coastal waters at the beginning of the new millennium. Having completed the task, I now feel free to choose topics for Section 1 of this book — some historical and some current — that qualify as "undersea horrors." Each of the next eight chapters will focus on specific problems: cholera, mercury, PCBs, polluted bathing beaches, algal toxins, anoxia, oil, and alien species (please note that other authors might emphasize different horrors, such as radioactive pollution or impacts of global warming — but I am comfortable with my choices). After a discussion of those eight specific problems, I propose to turn to general problems in Chapter 9, Chapter 10, and Chapter 11: sublethal effects of coastal pollution on marine animals, effects of coastal pollution on yields from fisheries and aquaculture, and effects of coastal pollution on mass mortalities of marine mammals. Then, in a series of chapters (Chapter 12, Chapter 13, and Chapter 14), I will try to review the totality of impacts of coastal pollution on the human species. After that, I'll attempt some obligatory general conclusions. This structure makes sense from my perspective as a long-term pollution watcher. Join me on the trip!

REFERENCES

Brown, L.R. 1991. The Aral Sea: Going, going .... *Worldwatch,* January/February, pp. 21–27.

Carson, R. 1963. Keynote address. Women's National Book Association Annual Meeting, Washington, DC.

Cederwall, H. and R. Elmgren. 1990. Biological effects of eutrophication in the Baltic Sea, particularly the coastal zone. *Ambio 19:* 109–112.

Green, E. 1993. Poisoned legacy: Environmental quality in the newly independent states. *Environ. Sci. Technol 27:* 590–595.

Greenberg, D.S. 2003. Bring back OTA — Congress' own think tank. *The Scientist* (May 19), p. 64.

Hansson, S. and L.G. Rudstam. 1990. Eutrophication and Baltic fish communities. *Ambio 19:* 123–125.

Mee, L.D. 1992. The Black Sea in crisis. A need for concerted international action. *Ambio 21:* 278–286.

Miller, A. 1994. Health of the North Sea. *Environ. Sci. Technol. 18:* 257A.

OTA (Office of Technology Assessment). 1987. Wastes in marine environments. U.S. Congress OTA-0-334. U.S. Government Printing Office, Washington, DC.

Pauly, D., V. Christensen, S. Guénette, T.J. Pitcher, U. Rasheed-Sumail, C.J. Walters, R. Watson, and D. Zeller. 2002. Towards sustainability in world fisheries. *Nature 418:* 689–695.

Pearce, F. 1995. Dead in the water. *New Scientist* (February 4), pp. 26–31.

Platt, A.E. 1995. Dying seas. *Worldwatch* (January/February), pp. 10–19.

Watson, R. and D. Pauly. 2001. Systematic distortions in world fisheries catch trends. *Nature 414:* 534–536.

Wulff, F. and Å. Niemi. 1992. Priorities for the restoration of the Baltic Sea — A scientific perspective. *Ambio 21:* 193–195.

Section I

Eight Specific Examples of Pollution-Related Undersea Horrors

It seems logical to begin the book with specific examples — some current and some historical — of pollution-associated events that can be described as "undersea horrors." I have picked eight such events for inclusion in Section 1 (from a much longer list). My choices are:

- The global reach and human impacts of cholera
- The persistent and increasing problem of mercury in seafood
- The worldwide distribution of PCBs and other toxic chlorinated hydrocarbons
- Microbial pollution of recreational waters
- The expanding occurrence of harmful algal blooms, with associated toxicity
- The growing influence of anoxia in coastal waters
- The localized impacts of oil spills in coastal waters
- Invasions of coastal ecosystems by alien species

An objective for this grouping of specific problem areas is to provide substance for claims of harm to public health, damage to marine populations, and negative impacts on ecosystems that can be consequences of coastal pollution.

1 Cholera

Cholera in the Western Hemisphere

The disease had not visited the western hemisphere for almost a hundred years — cholera in epidemic form — but here it was in the closing decade of the 20th century. It began during fiesta time, January 1991, in a tiny coastal town, north of Lima, Peru. Food, drink, and pleasure abounded, including one special treat, ceviche, always a favorite during any fiesta or special occasion. Prepared by marinating raw fish or shellfish for a few hours, this traditional dish had gained its popularity long ago, in a less complex time when people were not so numerous and coastal waters not so polluted. But, unfortunately, the risks of human disease from contaminated raw seafood have increased in proportion to population size, and the ceviche served in that Peruvian village acted as a minuscule but critical nucleus for catastrophic events that were to have effects far beyond the town.

As background information, it should be noted that an Asian freighter had been anchored in the harbor during the previous week, and that the ceviche for the fiesta had been prepared from raw shellfish harvested from that harbor. On the day following the celebration, many residents of the town became very ill, with vomiting, acute diarrhea, and extreme dehydration. Thirty-seven people died. A virulent Asiatic strain of a bacterial pathogen, Vibrio cholerae, was isolated from those stricken. The disease was diagnosed as cholera, a scourge from the dark ages that has never really disappeared from some parts of the world, but one that prospers where sanitary conditions are absolutely abominable.

The pathogen spread quickly to nearby towns, then to Lima itself, where thousands became ill and hundreds died. Drinking water in the poorer districts became contaminated with the pathogen, and primitive sanitary conditions added momentum to expansion of the disease. Within weeks, sporadic outbreaks were occurring in other coastal areas of Peru, and travelers soon carried the pathogen inland, and then to other South and Central American countries. Within three months the disease was pressing against the Mexican border of the United States. By the end of 1991, more than 3000 Peruvians had died from cholera, as had over 1000 from other countries in South and Central America. In the summer of

1992, the epidemic showed signs of lessened intensity, although 62,000 new cases had been diagnosed in the Americas in the first three months of that year. By February 1993, Brazil had become a focus of the disease, with 32,313 cases and 389 deaths reported. By December 1993, cases of cholera in Latin America and the Caribbean had reached 700,000, with an estimated 6400 deaths. The epidemic in the Americas diminished in intensity after 1995. Recorded cases during the period 1991–1995 (over one million) may well represent only a small fraction of the actual numbers of infections during those years.

The United States, with reasonable levels of sanitation, was spared most of the anguish and death caused by the disease. Less than 100 of its citizens acquired cholera, either during visits to South America, or by eating contaminated food transported home by travelers. Of this total, 65 were infected and one died from cholera after eating contaminated seafood salad served on a plane bound from Lima to Los Angeles.

The likelihood of a major cholera outbreak in the United States is considered to be slight, since the disease is associated with dreadful hygienic conditions not often found in this country. One exception might be among residents of slums along the border with Mexico — areas that lack public drinking water or sewage disposal systems.

From Field Notes of a Pollution Watcher
(C.J. Sindermann, 1997)

[It should be noted here that whereas the story of the *spread* of epidemic cholera throughout much of South and Central America beginning in early 1991 seems to agree with available statistics, there are conflicting interpretations of its *inception* in Peruvian coastal areas. An alternate version has the first reports of cholera coming in early January 1991 from the seaport city of Chancay, 36 mi north of Lima, with an almost simultaneous outbreak in Chimbote, 240 mi farther north, and a subsequent spread to the entire Peruvian coast (a total distance of 1200 mi) by early February 1991. In the same time period (one month), the disease had spread to towns up to 90 mi from the coast (Colwell 1996). A climate event, probably El Niño, was postulated as a trigger for the outbreak, rather than contamination from outside sources (Colwell 1996).]

A BRIEF HISTORY OF CHOLERA OUTBREAKS

We need to place this recent cholera outbreak in the western hemisphere in its proper historical perspective. The epidemic that began in Peru in 1991 was actually a later phase of a global pandemic that originated in Indonesia in 1961 and moved through India, Bangladesh, and Thailand in the 1960s, then to eastern and western Africa in the 1970s and 1980s, before reaching the western coast of South America in 1991. Furthermore, that outbreak was the seventh such pandemic in a series that began in 1816 — usually spreading from a reservoir in southeastern Asia, in the delta of the Ganges River. Cholera pandemics were recorded in 1816, 1829, 1852, 1863, 1881,

1889, and 1961. The disease had existed in India and other Asian countries well before 1800 — probably for a thousand years — but had not occurred in global pandemics until 1816 (Barua & Burrows 1974, Colwell 1984). Its early history will forever remain cloudy, though, because the causative organism, the bacterium *Vibrio cholerae*, was not isolated until 1883.

The epidemiology of cholera has yielded slowly to scientific examination, even after the finding that the disease agent was waterborne in fecally contaminated drinking water, and even after the disease agent was identified and characterized. This slow pace resulted from the reality that cholera epidemics in human populations are indicators of a complex global interactive system, involving such diverse elements as coastal plankton blooms, seasonal sea level heights and seawater temperatures, transport of the pathogen in ships' ballast water, and ingestion by humans of pathogenic bacteria with food or drinking water. Outbreaks have their origins in coastal areas and tidal portions of rivers, and relationships have been proposed with abundance of zooplankton (which may harbor or serve as substrates for the pathogens). Further spread to inland areas is fostered by fecal contamination of food and drinking water, in the large areas of the world where sanitation is inadequate or absent.

An obvious key to epidemic control is to follow procedures that prevent sustained transmission of the pathogen (Tauxe et al. 1995). Major routes of cholera transmission are:

1. Contaminated drinking water
2. Food contaminated in the market or home (or in farms that use fresh sewage for irrigation)
3. Seafood, cooked or uncooked (Animals may harbor the pathogens before harvesting, or they may be contaminated by water used in washing or processing.)

The first six pandemics (all in the 19th century) were caused by a so-called "classic" form (biotype) of *Vibrio cholerae*, characterized by high virulence but only modest survival in estuarine waters. The current (seventh) pandemic, which began in 1961 and has been responsible for millions of illnesses and thousands of deaths, was caused by a different biotype, called *V. cholerae* 01 El Tor, which has been reported from nearly 120 countries since 1991. The El Tor biotype survives longer in the environment, multiplies more rapidly in foods, is less virulent, and induces less immunity than the classical form.

But there is already in existence in India, Bangladesh, Thailand, and elsewhere in the world a new and more aggressive biotype, called *V. cholerae* 0139 Bengal, which has caused severe illness in thousands of people since 1992 and which could become the agent for a future eighth pandemic (Wachsmuth et al. 1993).

The disease itself is an acute intestinal infection with a short (1 to 5 d) incubation period. Toxin production leads to severe diarrhea, dehydration, and (without prompt treatment) death. It is important to note, though, that most people infected with *V. cholerae* do not become ill, or if they do, 90% of cases are of mild or moderate severity. The disease is self-limiting, with recovery in 3 to 6 d. Oral rehydration is the principal treatment (augmented by antibiotics).

Sporadic minor outbreaks of cholera have occurred in recent decades in a number of temperate zone countries — in Italy in 1973 and 1980, in Portugal in 1974, and in the United States (Louisiana) in 1978 (this was the first reported outbreak in the United States, with 11 cases, since 1911). Contaminated shellfish were implicated in each outbreak: mussels in Italy, cockles in Portugal, and crabs in Louisiana. Whereas *V. cholerae* may be a normal part of the brackish-water microflora, its potential for causing human disease seems to be enhanced in heavily polluted shellfish growing areas, especially if raw or improperly processed products are consumed, or if confirmed cases of cholera have been reported in the adjacent towns.

The recent history of cholera in the United States is generally comforting (relatively speaking). The disease was not reported during much of the 20th century until 1973, when a case was reported from Texas. Since then, sporadic small outbreaks have occurred (an "outbreak" consists of two or more cases) — 11 cases in Louisiana in 1978, 2 cases in Texas in 1981, 17 cases on a Texas oil rig in 1982, and 13 cases in Louisiana and Florida in 1986. Most of the cases were associated with eating contaminated shellfish. During the recent epidemic in Latin American countries that began in Peru in 1991, cases of cholera were diagnosed in Mexican cities near the U.S. border, and isolates of a *V. cholerae* strain identical to that found in Peru were recovered from oyster reefs in Alabama as early as September 1991, resulting in temporary closure of the beds. The source of the pathogens was not determined, but human carriers from South America were suspected. Isolated cases in the United States since then (numbering fewer than 100) have been associated indirectly with the South American epidemic, but no major outbreaks have resulted. Epidemic cholera, in the generally accepted usage of the term, has not occurred in the United States since the 19th century (Rosenberg 1987).

Hence, the development of understanding about this centuries-old pestilence has been slow, at least until the closing decades of the 20th century, when the rate of acquisition of new information accelerated enormously. Major events in the early history of the disease included:

- The demonstration by an English physician, John Snow, in 1849 to 1854 that localized cholera outbreaks in London were caused by a waterborne agent transmitted by drinking fecally contaminated water (Legend has it that the critical event in his classic study was removal of the handle of a pump connected to a contaminated well — an action that resulted in immediate subsidence of the local outbreaks.)
- Identification of the causative bacterial pathogen in 1883 by the great early microbiologist Robert Koch
- Recognition of the existence of different types and strains of *V. cholerae*, with varying pathogenicity, based on serological and biochemical tests, in the early years of the 20th century

Recent developments in understanding the disease, constituting the logarithmic growth phase of knowledge during the past several decades, include:

- The recognition and demonstration that *V. cholerae* is part of the indigenous bacterial populations of brackish and estuarine waters, and that it is usually associated with plankton organisms. The relationship of *V. cholerae* abundance in coastal or estuarine waters with zooplankton has been described. The bacteria, which are normal parts of the inshore flora, use the exoskeletons of crustaceans as growth substrates. The latest step in unraveling a complex ecological relationship is remote sensing of algal blooms, which can trigger subsequent increases in zooplankton that provide substrate for the vibrio.

- The recognition and demonstration that *V. cholerae* can exist in a viable but nonculturable state. Under unfavorable environmental conditions, the cholera organism — as well as certain other bacteria — can assume a quiescent nonculturable form associated with plankton organisms. This can result in failure to isolate the pathogen by standard culture methods, even when it may be present in a given sample. With a return of nutrient and temperature conditions favoring algal blooms, the vibrios revert to a culturable and infectious state. Plankton can thus be considered to form an "environmental reservoir" for *V. cholerae* (as was pointed out by Colwell & Huq 1994).

- The appearance in southern Asia in 1992 and the subsequent spread of a new virulent strain of *V. cholerae*. The new 0139 Bengal strain appears to have incorporated characteristics of both classic and El Tor *V. cholerae*. Its environmental survival is high, similar to type 01 El Tor, but its virulence is also high, similar to the classic type. Furthermore, its antigens are sufficiently different from type 01 El Tor that humans who survived earlier outbreaks are still susceptible to infection by type 0139 Bengal (Ramamurthy et al. 1993).

- The demonstration of transport of *V. cholerae* globally in ballast water of commercial vessels. As an example of the danger of international transfer of pathogens, ballast water of 15 ships entering Chesapeake Bay was examined recently for the presence of *V. cholerae* of types 01 El Tor and 0139 Bengal (Ruiz et al. 2000). The bacteria were found associated with plankton from all ships, and both types of the pathogen were isolated from 93% of the samples. The authors pointed out the likelihood of colonization of receiving waters in any country by viable nonindigenous pathogens such as the new 0139 Bengal type of *V. cholerae*.

- The description of the genome of *V. cholerae* 01 El Tor (Trucksis et al. 1998, Heidelberg et al. 2000, Lewis 2000) with findings that should lead to better understanding of how a normal inhabitant of estuaries can also be a severe pathogen of humans. The genome consists of two circular chromosomes (most bacteria other than vibrios have a single chromosome). DNA sequences of both chromosomes have been determined (Heidelberg et al. 2000).

Pathogenicity of *V. cholerae* depends on a combination of properties, and genotypes or chromosomal regions that code for virulence-associated factors (including

TABLE 1.1
**Virulence-Associated Factors (Genotypes or Chromosomal Regions) in
01 and 0139 Serogroups of *Vibrio cholerae***

Pathogenicity Factor (Genotype of Chromosomal Region)	Description Properties/Functions Coded for
ctxA	Cholera toxin (CT)
	Enterotoxin production
hlyA	Hemolysin
	El Tor-like hemolysin production
stn/sto	Non-01 heat stable enterotoxin
	Enterotoxin production
ompU	Outer membrane protein
	Outer membrane production
tcpA and tcpI	Colonization/adherence
	Ability to adhere to and colonize the small intestine
toxR	Tox R regulatory protein
	Regulatory protein production
zot	Zonula occludens toxin
	Toxin that increases permeability of the small intestine

Source: After Rivera, I.N.G., J. Chun, A. Huq, B. Sack, and R.R. Colwell. 2001. *Appl. Environ. Microbiol. 67:* 2421–2429.

toxin and hemolysin production and colonization capabilities) have been identified (see Table 1.1). Three major genetic elements, called pathogenicity islands or gene clusters, exist (Hacker et al. 1997, Rivera et al. 2001; see Table 1.2).

Nonpathogenic strains of *V. cholerae* (non-01 and non-0139) can acquire genes for toxin production from pathogenic strains (by transduction) and may become

TABLE 1.2
**Gene Clusters (Also Called Pathogenicity Islands or Larger
Genetic Elements) That Code for Major Virulence-Associated
Factors on the *Vibrio cholerae* Chromosome**

Genetic Element or Gene Cluster	Function
CTX	Toxin production. Reported to comprise the genome of a filamentous bacteriophage CTX
VPI	Large pathogenicity island that encodes toxin-coregulated pilus (TCP) gene cluster and a pilus that functions as an essential colonization factor
RTX toxin	Encodes cytotoxicity activity

Source: After Rivera, I.N.G., J. Chun, A. Huq, B. Sack, and R.R. Colwell. 2001. *Appl. Environ. Microbiol. 67:* 2421–2429.

sources of new epidemic clones (Lipp et al. 2003). Thus, the toxigenic 0139 sero-group was hypothesized to have arisen from recombination with toxigenic or non-toxigenic 01 strains (Faruque et al. 1998).

As an example of the widespread occurrence of *V. cholerae* in the environment, a recent paper by Chen et al. (2004) described the prevalences in seafood of toxigenic strains in Malaysia (those with ctx genes in the CTX genetic element). Of 97 strains isolated, 20 carried the gene and produced cholera toxin. The authors pointed out that cholera outbreaks occur periodically in Malaysia, and a public health problem exists when fresh market seafoods (especially molluscan shellfish and crustaceans) are con-taminated with toxin-producing strains of the pathogen. These findings from seafood and the isolation of toxigenic 0139 strains from seawater (Son et al. 1998) indicate that the aquatic environment serves as a reservoir for toxigenic *V. cholerae* in Malaysia.

In conclusion, then, cholera has been characterized aptly and memorably as "an ancient disease, associated with poverty, filth, and ignorance, and linked in the popular imagination with other dreaded killers such as bubonic plague and smallpox" (May 1985). It can be also identified, more hopefully, as a disease that is yielding to improvements in sanitation and in scientific understanding of its epidemiology and control, even though it is still a global threat to human populations with low sanitary standards, extreme poverty, and inadequate medical treatment, as well as human populations dislocated as a result of military action.

CHOLERA AND THE ENVIRONMENT

Remarkable progress has been made in attempts to understand *V. cholerae*'s inter-actions with its environment. The usual contaminants of coastal waters are derived from some land-based activity of humans and have greatest impacts where pollution levels are highest. *V. cholerae* is not one of those usual contaminants. With the cholera pathogen, we are dealing with a *normal inhabitant of estuarine waters* with abilities to survive and prosper as part of that ecosystem, relatively independent of human intrusions or contributions. Humans bumble into these self-contained estua-rine systems at their peril. They may ingest infective dosages of the pathogens initially with contaminated seafood and then become sources of enormous quantities of the organisms in their feces, as the self-limiting infection runs its course. Stool samples from acute cholera cases consist of almost pure cultures of *V. cholerae* in very high numbers (10^6 to 10^8 organisms/ml stool; Farmer et al. 1985). A critical phase in the development of the outbreak level of cholera is fecal contamination of drinking water supplies and food; that is when the organisms escape from their normal estuarine habitat and move inland.

The fascinating and still emerging story of environmental interactions of *V. cholerae* has been developed to a remarkable extent with the aid of research by Dr. Rita Colwell and her past and present associates, during more than three decades of highly focused work on the genus *Vibrio*. Significant milestones in that long-term research include these findings:

- The demonstration that *V. cholerae* is an autochthonous member (a natural inhabitant) of coastal ecosystems, and not merely a microbial contaminant.

Organic loading of coastal and estuarine waters can lead to increased abundance of many microbes, including *V. cholerae*.

- The discovery that *V. cholerae* (as well as other vibrios and other genera of marine bacteria) may enter a viable but nonculturable state — a counterpart in some ways to encystment — when environmental conditions are unfavorable for growth.
- The discovery that *V. cholerae* has an intimate relationship, described as commensal, with crustacean zooplankton, so that the epidemiology of cholera in humans must take into account seasonal phytoplankton and zooplankton abundances.
- The finding that even with the limited remote sensing technology available today, it is feasible to predict outbreaks of cholera in some countries where it is endemic (Bangladesh for example) with data on sea surface temperature and sea surface height. The expected near-term inclusion of ocean color data (for detection of phytoplankton blooms) is expected to enhance reliability of those predictions.
- The development and demonstration of the validity of a hypothesis that the natural aquatic environment serves as a reservoir for *V. cholerae*, and that it may play a critical role in pandemics (Colwell 1996, Rivera et al. 2001).

Many other researchers have of course contributed to understanding of *V. cholerae*, especially its genetics and epidemiology. As examples, the emergence of new pathogenic strains by horizontal gene transfer has been proposed and confirmed by Waldor and Mekalanos (1996) and Faruque et al. (2002). The genetic basis for virulence of *V. cholerae* has been examined by several research groups, including Rivera et al. (2001).

Cholera, then, is a disease of humans that has its origins in coastal waters of tropical countries, especially those that are overcrowded and poor and have little or no sanitary infrastructure. Epidemic outbreaks begin in coastal communities of those countries but quickly spread to inland areas, primarily via contaminated drinking water or contaminated shellfish. Such outbreaks may reach pandemic proportions, as the pathogen is gradually freed from dependence on coastal or estuarine environments for survival, growth, and replication. Diseased humans increase the infective populations of the microorganisms enormously when any aquatic environment is contaminated.

REFERENCES

Barua, D. and W. Burrows, editors. 1974. *Cholera.* W.B. Saunders, Philadelphia, PA.

Chen, C.-H., T. Shimada, E. Nasreldin, S. Radu, and M. Nishibuchi. 2004. Phenotypic and genotypic characteristics and epidemiological significance of *ctx*[+] strains of *Vibrio cholerae* isolated from seafood in Malaysia. *Appl. Environ. Microbiol.* 70: 1964–1972.

Colwell, R.R., editor. 1984. *Vibrios in the Environment.* John Wiley & Sons, New York.

Colwell, R.R. 1996. Global climate and infectious disease: The cholera paradigm. *Science* 274: 2025–2031.

Colwell, R.R. and A. Huq. 1994. Environmental reservoir of *Vibrio cholerae*. *Ann. N.Y. Acad. Sci. 740:* 44–54.

Farmer, J.J., F.W. Hickman-Brenner, and M.T. Kelly. 1985. Vibrio. *Manual of Clinical Microbiology.* 4th ed., pp. 282–301. In E.H. Lennette, A. Ballows, W.J. Hausler, Jr., and H.J. Shadomy (eds.), Washington D.C. American Society for Microbiology.

Faruque, S.M., M.J. Albert, and J.J. Mekalanos. 1998. Epidemiology, genetics and ecology of toxigenic *Vibrio cholerae*. *Microbiol. Mol. Biol. Rev. 62:* 1301–1314.

Faruque, S., M. Asadulghani, M. Kamruzzaman, R.K. Nandi, A.N. Ghosh, G. Balakrish Nair, J. Mekalanos, and D.A. Sack. 2002. RS1 element of *Vibrio cholerae* can propagate horizontally as a filamentous phage exploiting the morphogenesis genes of CTX. *Infect. Immunol. 70:* 163–170.

Hacker, J., G. Blum-Oehler, I. Mühldorfer, and H. Tschäpe. 1997. Pathogenicity islands of virulent bacteria: Structure, function and impact on microbial evolution. *Mol. Microbiol. 23:* 1089–1097.

Heidelberg, J.F., J.A. Eisen, W.C. Nelson, R.A. Clayton, M.L. Gwinn, R.J. Dodson, D.H. Haft, E.K. Hickey, J.D. Peterson, L. Umayam, S.R. Gill, K.E. Nelson, T.D. Read, H. Tettelin, D. Richardson, M.D. Ermolaeva, J. Vamathevan, S. Bass, H. Qin, I. Dragoi, P. Sellers, L. McDonald, T. Utterback, R.D. Fleishmann, W.C. Nierman, O. White, S.L. Salzberg, H.O. Smith, R.R. Colwell, J.J. Mekalanos, J.C. Venter, and C.M. Fraser. 2000. DNA sequence of both chromosomes of the cholera pathogen *Vibrio cholerae*. *Nature 406:* 477–484.

Lewis, R. 2000. TIGR [The Institute for Genomic Research, Rockville, MD] introduces *Vibrio cholerae* genome. *The Scientist 14*(16): 8.

Lipp, E.K., I.N.G. Rivera, A.I. Gil, E.M. Espeland, N. Choopun, V.R. Louis, E. Russek-Cohen, A. Huq, and R.R. Colwell. 2003. Direction detection of *Vibrio cholerae* and *ctxA* in Peruvian coastal water and plankton by PCR. *Appl. Environ. Microbiol. 69:* 3676–3680.

May, C.D. 1985. An ancient disease advances. *New York Times* (August 18), p. E-5.

Ramamurthy, T., S. Garg, R. Sharma, S.K. Bhattacharya, G.B. Nair, and T. Shimada. 1993. Emergence of a novel strain of *Vibrio cholerae* with epidemic potential in southern and eastern India. *Lancet 341:* 703–704.

Rivera, I.N.G., J. Chun, A. Huq, B. Sack, and R.R. Colwell. 2001. Genotypes associated with virulence in environmental isolates of *Vibrio cholerae*. *Appl. Environ. Microbiol. 67:* 2421–2429.

Rosenberg, C.E. 1987. *The Cholera Years: The United States in 1832, 1849, and 1866*. UNIV. Chicago Press, Chicago, IT

Ruiz, G.M., T.K. Rawlings, F.C. Dobbs, L.A. Drake, T. Mullady, A. Huq, and R.R. Colwell. 2000. Global spread of microorganisms by ships. *Nature 408:* 49–50.

Son, R., G. Rusul, L. Samuel, S. Yuherman, A. Senthil, A. Rasip, E.H. Nasreldin, and M. Nishibuchi. 1998. Characterization of *Vibrio cholerae* 0139 Bengal isolated from water in Malaysia. *J. Appl. Microbiol. 85:* 1073–1077.

Tauxe, R.V., E.D. Mintz, and R.E. Quick. 1995. Epidemic cholera in the New World: Translating field epidemiology into new prevention strategies. *Epidemiol. Infect. Dis. 1*(4): 1–9.

Trucksis, M., J. Michaelski, Y.K. Deng, and J.B. Kaper. 1998. The *Vibrio cholerae* genome contains two unique circular chromosomes. *Proc. Natl. Acad. Sci. USA 95:* 14464–14469.

Wachsmuth, I.K., G.M. Evins, P.I. Fields, O. Olsvik, T. Popovic, and C.A. Bopp. 1993. The molecular epidemiology of cholera in Latin America. *J. Infect. Dis. 167:* 621–626.

Waldor, M.K. and J.J. Mekalanos. 1996. Lysogenic conversion by a filamentous phage encoding cholera toxin. *Science 272:* 1910–1914.

2 Minamata Disease

The "Dancing Cat Disease" of Minamata

Sometime in the early 1950s, weird behavior began to be noticed in some of the nonhuman inhabitants of small fishing towns near the industrial city of Minamata, on the island of Kyushu in southern Japan. Birds tumbled from trees, or flew erratically, occasionally bumping into houses in their flight. Cats walked with a peculiar lopsided gait, or ran in tight circles, eventually becoming totally disoriented and dying. Locals called the condition "dancing cat disease." Not long after these strange happenings were first observed, abnormalities began to occur in humans as well — abnormalities of increasingly frightening proportions.

Fishermen and members of their families began to experience neurological disabilities: tremors, numbness of face and limbs, paralysis, visual disturbances (especially constricted vision), mental disorientation, and speech disorders. Advanced cases lost control of body functions, became bedridden, and died. Mortality was about 40% of affected individuals. The full horror of the disease was still ahead — in the birth of blind, dreadfully deformed, and mentally impaired children (Smith and Smith 1975). The affliction was first called "the strange disease" and later "Minamata Disease." By the end of 1956, 52 victims had been identified in the small fishing communities surrounding Minamata Bay.

The search for a cause was painfully slow, and was impeded by government/industry foot-dragging and denials by both parties that a problem existed. Mercury contamination of Minamata Bay and its fish and shellfish populations by the effluents of a chemical production company was suspected as the cause, and was reported as such in the scientific literature in 1959. It was not until 1968, however, that the Japanese government stated officially that organic mercury contamination of fish and shellfish was the cause of the disease in human consumers, and that the chemical company (a part of the Chisso Corporation) was the source. Later investigations disclosed that during the period 1932 to 1965 Chisso had dumped 80 tons of organic mercury into the bay. The company denied legal responsibility until 1973, although beginning in 1959 it had begun offering poisoning victims and their families minuscule compensations (so-called "solatiums," which acknowledged corporate concerns but were not to be considered admissions of guilt). For all the suffering, disfigurement, and death that this company had caused, it paid "solatiums" of only $800 for a death, $280 per year for adult victims, and $83 per year for afflicted children! (These amounts were increased in the 1970s after public outrage and legal decisions in civil suits forced action by Chisso.) Present compensation consists of a lump sum payment of about 24 million yen (about 240,000 U.S. dollars).

By 1975, 793 victims had been designated officially, but about 2700 other people living in the polluted zone around Minamata City had symptoms of mercury poisoning, and an additional 10,000 residents were considered latent victims. [According to the most recent government information (March 1997), 2762 victims have been certified in the Minamata area, although many other cases are still being contested in courts, or have gone unobserved, or have been concealed because of family pride, during the long interval from the early 1950s to the present.]

But the Japanese mercury poisoning story was not confined to Minamata. In 1964, another outbreak of the "strange disease" due to mercury poisoning was discovered on the west coast of Japan, near the city of Niigata. The source of contamination was traced to a factory of the Showa Denko Corporation that was using the same acetaldehyde production process (with a mercury sulfate catalyst) as that used by Chisso in Minamata — with mercury being dumped into adjacent river waters. Cases were mostly confined to fishermen and their families, for whom fish was a dietary staple. Fortunately, the scale of contamination was less than that of Minamata Bay, with fewer deaths and disabilities. To the present time, 690 cases of Minamata Disease have been certified officially from the Niigata area.

Then in 1973, a third outbreak of the disease, involving only 10 cases, was reported from the coastal area bordering the Ariake Sea, 40 km north of Minamata Bay. Announcement of that outbreak touched off near-panic in all of Japan, accompanied by short-term but widespread incidents of civil disorder, aimed at forcing reduction of pollution by large industries and increasing government regulatory activities at prefectural and national levels.

From Field Notes of a Pollution Watcher
(C.J. Sindermann, 1983)

CHRONOLOGY OF EVENTS RELATED TO MINAMATA DISEASE

The Minamata mercury poisoning episode has become an important milestone in global environmental thinking, and, because components of such events become obscured with the passage of time, it seems useful here to summarize principal events on a time scale. It also seems relevant to consider those events from several perspectives: those of the victims, the scientific community, and the polluting industries and the regulatory agencies.

All the elements of high drama are here: the human pain and suffering, the struggles by scientists for understanding, the uncaring polluting industry, the reluctant governmental regulatory bodies, and the endless legal maneuvering to assign guilt and to secure adequate compensation for victims. Each group has its own perspective on the happenings. To minimize confusion, a time and event summary for each of the principal entities involved follows, with their perspectives on the unfolding story of Minamata disease in Japan.

Perspective of the Victims

Early 1950s — Fishermen and their families began exhibiting signs of an unknown illness characterized by trembling, numbness of hands and feet, paralysis, and loss of control over body functions. It was first thought to be an infectious disease, and signs of the affliction resulted in social ostracism of the victims. The Chisso factory was a known polluter of the bay, but it was also essential to the economy of the region, so when the disease became associated with eating contaminated fish, the local business community and the local government refused to confront the company directly.

1958 — Disease victims and their families formed a mutual assistance society to negotiate with Chisso for compensation but were rebuffed (Huddle et al. 1975).

1959 — With the publication of a university scientific report implicating Chisso in polluting the bay, the demands of the fishermen's union for compensation for reduced catches became more violent. Repeated demonstrations were held at the factory. A mediation committee formed of local politicians proposed a minimum solatium that specifically ignored the matter of compensation to disease victims. More than 2000 fishermen rioted, stormed the factory gates, and caused extensive damage. The riots brought national attention; disease victims conducted a sit-in, demanding that medical compensation be included in any settlement. An agreement was reached at year's end, even though the victims' mutual assistance society was extremely unhappy with the ridiculously small amounts of the solatiums ($800 for a death, $280/yr for adult victims, $83/yr for children, and $56 for funeral expenses). The fishermen's union was the principal negotiator of the agreement (Huddle et al. 1975).

1960–1967 — The insultingly small solatiums to victims paid by Chisso subdued protests for a number of years, even though the victims continued to live in poverty and many received no treatment. Late in 1964, a second outbreak of Minamata disease occurred in the area of the Agano River near the city of Niigata. The sequence of events followed the course of the original outbreak in Minamata: illness and death among poor people with a fish diet, denial of responsibility by the polluting company, suppression of academic research findings by government agencies, and demonstrations and legal actions by victims' associations. By 1967, victims from Niigata were cooperating with those from Minamata with joint protests and legal claims.

1968 — The national government finally and belatedly agreed that organic mercury caused Minamata disease and that its source in the Minamata area was the Chisso factory, and in the Niigata area the Showa Denko factory. It also passed in December 1968 the Pollution Victims Relief Law (later called the Pollution-Related Health Damage Compensation Law), in which certified victims of mercury poisoning received a lump sum payment. Chisso claimed inability to pay the compensations, so the national government has "assisted" (National Institute for Minamata Disease 1997).

1973 — A prefectural court found Chisso guilty of gross negligence, and the company finally admitted its responsibility for creating conditions leading to organic mercury poisoning — a clear victory for Minamata victims. Chisso also agreed to pay compensation of 16 to 18 million yen (US$51,000 to US$59,000) to victims, but by 1975 Chisso was already requesting government loans.

1992 — In addition to compensation to certified victims, the Japanese government began providing financial or medical support to residents of contaminated areas, to reduce health-related anxieties.

2000 — Almost 3000 victims of organic mercury poisoning have been certified for compensation under the terms of the Victims Relief Law. Victims are now eligible for lump sum payments of about 24 million yen (roughly US$240,000 at the present rate of exchange). Chisso has claimed inability to pay the cost, so the national government has "assisted" with its payments.

2001 — The Japanese Supreme Court awarded an additional total sum of US$2.8 million to 51 plaintiffs who claim illness from mercury poisoning.

The drama is not over. Every year, 400 to 700 people apply for certification and compensation — individuals who did not apply in the past, for various reasons, or those who were rejected earlier.

A Scientific Perspective

1956 — The "strange disease" was first officially described by the director of Minamata's Chisso Hospital as "severe damage to the central nervous system of humans." In that same year, a scientific/medical group (including Kumamoto University researchers) was designated by the prefectural government to search for the cause. The first report of that group suggested very tentatively that the cause might be "heavy metal intoxication from eating contaminated fish."

1957 — The university group reproduced the disease signs seen in cats by feeding experimental animals fish caught in Minamata Bay. The group recommended a ban on fishing in the bay, but no action resulted. Local fishermen were urged to limit fishing in the bay, but no compensation was provided.

1958 — After many inconclusive tests of various pollutants, organic mercury became a prime suspect, principally because a British neurologist, D. McAlpine, visited Minamata for 2 days and subsequently published a paper in *The Lancet* (September 20, 1958) suggesting that concentration of organic mercury in nerve tissue could produce the symptoms seen in Minamata patients.

1959 — Despite delays imposed by the Kumamoto Prefecture government, a university research report was published concluding that the causative agent of Minamata disease was organic mercury. The report identified Chisso effluents as "the likely source" of organic mercury compounds found in the environment (Huddle et al. 1975). Minamata thus became the first recorded episode of mass mercury intoxication from *indirect* transmission from a contaminated coastal environment.

1960–1967 — Charges by Chisso scientists resulted in the termination in 1960 of research grants to Kumamoto University scientists studying Minamata disease. Grant support was assumed for 3 yr by the U.S. National Institutes of Health (NIH). Organic mercury was found in the factory effluents in 1962, and it was found to be concentrated as it moved up estuarine food chains. The chemistry of conversion of inorganic mercury used as a catalyst for acetaldehyde production to methylmercury was also described in 1962. The prefectural and national governments, particularly the National Ministry of Health and Welfare, repeatedly suppressed or delayed scientific reports on Minamata disease. This was in part a result of sometimes obstructive cooperation between government ministries and industry (Huddle et al. 1975).

1973 — The national government of Japan announced that during 33 yr of acetaldehyde production at Minamata (1932–1965), Chisso had discharged an estimated 80 tons of methylmercury into the bay.

1978 — The National Institute for Minamata Disease was established to carry out medical studies of the disease and to conduct medical examinations of possible victims.

1996 — The institute's role was expanded to include international chemical and environmental studies of mercury poisoning in humans.

1997 — Monitoring of methylmercury concentrations in fish and shellfish continued in the Minamata area, as did other investigations that led to the conclusion that "continuous methyl mercury exposure at the level which can cause Minamata Disease existed until no later than 1968 in the Minamata area.... and after that there has not been such exposure" (National Institute for Minamata Disease 1997).

2001 — Medical scientists disagree about the validity of neurological tests used to identify and certify victims of Minamata disease.

Industry/Regulatory Perspective

Before 1950 — The huge Chisso Corporation, located on the shore of Minamata Bay in southern Japan and established in 1908, became a major producer of polyvinyl chloride, acetaldehyde, and other synthetic chemicals used in plastics manufacturing. Increased production for a booming economy, without concern for environmental consequences, was the prevailing philosophy of the time, and this philosophy was reflected in very tolerant government regulatory attitudes toward the actions of large polluting industries like Chisso.

1950 — Because of conspicuous postwar pollution damage to the fisheries, the local fishermen's union appealed to Chisso for compensation. The company, citing lack of scientific evidence, denied responsibility for any deterioration of the bay.

1953 — The first cases of what came to be called the "strange disease" were reported to local authorities in Minamata City.

1954 — Chisso purchased land and fishing rights from the fishermen's union to compensate for past and future damages, with the provision that "even if further damages occurred, no new claims would be made." The company promised to test its effluents, install treatment equipment, and carry out pollution control measures (none of these steps was implemented, and fish production continued to decrease).

1959 — A Kumamoto University report identified organic mercury as the cause of the "strange disease" (which became known as "Minamata disease"). The report implied that Chisso was the source of the mercury, and its publication caused intensified efforts by fishermen and victims to receive compensation, as well as efforts by the company to deny responsibility and to discredit the organic mercury theory. Also in 1959, a scientist employed by Chisso reported to company management his experimental results with cats, finding that the factory effluents produced signs of the condition then known as "dancing cat disease." Chisso suppressed the report until 1968, and the experiments were terminated (Ui 1968).

1968 — The Japanese government stated officially that organic mercury compounds in industrial effluents caused Minamata disease.

1970 — The Water Pollution Control Act was enacted to reduce or eliminate industrial mercury and cadmium discharges in all of Japan.

1973 — Chisso admitted its responsibility for the mercury pollution and agreed to pay compensation to present and future victims. It also agreed to cooperate with the government in the cleanup of the bay.

1975 — Chisso applied for a government loan of US$33 million, in an attempt to shift responsibility for compensation payments, and began spinning off its more profitable subsidiaries as independent companies.

1976 — Kumamoto Prefecture began removing 1.5×10^6 m³ of contaminated bottom sediments from Minamata Bay. Chisso was to pay 60% of the costs of removal. Work was completed in 1990.

1977 — Nets that had been installed in 1959 to reduce movement of fish into or out of Minamata Bay were removed, with the assumption that the methylmercury problem had been resolved.

1997 — Testing for methylmercury contamination in fish continued, as did other investigations by the National Institute for Minamata Disease.

2001 — Government and industry appealed the Japanese Supreme Court award of US$2.8 million to Minamata disease victims.

Table 2.1 is a further condensation of events from the perspectives of victims, scientific, and industry/regulatory bodies.

WHAT HAS BEEN LEARNED FROM THE MINAMATA EXPERIENCE?

The sheer horror and accompanying public outrage over the Minamata poisonings marked an important turning point in national and eventually international attitudes about industrial development and environmental abuse. The phenomenal economic growth of Japan after World War II was accomplished at a huge cost in environmental degradation and human suffering. Minamata continues as a stark illustration of that reality, and one that the Japanese people recognized and began changing in the 1970s.

Events during the course of massive and culpable polluting episodes like this one seem to follow an almost predictable course. The polluting industry, deliberately or through negligence, dumps toxic wastes into rivers or estuaries. Fish and shellfish may be affected, in that abnormalities appear and populations may decline. Fish-eating birds and mammals (including humans) may be affected by ingested toxic chemicals, so that sickness and death may result. Medical scientists address the problem after growing public concern results in release of research funds. Investigations focus on suspect chemicals, and evidence accumulates about a specific contaminant and its likely source.

During this period of initial uncertainty, the polluting company vehemently denies that its effluents can maim or kill, even when it has evidence from studies by its own

TABLE 2.1
Important Milestones in the History of Minamata Disease

Victims	Science	Industry/Government
Fishermen of Minamata encounter reduced catches and blame pollution from Chisso factory, beginning in 1926		
Fishermen's union appeals to Chisso for compensation for lost catches in 1926, 1943, and 1950		
		Chisso denies responsibility for reduced fish catches, 1950
First cases of "strange disease" reported during early 1950s		
	A university scientific/medical group is designated by the prefectural government to search for the cause of the "strange disease," 1956	Chisso factory management refuses to allow scientists access to its effluents, 1956
Disease victims form mutual assistance society to negotiate with Chisso, 1958	First report of the scientific investigation of the "strange disease" suggests heavy metal intoxication as a cause, 1956	
After demonstrations and riots outside Chisso factory, an agreement was reached at the end of 1959 in which the company paid tiny "solatiums" to disease victims and small compensations to the fishermen's union		
	The first scientific report implicating organic mercury as the probable cause of Minamata disease was published in 1959	Chisso begins paying "solatiums" but continues to deny responsibility, 1959
	The U.S. National Institutes of Health provided principal financial support to Kumamoto University research on Minamata disease, 1960–1963	
A second outbreak of Minamata disease occurred at Niigata, near the Showa Denko factory in 1964	Niigata University designated by the Japanese government to do research on Minamata disease, 1964	

TABLE 2.1 (Continued)
Important Milestones in the History of Minamata Disease

Victims	Science	Industry/Government
	First report of Niigata research scientists points to mercury pollution from the Showa Denko factory as the cause of the disease outbreak, 1966	
Victims from Minamata joined with those from Niigata in legal actions for compensation in 1967		
Production of acetaldehyde by Chisso discontinued at Minamata in 1968		
National government passes pollution-related Health Damage Compensation Law, providing compensation to victims, 1968		Japanese government states officially that Minamata disease was caused by organic mercury from the Chisso factory, 1968
		National Water Pollution Control Law enacted, 1970
Chisso agrees to pay victims compensation of US$51,000 to US$59,000, 1973		
		Chisso requests loan of US$33 million from national government to help pay victims' compensation costs, 1975
		Kumamoto Prefecture begins dredging 1.5×10^6 m^3 of contaminated bottom sediment from Minamata Bay, 1976
		Nets preventing fish from moving into or out of Minamata Bay removed, with the assumption that the mercury problem had been eliminated, 1977
	National Institute for Minamata Disease established, 1978	
		National government begins providing financial and medical support to residents of contaminated areas, 1992

TABLE 2.1 (Continued)
Important Milestones in the History of Minamata Disease

Victims	Science	Industry/Government
	Role of National Institute for Minamata Disease expanded to include environmental and international studies, 1996	
Lump sum payments underwritten by the Japanese government are specified at 24 million yen (US$240,000), 1997		
Japanese Supreme Court awards an additional total sum of US$2.8 million to 51 plaintiffs who claim illness from mercury poisoning, 2001	Medical scientists disagree about the validity of neurological tests used to identify and certify Minamata disease, 2001	Industry/government appeal Supreme Court award of US$2.8 million to Minamata disease victims, 2001

captive scientists that the company's effluents are responsible (as did the Chisso management early in the Minamata incident). Scientists in the employ of the polluting industry conduct their own investigations, but they never release their findings (except to the CEO and the board of directors). Governments and their regulatory agencies, during this phase of denial, are reluctant to act against the industry in the absence of truly overwhelming data and vigorous public outcry against the polluters. Civil suits instituted by individuals or groups may eventually resolve the impasse, if decided in favor of the victims, and if endless appeals by the polluting industry are denied. The proffered settlements are invariably far too small to compensate for damages; the polluter then pleads poverty and begs for a government bailout. The final stage is threatened or actual declaration of bankruptcy, after the company has spun off the more profitable of its subsidiaries as independent companies.

The sequence of events in these major pollution episodes, often dragging on over several decades, seems to follow this common pattern — almost a formula — augmented occasionally by new legal devices designed to avoid admission of responsibility for environmental degradation or damage to resources and people. One of the few positive aspects of this prolonged struggle of victims against polluting industries has been the reinforcement of an important legal concept that had been developed during repeated earlier confrontations with polluters of Japanese waters, which states: "In the absence of conflicting clinical or pathological evidence, *epidemiological* proof of causation [of human disease] suffices as *legal* proof of causation." This hard-won concept — which has become known as the "Minamata Principle" — will be important in any future major contamination episode, anywhere in the world, that results in human illness.

Today there is widespread recognition in Japan of the reality that post–World War II industrial development succeeded at a large cost in environmental damage and in damage to human health. Environmental protection has received increasing attention there since the late 1960s, when the consequences of earlier polluting practices began to be identified. Pollution prevention and control policies have been implemented throughout the country and in adjacent waters.

The Minamata experience, partly because of widespread media coverage and public demonstrations, was an important factor in developing awareness of a national pollution problem and in gaining support for preventive and remedial actions (Ui 1968).

REFERENCES

Huddle, M., N. Reich, and N. Stiskin. 1975. *Island of Dreams: Environmental Crisis in Japan.* Autumn Press, New York. 225 pp.

National Institute for Minamata Disease. 1997. Our intensive efforts to overcome the tragic history of Minamata Disease. Spec. Rep., NIMD. 23 pp. (In English)

Smith, W.E. and A.M. Smith. 1975. *Minamata.* Holt, Rinehart and Winston, New York. 220 pp.

Ui, J. 1968. *The Politics of Pollution.* Sanseido, Tokyo. (In Japanese)

3 PCBs and Related Chemicals

Early Experiences with PCBs in New England

*From high school to a production job in the Sprague Electric Company —
that was a common career path for many 18-year-olds in the western Massachusetts mill town where I was born too many years ago. The company was
one of the early users of PCBs, a versatile "oil" that filled the condensers and
capacitors for a primitive pre-transistor electronic industry. I worked the night
shift, sometimes up to my elbows in that magic fluid, filling thousands of metal
containers before riveting their terminals and sending them along for "degreasing" to a cohort standing over a fuming vat of solvent. It was like a scene from
Hades, and some of my cohorts didn't tolerate the "oil" too well. Skin rashes,
headaches, dizziness, bronchitis, eye irritations, asthma, and nausea drove some
back to other less stressful departments, with lower pay scales. Outside the
walls, residues of the oil were part of the effluent that emptied into the nearby
Hoosac River, a terribly abused slimy tributary of a similarly degraded major
river, the mighty Hudson. [Later it was discovered that the company had been
also, quite legally at the time, burying drums of PCBs on the shores of the
Hoosac. Burying toxic wastes in vulnerable locations did not become illegal
until enactment of the federal Toxic Substances Control Act of 1979.]*

*All this action took place in post-Neanderthal times — the very early 1940s
— long before the first faint stirrings of any kind of environmental ethic. We
knew nothing in those days about ecology or toxicology; PCBs were not recognized as a serious public health problem until the late 1960s. Evidence of harmful
effects on humans and other animals has accumulated slowly since that time,
especially on effects of PCBs on reproductive and developmental physiology.*

*Unknown to most of us working the night shift in that decaying factory building,
a remarkably similar scenario was unfolding in the basement rooms of Sprague's
arch-competitor — a foreign-based company located in New Bedford, Massachusetts, with an improbable name, Cornell-Dubilier Corporation. The one significant
and critical difference between the two companies was that in New Bedford the
manufacturing wastes, including PCBs, were dumped directly into the harbor,
where even now, more than half a century later, commercial fishing is restricted
and PCB levels in some resource species are still greater than action levels.*

*I left that Massachusetts mill town forever soon after World War II, but I have
often wondered about what happened to all of those innocent co-workers, male*

and female, whose early life histories placed them at extreme risk from long-term exposure to toxic environmental levels of PCBs. Did they have normal kids? What about their cancer histories? What have been their survival rates? That no-doubt fascinating story has never been told; the appropriate detailed research was not done, and the consequences of those early and prolonged contacts with a toxic industrial chemical are now rapidly receding from living memory.

From Field Notes of a Pollution Watcher
(C.J. Sindermann, 1990)

Addendum, 2003:

Several small shreds of technical information have survived from the PCB episode at Sprague Electric Company. One is a 1990 report of a limited survey of cancer prevalences throughout all of Massachusetts, conducted from 1982 to 1986. Findings were "higher than expected rates" of certain types of cancer (including stomach and bladder cancer) in residents of my hometown. No subsequent study has been made, to my knowledge, despite the brief flurry of local publicity that the report engendered.

New Bedford — our rival municipality for early PCB contamination of Massachusetts waters — was more aggressive. It was able to promote two studies. In 1980 and 1981, a determination was made of serum PCB levels in volunteers who reported eating locally caught fish. More than 35% (n = 72) had levels greater than or equal to 30 ppb. Armed with these preliminary findings of elevated body burdens of PCBs, the city then convinced the federal Centers for Disease Control in Atlanta to cooperate in a 4-year study from 1984 to 1988 to assess the prevalences of elevated PCB levels in a larger sample of 840 residents — levels that might reflect known contamination of New Bedford Harbor and adjacent estuaries by Cornell-Dubilier. A report, published in 1991, did not disclose unusual serum levels of PCBs, as compared with other U.S. urban populations. Health problems were (unfortunately) not examined. The study and its findings were severely compromised by the fact that the survey was done too late: the sampling did not occur until 8 to 12 years after the 1979 EPA ban on use of PCBs, and the median half-life of PCBs in only 4 years in individuals with an initial 30- to 100-ppb serum level. At any rate, a planned follow-up study of health effects was not conducted, to my knowledge.

Results of one final related study were published only 6 years ago (Kimbrough, Doemland, & LeVois 1999). General Electric Company workers from northeastern states, some of whom had occupational contacts with PCBs, were surveyed for unusually high cancer prevalences or differential mortality rates. The survey disclosed no elevated cancer occurrences, but many questions about the validity of the sampling have been raised.

My personal story of PCB use and misuse in two small New England towns is only a tiny fragment of a much larger environmental horror, because since those earlier times, PCBs have permeated all the world's oceans, principally by airborne transport and (to a lesser extent) by riverine transport to coastal waters in association with sediment particles — followed by deposition and subsequent movement through aquatic food chains. The primary route to human exposure to PCBs is through eating contaminated fish and marine mammals (the latter in Arctic and Subarctic areas).

I will discuss next some examples of PCB contamination of humans from many regions of the planet — emphasizing the emerging evidence for illnesses and disorders that can be traced to that source. Possible dangers to humans include reproductive/developmental impairment, endocrine disruption, carcinogenicity, and immunotoxicity (Kimbrough 1974). Except for instances of gross accidental contamination of food by high PCB concentrations, health effects are usually subtle, and they are often extrapolated from results of experiments with laboratory mammals (Cordle, Locke, & Springer 1982) or are based on correlations with elevated levels in blood or body fat. Some findings are controversial or inconclusive, but there is enough evidence, from enough studies, to indicate a role for PCBs in causing human abnormalities and diseases. This role can be examined under the major dangers listed above: reproductive/developmental disorders, endocrine disruption, carcinogenicity, and immunotoxicity.

REPRODUCTIVE/DEVELOPMENTAL DISORDERS

Effects of PCBs on humans were demonstrated dramatically in Japan in 1968 and in Taiwan in 1978 to 1979 by several unfortunate outbreaks of accidental poisonings through contamination of cooking oil by PCBs and their breakdown products (dioxins, furans, and quaterphenyls). Originally called Yu-sho disease in Japan and Yu-cheng disease in Taiwan, PCB intoxication caused skin eruptions, increased skin pigmentation, and disturbed liver function (Rogan et al. 1988, Rogan 1989). Fetuses suffered the most obvious damage. They were undersized before and after birth, had persistent skin rashes, and later had neurocognitive deficiencies (Chen et al. 1997) — some of which persisted into adulthood (Brouwer et al. 1999). The "Yu-cheng children," exposed prenatally to PCBs, have been examined periodically as they have grown and have suffered from a spectrum of physical, neurological, and behavioral problems (Chen 1994). Additionally, women from that exposure cohort continued to report significantly higher occurrences of stillbirths and childhood deaths than controls (Yu et al. 2000).

These examples of short- and long-term effects of dietary exposure to PCBs at high dosages provide good evidence for dangers to human health, clouded only by the possible role of bioactive contaminants other than PCBs.

Probably the best demonstration of human health effects of PCBs at *environmental* levels has emerged from long-term studies in the Great Lakes states of the effects on offspring of a maternal diet that included significant intake of PCB-contaminated fish. Most species of fish at the time of the investigation — the early 1970s — contained detectable levels of PCBs, with trout and salmon being highly contaminated (Humphrey, Price, & Budd 1976). Mean PCB levels were 3 to 11 ppm. The testing focused on mothers and newborn infants who had been chronically exposed to PCBs through maternal consumption of fish from Lake Michigan. Maternal health effects included tendencies toward increases in anemia, edema, and susceptibility to infections. Effects on infants via transplacental transmission of PCBs and PCB contamination of maternal breast milk included delays in development, decreased responses to stimuli, and decreased visual recognition memory. The startling estimation was made by the investigators that if the input of PCBs into humans were to be stopped in a given maternal generation, the contaminants would still be transmitted in measurable concentrations ... for at least five generations.

Follow-up studies were done on some of the children whose mothers had eaten PCB-contaminated fish from Lake Michigan. At age 11, children with highest fetal exposure were three times as likely as nonexposed children to have relatively low IQ scores and twice as likely to lag at least 2 yr in reading comprehension (Jacobson & Jacobson 1996). Highly exposed children were also found to have memory, word comprehension, and attention span deficiencies.

Disagreements have been raised with some of the conclusions drawn from the Great Lakes studies — regarding everything from inadequate sample size to the possible existence of other environmental confounding influences (such as maternal alcohol consumption or presence of methylmercury or pesticides in the fish consumed by mothers, and the small differences in blood PCB levels of mothers who consumed or did not consume fish; Flynn 1997).

As a supplement to the Great Lakes studies of possible effects of PCBs on humans, experiments with monkeys have provided further evidence of PCB-induced reproductive dysfunction (Allen & Norback 1976). Of eight animals fed diets containing 5 ppm PCBs for 6 months, two did not conceive, four had early abortions, and one produced a stillborn infant. Of eight females fed 2.5 ppm for 6 months, only five were able to carry infants to term, whereas three aborted. At both dietary PCB levels, infants were small at birth and had detectable PCB levels in their skin.

Other studies in other areas have examined effects of occupational exposure to PCBs on neural functions in adults. One report (Kilburn 2000) found statistically significant deficits in an array of visual and other neural activities to result from such exposures in the workplace. Testing focused on visual performance, reaction time, memory, color determination, and other variables.

ENDOCRINE DISRUPTION

The reproductive/developmental disorders just considered are, of course, intimately associated with endocrine (hormonal) activity. During the past 3 decades there has been increasing concern about the possible roles of some environmental pollutants, including PCBs, as endocrine disruptors. Some evidence has accumulated about possible suppression by PCBs of thyroid hormone production and about a role for PCBs and related pollutants as sex hormone mimics. Studies in laboratory mammals indicate that fetal exposure to PCBs results in lower postnatal thyroxin production, and, in a field experiment, seals fed for several years on PCB-contaminated fish from the Baltic Sea were found to have lower levels of thyroxin than seals from other less contaminated waters (Reijinders 1986). Additionally, reproductive success of the seals receiving highest PCB exposure was significantly decreased, due apparently to hormonal disturbance, possibly caused by disruption of the steroid synthetic pathway. Estradiol levels were lower in all PCB-exposed seals when compared with controls.

Disturbed thyroid hormone production may also result in abnormal brain development in laboratory animals. Subtle effects occur when PCBs and other environmental contaminants inhibit or mimic actions of thyroid hormones — especially when exposure occurs during the prenatal period (Porterfield 1994).

In humans, a direct correlation of PCB ingestion and thyroid hormone levels was found in a study of Native Americans of the Akwesane tribe living in northern New York State, near the Mohawk River. PCB pollution from General Motors factories resulted in high levels in fish and in tribeswomen and their offspring until 1990, when consumption of fish stopped. Children 10 to 15 yrs old at that time had PCB levels up to 5 ppm, and a direct relationship was seen between blood PCB levels and thyroid hormone levels (Lewis 2001).

CARCINOGENICITY

The cancer/PCB relationship in humans is still described as "problematic" and is based on animal experiments, where liver and kidney tumors have been induced. Liver carcinogenetic effects were first reported in rats and mice exposed to relatively high dosages (100 to 300 ppm) of PCBs in their diets (Kimbrough 1974, Kimbrough et al. 1975). Adult victims of mass PCB poisonings in Japan (Yu-sho disease) and in Taiwan (Yu-cheng disease) were examined for differential cancer occurrence, with inconclusive findings. Causes of deaths among 1761 Yu-sho patients through 1983 disclosed excess mortality due to liver cancer in one geographic sector of the exposure area but not in another (Kuratsune 1989). After the later (1978 to 1979) Yu-cheng poisoning in Taiwan, elevated mortality due to liver diseases other than cancer was observed within 3 yr (Hsieh et al. 1996).

Studies of humans exposed to high environmental levels of PCBs have produced equivocal results. One of the largest, using a population of more than 7000 capacitor workers at General Electric factories (and funded by General Electric) disclosed no increased mortalities due to cancer (Brown & Jones 1981; Kimbrough, Doemland, & LeVois 1999).

PCBs are clearly carcinogens in nonhuman mammalian models, and it is generally accepted that a proven animal carcinogen is a potential human carcinogen (Lewis 2001), but a direct causal link between PCB exposure and human cancer has not yet been demonstrated.

IMMUNOTOXICITY

Evidence is accumulating, from studies of victims of PCB poisoning events and from examinations of human groups whose everyday diets include high levels of PCB contaminants, for a role of this ubiquitous pollutant in immunosuppression — in reduced responses to disease agents and other foreign proteins.

Studies of victims of the Japanese (1968) and Taiwanese (1978 to 1979) PCB poisoning episodes disclosed evidence of impaired humoral and cellular immunity in the form of decreased immunoglobulins, reduced skin responses to injected antigens, and decrease in numbers of total T-cells and active T-lymphocytes (Kuratsune et al. 1969, Lü & Wong 1984).

Several regional studies of human populations have provided additional evidence of impaired immune functions as a result of long-term consumption of food with high environmental levels of PCBs. One study of environmental contamination in the Canadian Arctic (Dewailly et al. 1989) found mean PCB levels in breast milk of Inuit women from northern Quebec to reach 111 µg/l — among the highest ever reported (and compared with a mean of only 28 µg/l for Caucasian women from southern Quebec). Major contributors to the Inuit diet and to PCB intake were marine mammals (narwhal, seal, and walrus) and fish. Lower disease resistance in children of the study region was suggested by observed infection rates that were 10- to 15-fold higher than those observed in the southern control region (Reece 1987). Impairment of immunity was also reported in another study of Inuits, this time from Iqualuit (formerly Frobisher Bay), the capital of Nunavut Territory in northern Canada. A sample of 40 otherwise healthy infants was found to have lower percentages of several kinds of cells (T-cells and T-helper cells) critical to disease resistance.

Another study, this one in the Netherlands, of common infections in preschool children found that prenatal exposure to PCBs correlated with decreased levels of antibodies to certain childhood diseases (measles, mumps), decreased allergies, and increased ear infections (Weisglas-Kuperus et al. 2000).

Other studies of PCBs in high-latitude populations of marine mammals and humans have been done in Alaska, where subsistence foods include beluga whales, walrus, and bearded and ringed seals — all of which have relatively high concentrations of persistent organic contaminants in their blubber — with consequent risks to human health (Krahn et al. 1999).

A pattern of higher PCBs and chlorinated pesticides in adult males than in females has been found in marine mammals, because some of the contaminant load in the female is transferred to the offspring during pregnancy and lactation. Atmospheric transport of contaminants from industrialized regions at lower latitudes is an important source of dietary exposure for high-latitude animals such as whales — as well as transport of contaminants by ocean currents.

CONCLUSIONS

PCBs now have a truly global distribution — in every environment, including the oceans. Coastal waters are recipients of PCBs from airborne transport and from riverine sources, with concentrations in certain coastal resource species in excess of levels considered to be harmful to humans (New Bedford Harbor, Hudson River) and with concentrations in fish and marine mammals sufficient to be of public health concern (Great Lakes, Arctic, and Subarctic waters).

PCBs and related chemicals have been examined from four principal public health perspectives: reproductive/developmental disorders, endocrine disruption, carcinogenicity, and immunotoxicity. The best evidence for effects has been found in studies of mass accidental poisonings of local populations of humans and through exposures of experimental mammals — usually to dosages in excess of environmental levels. Results of some studies have been compromised by the presence of toxic degradation products of PCBs (dioxins, furans, quaterphenyls) or by other contaminants, and results can be distorted by biased media reporting or statistical manipulation. Other problems with the data are the following:

- Extrapolation of effects found in animals to humans is not considered to provide definitive evidence.
- There is difficulty in linking hormonal effects with specific pollutants.
- There is difficulty in correlating elevated PCB levels in blood or body fat with health effects, which are subtle.
- There is difficulty, in field studies, in distinguishing effects of exposure to PCBs vs. exposure to other bioactive contaminants or their degradation products.
- There is difficulty in equating occupational exposure (usually external) to dietary exposure (usually from eating contaminated fish).

Despite these and other shortcomings, a growing body of evidence points to potentially damaging effects of PCBs on human health. This evidence will be discussed further in Chapter 12, "Effects of Coastal Pollution on Public Health."

REFERENCES

Allen, J.R. and D.H. Norback. 1976. Pathobiological responses of primates to polychlorinated biphenyl exposure. In: National Conference on Polychlorinated Biphenyls, Chicago, IL. EPA Publ. No. 560/6-75-004, Environmental Protection Agency, Washington, DC.

Brouwer, A., M.P. Longnecker, L.S. Birnbaum, J. Cogliano, P. Kostyniak, J. Moore, S. Schantz, and G. Winneke. 1999. Characterization of potential endocrine-related health effects at low-dose levels of exposure to PCBs. *Environ. Health Perspect. 107:* 639–649.

Brown, D.P. and M. Jones. 1981. Mortality and industrial hygiene study of workers exposed to polychlorinated biphenyls. *Arch. Environ. Health. 36:* 120–129.

Chen, Y.-C. 1994. A 6-year follow-up of behavior and activity disorders in the Taiwan Yu-cheng children. *Am. J. Public. Health 84:* 415–421

Chen, Y.C., Y.L. Guo, C.C. Hsu, and W.J. Rogan. 1997. Cognitive development of Yu-cheng ("oil disease") children prenatally exposed to heat-degraded PCBs. *J.A.M.A. 268:* 3213–3218.

Cordle, F., R. Locke, and J. Springer. 1982. Risk assessment in a federal regulatory agency: An assessment of risk associated with the human consumption of some species of fish contaminated with polychlorinated biphenyls (PCBs). *Environ. Health Perspect. 45:* 171–182.

Dewailly, E., A. Nantel, J.-P. Weber, and F. Meyer. 1989. High levels of PCBs in breast milk of Inuit women from Arctic Quebec. *Bull. Environ. Contam. Toxicol. 43:* 641–646.

Flynn, L.T. 1997. Public health concerns about environmental polychlorinated biphenyls: A position paper of the American Council on Science and Health. Academic Press, New York.

Hsieh, S.-F., Y.-Y. Yen, S.-J. Lan, C.-C. Hsieh, C.-H. Lee, and Y.-C. Ko. 1996. A cohort study on mortality and exposure to polychlorinated biphenyls. *Arch. Environ. Health 51:* 417–424.

Humphrey, H.E.B., H.A. Price, and M.L. Budd. 1976. Evaluation of changes of the level of polychlorinated biphenyls (PCBs) in human tissue. Final Rep., FDA Contract No. 223-73-2209.

Jacobson, J.L. and S.W. Jacobson. 1996. Intellectual impairment in children exposed to polychlorinated biphenyls in utero. *N. Engl. J. Med. 335:* 783–789.

Kilburn, K. 2000. Visual and neurobehavioral impairment associated with polychlorinated biphenyls. *Neurotoxicology 21:* 489–499.

Kimbrough, R.D., R.A. Squire, R.E. Linder, J.D. Strandbest, J.J. Mondali, and V.W. Bubse. 1975. Induction of liver tumors in Sherman strain female rats by polychlorinated biphenyl Aroclor 1260. *J. Natl. Cancer Inst. 55:* 1453–1459.

Kimbrough, R.D. 1974. Toxicity of polychlorinated polycyclic compounds and related chemicals. *Crit. Rev. Toxicol. 2:* 445–448.

Kimbrough, R.D., M.L. Doemland, and M.E. LeVois. 1999. Mortality in male and female capacitor workers exposed to polychlorinated biphenyls. *J. Occup. Environ. Med. 41*(3): 161–171.

Krahn, M.M., D.G. Burrows, J.E. Stein, P.R. Becker, M.M. Schantz, D.C.G. Muir, T.M. O'Hara, and T. Rowles. 1999. White whales (*Delphinapterus leucas*) from three Alaskan stocks: Concentrations and patterns of persistent organochlorine contaminants in blubber. *J. Cetacean Res. Manage. 1:* 239–249.

Kuratsune, M. 1989. Yusho, with reference to Yu-cheng, pp. 381–400. In: R.D. Kimbrough and A.A. Jensen (eds.), *Halogenated Biphenyls, Terphenyls, Naphthalene Dibenzodioxins and Related Products*. Elsevier, Amsterdam.

Kuratsune, M., Y. Morikawa, T. Hirohata, M. Nishizumi, S. Kohchi, T. Yoshimura, J. Matsuzaki, A. Yamaguchi, N. Saruta, N. Ishinishi, E. Kunitake, O. Shimono, K. Takigawa, K. Oki, M. Sonoda, T. Ueda, and M. Ogata. 1969. An epidemiologic study on "Yu-sho" or chlorobiphenyls poisoning. *Fukuoka Acta Med. 60:* 513–532.

Lewis, R. 2001. Evaluating the evidence. *The Scientist 15*(6): 15–17.

Lü, Y.C. and P.N. Wong. 1984. Dermatological, medical and laboratory findings of patients in Taiwan and their treatments. *Am. J. Indus. Med. 5:* 81–117.

Porterfield, S.P. 1994. Vulnerability of the developing brain to thyroid abnormalities: Environmental insults to the thyroid system. *Environ. Health Perspect. 125* (Suppl. 2).

Reece, E.R. 1987. Ontogeny of immunity in Inuit infants. *Communication of 7th International Congress on Circumpolar Health. Arctic Med. Res. 45:* 62.

Reijnders, P.-J. 1986. Reproductive failure in common seals feeding on fish from polluted waters *Nature 324:* 456–431.

Rogan, W.J. 1989. Yu-cheng, pp. 401–415. In: R.D. Kimbrough and A.A. Jensen (eds.), *Halogenated Biphenyls, Terphenyls, Naphthalenes, Dibenzodioxins, and Related Products*. Elsevier, Amsterdam.

Rogan, W.J., B.C. Gladen, J.D. McKinney, N. Carreras, P. Hardy, J. Thullen, J. Tinglestad, and M. Tully. 1988. Congenital poisoning by polychlorinated biphenyls and their contaminants in Taiwan. *Science 241:* 334–336.

Weisglas-Kuperus, N., S. Patandin, G.A.M. Berbers, T.C.J. Sas, P.G.H. Mulder, P.J.J. Sauer, and H. Hoojkaas. 2000. Immunologic effects of background exposure to polychlorinated biphenyls and dioxins in Dutch preschool children. *Environ. Health Perspect. 108:* 1203–1207.

Yu, M.-L., Y.L. Guo, C.-C. Hsu, and W.J. Rogan. 2000. Menstruation and reproduction in women with polychlorinated biphenyl (PCB) poisoning: Long term follow-up interviews of the women from the Taiwan Yu-cheng cohort. *Int. J. Epidemiol. 29:* 672–677.

4 Microbial Pollution of Recreational Waters

A Perspective from the Beach at Sandy Hook, New Jersey

A significant part of my career in marine biology was spent at the Sandy Hook Marine Laboratory, a federal facility located on the New Jersey coast, on a sandspit pointed directly at the New York skyline, dimly visible through the almost daily haze of air pollutants there. The laboratory and its grounds were and are parts of a national seashore. People pay fees to use the park's badly despoiled beaches — desperate for some contact with even this sad example of what could be a superb public resource.

The primary objective of the federal EPA and the New Jersey Department of Environmental Protection was then and still is regulation — such as ensuring that recreational waters meet arbitrary standards if humans are to be allowed entry. Fecal contamination was and is a persistent problem in any recreational waters near large human populations, and the New York/New Jersey beaches are excellent examples of an almost overwhelming public health and regulatory problem.

*Despite efforts of local, state and federal regulatory agencies, things happen: a sewage treatment plant malfunctions and raw sewage sometimes mixed with medical wastes is dumped into shallow waters; a cruise ship illegally disposes of wastes near land; a retention dike of a waste collection pond of a pig production facility collapses and pollutes an adjacent estuary. These horrific events can combine with the usual continuing pressures on coastal environments to make inshore waters potentially dangerous temporary habitats for people. Microbial problems increase in proportion to increasing density of human populations in coastal zones, even with sincere efforts to alleviate them. Some relief can come from constant monitoring, restrictive regulations, and strict enforcement, but recreational use of coastal waters will probably always be accompanied by a certain degree of disease risk, despite the sensibility of an admonition that I used to give my kids in those early days at Sandy Hook: "Always keep your mouth closed while you are in the water!"**

From Field Notes of a Pollution Watcher
(C.J. Sindermann, 2000)

---

* During those early days at Sandy Hook, our kids, quite independently of any scientific advice, had developed their own identifier of the better places to swim in those degraded waters. They routinely looked for a so-called "scum line" off the beaches and spent their time in the clearer water beyond that line. They survived!

Enjoy the Beach, But Don't Go in the Water

I disclosed some details about my love–hate relationship with polluted beaches of the New York Bight in the prologue of this book [Field Notes of a Pollution Watcher]. The "sludge monster" — a media creation — was a figurative aspect of the environmental problem, but not the most serious part. In 1976 and 1988 inshore pollution was most often on the minds of coastal residents of New York and New Jersey, and in their newspapers, and on their television sets. Let me review more of the history of that dark period, and then consider one of the really important consequences of increasing environmental degradation.

Beginning in early summer of 1976, major ocean pollution events occurred. "Washups" of foreign material (sewage, tar balls, and other repulsive trash) became common just before the great bicentennial extravaganza on July 4, 1976, in New York Harbor (the federally-funded Job Corps was mobilized for cleanup duty just before the event was to take place).

An extensive and unusual algal bloom invaded the New York Bight, with accompanying widespread anoxia and mortalities of fish and shellfish on the New Jersey coast. [This event is discussed from a resource perspective in Chapter 6.]

For the next decade, sporadic but persistent shoreline pollution occurred, accompanied by frequent beach closures. In 1987, though, the frequency of shoreline "washup events" increased dramatically, reaching a peak in the following year. In the summer of 1988, contamination of New Jersey beaches with medical wastes, including syringes containing HIV-positive blood, caused a wave of concern and revulsion, and received nationwide publicity. The two largest national news magazines — Time and Newsweek — both ran extensive feature stories on the reality and consequences of ocean pollution. Public sensitivities were temporarily aroused, and vacationers avoided those fouled beaches (and all other area beaches) in great numbers. But public memory is transient; even though washups and beach closings have occurred since 1988, no prolonged public outcry has resulted. It seems that urban beach visitors have high tolerances for the presence of trash and other suspicious debris. Unless shoreline conditions become absolutely disgusting, they are considered a "normal" part of everyday environmental experiences in this much-abused overpopulated sector of the coastline.

From Field Notes of a Pollution Watcher
(C.J. Sindermann, 2000)

Public patience is most severely stressed, though, when reports appear about actual illnesses caused by recreational contact with polluted coastal waters — swimming, diving, surfing, snorkeling. Esthetic concerns are justifiable enough in their own right, but the added possibilities of sickness (gastrointestinal, eye, ear, or skin infections in particular) caused by merely entering those contaminated waters approach the totally unacceptable. This chapter contains a brief review of past and

current experiences with what is a genuine problem: diseases caused by environmental exposure to even marginally polluted inshore waters — with emphasis on
bacterial and viral infections that can result.

THE EMERGENCE OF KNOWLEDGE ABOUT RISKS OF
DISEASE FROM RECREATIONAL CONTACT WITH
POLLUTED COASTAL WATERS

Understanding of human diseases resulting from recreational use of polluted coastal
waters (for swimming and diving in particular) has been achieved remarkably
slowly, even though the subject should be important to the hordes of people who
put themselves at risk by participating in such activities. Definitive scientific information did not become available until the 1980s, when extensive studies of illnesses
among bathers at coastal beaches were reported from the U.S. east and gulf coasts,
and from several countries bordering the severely polluted Mediterranean Sea
(Cabelli 1983).

The history of scientific investigations of the role of coastal waters in transmitting
human diseases is a fascinating one of inadequately controlled studies and of regulation in the absence of detailed information. Two large-scale research programs
were instituted in the early 1950s, ostensibly to provide a better database for actions
to protect human health. One was conducted by the U.S. Public Health Service
(Stevenson 1953), and one — a massive investigation conducted during most of the
decade of the 1950s — by the British Public Health Laboratory Services (Moore
1959). The studies resulted in controversy that persisted until the 1970s. The U.S.
findings were that swimming resulted in a higher rate of illnesses, whereas the British
study concluded, amazingly, that "swimming in sewage-polluted seawater carried
only a negligible risk to health, even on beaches that are aesthetically unsatisfactory"
(a delightful British euphemism for the presence of "intact aggregates of fecal
material"; Moore, p. 465). Both the U.S. and the British investigations had flaws in
experimental design, statistical treatment, and analyses; the ensuing controversy led
to recognition of a need for a carefully planned large-scale global epidemiological
research program. That need was satisfied in the 1970s and 1980s by international
efforts focusing on correlating health regulations with densities of microbial indicators of fecal pollution.

Beginning in 1972, studies in the United States by Cabelli and his associates
(Cabelli et al. 1983) focused on Boston Harbor, Long Island beaches near New York
City, and Lake Pontchartrain near New Orleans. Principal findings were that the risk
of gastrointestinal disease associated with swimming in marine waters impacted
with municipal wastewaters is related to quality of the water as indexed by mean
enterococcus density. That risk was detectable even at low pollution levels.

Methodology built on beach surveys developed in the U.S. studies during the
1970s was used in Egypt in investigations of heavily polluted beaches near Alexandria. Disease rates in residents and visitors were compared. One report
(El-Sharkawi & Hassan 1982) cited by Shuval (1986) found a significant risk of
contracting typhoid fever from bathing in polluted seawater near the city, especially

among children (who had, presumably, a lower level of acquired immunity than that of members of the adult population).

With the impetus of consistent findings of public health dangers, these definitive early studies by Dr. Cabelli and his associates were broadened by subsequent investigations during the 1980s that included beaches in Israel, Spain, and France. The principal finding in Israel was a "significant excess of gastrointestinal symptoms in children under 4 years old, in waters with high densities of bacteria known as enterococci" (Fattal et al. 1987).

The Spanish study of 24 beaches in Malaga and Tarragona confirmed the existence of public health risks associated with swimming in coastal waters with poor microbial quality, but it identified the most frequent ailments as skin infections, followed by eye and ear infections. Intestinal disorders in that study area were not the principal public health concern (Mujeriego, Bravo, & Feliu 1982). At about the same time, a study of five beaches in France indicated higher prevalence of conjunctivitis and skin infections in bathers when compared with nonbathers, but it did not find marked differences between polluted and relatively nonpolluted beaches (Foulon et al. 1983).

An excellent review of all available data from this major international effort, published by the United Nations Environmental Program (Shuval 1986), concluded with the statement: "There is finally a vast amount of firm data providing strong evidence that bathing in sewage-polluted seawater ... can cause a significant excess of credible gastrointestinal disease" (p. 37).

Although more investigations are still needed on a worldwide scale (and are being conducted sporadically), there is, beginning with the general conclusion of Shuval, a reasonable core of information about health risks from recreational contact with polluted coastal waters. Some elements of that core can be found in these statements:

- Swimming in seawater carries increased risk of illness (eye, ear, nose, throat, and gastrointestinal ailments) in swimmers than nonswimmers — a result attributable to immersion during water recreation and not the quality of the water (Stevenson 1953).
- Increased risk of gastroenteritis is associated with swimming in waters polluted with wastewater — even at beaches that meet existing state and federal guidelines (Cabelli 1983).
- The most probable etiologic agents of gastrointestinal infections associated with bathing (in U.S. studies) are human rotaviruses and the Norwalk-like viruses (now also called noroviruses; Cabelli 1983).
- Typhoid fever was in earlier times the most severe bacterial disease associated with swimming in contaminated seawater. The relationship has been known for more than a century (probably beginning with a report in 1888 of an outbreak resulting from swimming in the heavily polluted Elbe River in Germany). Subsequent reports of outbreaks from England, United States, Australia, and Egypt, mostly during the first half of the 20th century but a few subsequent to that, implicated bathing in polluted waters (Reece 1909, Winslow & Moxon 1928, Snow 1959, Shuval 1974, Cabelli 1983).

- The classification of bathing waters depends principally on determining the numbers of coliform bacteria as indicators of quality — a monitoring tool first proposed in 1943 and now used (with some variations) in most states of the United States. Classification usually functions with the premise that total coliform densities less than 1000 per 100 ml are acceptable for swimming, as a general value that can be used to distinguish high-quality from low-quality recreational waters, and one that can provide an estimate of the degree of disease risk from pollution.
- During the past 2 decades, however, research has focused on finding a more reliable indicator of fecal contamination of bathing waters, and progress has been made in the use of enterococci as better bacterial indicators of fecal pollution — often in conjunction with total or fecal coliform counts (Godfree, Jones, & Kay 1990). Numbers of coliform bacteria in a sample still form the standard basis for water quality determinations, despite the fact that some studies (for example, Cabelli et al. 1983) disclose that densities of enterococci can serve as the best indicators of water quality, insofar as risk of gastrointestinal illnesses is concerned.

Looking more broadly at disease risks associated with recreational use of sewage-polluted coastal waters, we should keep in mind that humans, as terrestrial animals, are adapted to a life surrounded by air. When they invade polluted coastal waters, usually as swimmers or divers, they place themselves at risk from infections by a wide spectrum of waterborne microbial pathogens. Invasions may occur through pre-existing skin lesions, through upper body orifices (ear, nose, mouth), or through membranes of the eye. The most common illnesses resulting from exposure to polluted water are gastrointestinal — nausea, vomiting, diarrhea — followed at some distance by eye, ear, nose, and throat disorders.

Most gastrointestinal illnesses probably result from swallowing sewage-polluted seawater containing enteric viruses (especially noroviruses and rotaviruses) or members of several bacterial groups (especially the enterococci, whose densities have a strong relationship to prevalences of swimming associated diseases; Dufour 1984).

CURRENT LEVELS OF VIRAL DISEASE RISKS FROM RECREATIONAL CONTACTS WITH POLLUTED COASTAL WATERS

Coastal pollution studies in the 1980s and 1990s demonstrated the important role of viruses of human origin in causing environmentally related diseases of humans. For example, it has been estimated that the aquatic environment may be contaminated by more than 140 serotypes of viruses via wastewater effluents (Gantzer et al. 1998). Of these, hepatitis A virus, caliciviruses (noroviruses), adenoviruses, rotaviruses, and enteroviruses can cause public health problems, and epidemics have been traced to environmental occurrences of hepatitis A virus and noroviruses (Murphy et al. 1979; Bosch et al. 1991; Le Guyader et al. 1996; Sugieda, Nakajima, & Nakajima 1996; Nishida et al. 2003). Adenoviruses are ubiquitous and are common worldwide.

Some serotypes are potentially important waterborne viruses, as they are relatively resistant to sewage treatment. They have been identified as the second most common agents of gastroenteritis in children, next to rotaviruses, in many studies (Ko, Cromeans, & Sobsey 2003).

In spite of clear public health problems, the contamination indicators currently used for regulatory purposes (such as closure of recreational beaches) are based on tests for the presence and abundance of coliforms, fecal coliforms, and fecal streptococci. These indicators are dependent, obviously, on *bacteriological* findings, and a number of studies have shown that they do not provide adequate information about *viral* contamination (Geldenhuys & Pretorius 1989, Gantzer et al. 1998).

Research on human infections related to recreational contact with coastal waters has, since the 1980s, emphasized the role of viruses as indicators of fecal contamination. Using recently developed molecular techniques, it is now possible to determine risks of fecal virus pollution with increased precision. A procedure known as PCR (polymerase chain reaction), the amplification of viral DNA and cDNA, accompanied by reverse transcriptase RNA detection (RT-PCR), has indicated high prevalences of human viruses in coastal waters that would not have been detected by classical methods (Pina et al. 1998). Distributions of enteroviruses, adenoviruses, and hepatitis A viruses, using this technique, were compared with findings from cell culture of enteroviruses and with fecal coliform tests. Adenoviruses were more frequently detected throughout the year, suggesting that this viral group could be used as an index of human viruses in the environment. As a specific example, in a recent study in coastal waters of southern California, adenoviruses were detected by PCR in one-third of the beach locations sampled. As in other studies, bacterial indicators (total coliforms, fecal coliforms, and enterococci) did not correlate with the presence of adenoviruses (Jiang, Noble, & Chu 2001). Additionally, occurrences of bacteriophages of *Bacteroides fragilis* (indicators of fecal contamination), detected by PCR, correlated well with the presence of human viruses and have also been suggested as indicators of viral pollution (Gantzer et al. 1998).

Viruses of human origin — enteroviruses, noroviruses, adenoviruses, rotaviruses, and hepatitis A viruses — in fecally contaminated seawater can cause a variety of illnesses when ingested, even in low concentrations. The illnesses include epidemic vomiting and diarrhea, infectious hepatitis, paralysis, meningitis, respiratory disease, myocarditis, congenital heart anomalies, and eye infections (Pina et al. 1998). Unfortunately, the presence and numbers of fecal coliforms provide only poor correlations with numbers of viruses present (Geldenhuys & Pretorius 1989; Metcalf, Melnick, & Estes 1995).

All the fecal viruses discussed here — as environmental contaminants that can cause disease problems for bathers and divers in recreational areas — assume even greater significance when they are concentrated by molluscan shellfish, which are too often eaten raw or undercooked or poorly processed by misguided or risk-seeking humans. The greatest impacts of the enteric viruses are therefore caused by *consumption* rather than mere environmental exposure. Public health problems created by ingestion of fecally contaminated shellfish will be explored much further in Chapter 12. For purposes of the present chapter on recreational exposure to disease agents, microbial infections related to polluted water constitute a clear danger, with viruses currently in the forefront.

DISEASES OF HUMANS TRANSMITTED PASSIVELY BY MARINE FISH

Microorganisms that are opportunistically pathogenic to humans may be acquired by environmental exposure to contaminated seawater or by handling fish that carry the organisms (either passively or as infections). Humans particularly at risk are those with pre-existing skin lesions or with lacerations or puncture wounds resulting from contact with living or dead fish. A number of pathogens cluster beneath the commonly used term "fish handler's disease." Examples that have been described repeatedly in the scientific literature include *Erysipelothrix rhusiopathiae*, *Mycobacterium marinum*, and *Vibrio vulnificus*.

ERYSIPELOTHRIX RHUSIOPATHIAE (ALSO KNOWN AS E. INSIDIOSA)

Erysipelas has probably the longest history in this category of diseases of humans transmitted passively or accidentally by contact with marine fish. My first introduction to the disease was through an Australian report published over half a century ago (Sheard & Dicks 1949). This was a medical study of fish handlers from a tiny fishing community with the unforgettable name of Houtman's Abrolhos, on the west coast south of Perth. The disease was found to be common among dockside workers, usually as infections resulting from minor skin abrasions, punctures, or other lesions resulting from contact with fish spines.

The infections, which result in severe inflammatory responses, are caused by the halophilic bacterium *Erysipelothrix insidiosa* (also known as *E. rhusiopathiae*; Sneath, Abbott, & Cunliffe 1951; Langford & Hansen 1954). The organisms responsible have been isolated repeatedly from fish slime (Sheard & Dicks 1949; Wellmann 1950, 1957) but apparently are not pathogenic to the fish that carry them. Processing plant employees are often temporarily incapacitated by the infections, known as "erysipeloid or fish rose," which are particularly common after injury by spines of such fish as sea robins or redfish (Price & Bennett 1951).

MYCOBACTERIUM MARINUM

Skin granulomas in humans, due to infection by *M. marinum*, were first reported in 1954 (Linell & Norden 1954) and have been increasing since then — particularly reports of infections acquired in swimming pools (fresh and saltwater) or in tropical fish aquaria (Adams et al. 1970, Kelly 1976, Giavenni 1979). Scattered through the medical literature are reports of often extremely resistant infections with *M. marinum* in skin lesions. The lesions were either acquired in or were exposed to water containing the pathogens in aquaria, pools, or estuaries (Black, Rush-Munro, & Woods 1971; Zeligman 1972; Williams and Riordan 1973; Wilson 1976). In some cases, treatment has included surgery and even amputation, although other cases responded to chemotherapy and/or X-ray therapy — or healed spontaneously.

Mycobacterium marinum was first described and characterized by Aronson (1926). Dead and dying fish with mycobacterial infections (from the Philadelphia Aquarium) included Atlantic croaker (*Micropogon undulatus*) and black sea bass (*Centropristes striatus*). Later reviews of the disease, known as "fish tuberculosis,"* reported the disease, principally as visceral granulomatous lesions, in 120 species of marine and freshwater teleosts throughout the world (Vogel 1958, Nigrelli and Vogel 1963).

Mycobacteriosis at high prevalences was reported in the 1980s from wild and cultured striped bass (*Morone saxatilis*) sampled on the Pacific coast of the United States. Prevalences of up to 68% were reported in wild fish, and an epizootic among cultured juveniles killed half the population being reared (Sakanari, Reilly, & Moser 1983; Hedrick, McDowell, & Groff 1987). Most of the cultured fish had gross signs of systemic mycobacterial infections, with numerous nodules in the kidney, liver, and spleen. Similar infections have been reported in high prevalences (up to 76%) in resurgent (since 1990) striped bass populations of the Chesapeake Bay on the Atlantic coast. Mortality rates for the species have increased significantly since 1997, and chronic but eventually lethal infections with *M. marinum* have been postulated to be caused, at least in part, by the pathogen.

The news media, always ready to leap to weakly supported conclusions, have suggested that mycobacterial skin infections in recreational fishermen and party boat captains may be related to the high prevalences of infections in fish (see, for example, an article headlined "Watermen, Scientists Warn of 'Fish Handler's Disease,'" *Star Democrat*, Easton, Maryland, April 26, 2004).

The presence of *M. marinum* infections among many categories of "fish handlers" is real, and it seems to be a continuing problem for people who live on the Chesapeake Bay, according to the Curtis Hand Center in Baltimore, Maryland. A logical assumption might be that high infection levels of *M. marinum* in fish should result in increased populations of the potentially pathogenic bacterium in waters of the bay. However, no studies have been conducted that would demonstrate the validity of a quantitative relationship, to my knowledge.

Here, then, is a pathogen of fish — *M. marinum* — that can on occasion cause localized infection in humans. It can also enter symphytically in a broad range of habitats: freshwater and marine aquaria or pools, or open coastal/estuarine waters.†

VIBRIO VULNIFICUS WOUND INFECTIONS

Of the many bacteria in coastal waters that may become opportunistic pathogens of humans by infecting pre-existing skin lesions or invading lacerations or punctures resulting from contact with hard parts of marine animals — spines of crustaceans or of fish, most notably — few have greater lethality than *V. vulnificus*. This is a

---

* Tubercular lesions consist of a central core of necrotic tissue, acid-fast bacilli, and macrophages, surrounded by a fibrous capsule. Such lesions may be confined to a single location, or they may be dispersed throughout the viscera (especially in the liver, kidney, and spleen).

† Recently, a *Mycobacterium* pathogenic for striped bass in Chesapeake Bay has been proposed as a new species, *M. chesapeaki*, based on DNA sequencing data (Heckert et al. 2001). The isolate is biochemically similar to *M. marinum*, so its specific status requires further studies (as was pointed out by the authors).

widely distributed autochthonous marine bacterium, with effects on humans that are temperature dependent; infections occur when water temperatures exceed 15°C.

The mortality rate from wound infections with secondary septicemia is estimated at about 20%. However, *V. vulnificus* may also cause primary septicemia and death in immunocompromised humans who acquire infections from eating raw shellfish, oysters in particular. The mortality rate for this type of infection is high — an estimated 76%.

We will revisit *V. vulnificus* in more detail in Chapter 12, as the pathogen has now achieved status as the leading cause of death in the United States associated with consumption of seafood (DePaola et al. 2003). Let it be said here, though, that part of the ominous reputation of *V. vulnificus* is based on the consequences of wound infections, which can become systemic and often fatal in otherwise normal individuals as well as those with liver problems or immune deficiencies.

The problem of wound infections resulting from exposure to opportunistic microbial pathogens in coastal waters is not, of course, confined to these few examples that I have chosen. Many other genera of bacteria with invasive capabilities may produce infections in vulnerable humans who have contacts, recreational or otherwise, with coastal waters and their inhabitants. Pollution, especially that part of it that increases organic loading of coastal/estuarine environments, may act to augment the problem.

CONCLUSIONS

Microbial pollution of coastal/estuarine waters has become an ever-increasing problem during the past half century, as human populations on adjacent shores expand and as their effluents make corresponding impacts on the edges of the seas. Recreational use of shorelines (swimming and diving in particular) puts greater and greater numbers of people at risk from viral and bacterial pathogens, especially in waters near population concentrations. This obvious sequence of events has become established scientifically only by slow accretion of information, and, even today, uncertainties exist about the validity of indicators of the level of risk for any given coastal area.

Early concerns emphasized bacterial contaminants, especially those responsible for typhoid fever and gastrointestinal disorders. Bacterial indicators (coliforms and enterococci) are still, in the 21st century, standard tests for acceptability of recreational waters. In the past 2 decades, though, the relative importance of enteric viruses has become apparent. Epidemics of gastroenteritis have been traced to environmental exposure to adenoviruses, rotaviruses, and noroviruses — all of which have been found to be ubiquitous in polluted coastal waters. Hepatitis A virus has also been identified frequently in water samples taken near urban centers.

Enteric viruses are also transmitted to humans through ingestion of raw molluscan shellfish, and consumption rather than mere environmental exposure can be the principal route of infection. That aspect of exposure to microbial pathogens will be considered later, in Chapter 12.

In a separate category are those microorganisms that can be opportunistically pathogenic to humans and are transmitted by handling fish, during either recreation or commercial operations. Several bacterial agents are known to infect pre-existing lesions or can invade lacerations caused by contact with spines of fish, causing a condition known as "fish handler's disease." Infections are often resistant to treatment and, in the case of one bacterial pathogen (*Vibrio vulnificus*), may cause fatal septicemia.

It seems safe to conclude from the foregoing that recreational contact with polluted marine waters and their inhabitants is an exercise in risk-taking. A robust body of data now indicates that waters near population centers are particularly likely to contain bacteria and viruses that can infect innocent recreational users. Enjoy!

REFERENCES

Adams, R.M., J.S. Remington, J. Steinberg, and J.S. Seibert. 1970. Tropical fish aquariums: A source of *Mycobacterium marinum* infections resembling sporotrichosis. *J. Am. Med. Assoc. 211:* 457–461.

Aronson, J.D. 1926. Tuberculosis of cold blooded animals. *Am. Assoc. Adv. Sci. Symp. Ser. 1:* 80–86.

Black, H., F.M. Rush-Munro, and G. Woods. 1971. *Mycobacterium marinum* infections acquired from tropical fish tanks. *Aust. J. Dermatol. 12:* 155–162.

Bosch, A., F. Lucena, J.M. Diez, R. Gajardo, M. Blasi, and J. Jofre. 1991. Waterborne viruses associated with hepatitis outbreak. *Res. Technol. Manage. 3:* 80–83.

Cabelli, V.J. 1983. Health effects criteria for marine recreational waters. R&D Rep. No. EPA-600/1-80-031, U.S. Environmental Protection Agency, Research Triangle Park, NC.

Cabelli, V.J., A.P. Dufour, L.T. McCabe, and M.A. Levin. 1983. A marine recreational water quality criterion consistent with indicator concepts and risk analysis. *J. Water Pollut. Control Fed. 55:* 1306–1314.

DePaola, A., J.I. Nordstrom, A. Dalsgaard, A. Forslund, J. Oliver, T. Bates, K.L. Bourdage, and P.A. Gulig. 2003. Analysis of *Vibrio vulnificus* from market oysters and septicemia cases for virulence markers. *Appl. Environ. Microbiol. 69:* 4006–4011.

Dufour, A.P. 1984. Bacterial indicators of recreational water quality. *Can. J. Publ. Health 75:* 49–56.

El-Sharkawi, F. and M.N.E.R. Hassan. 1982. The relation between the state of pollution in Alexandria swimming beaches and the occurrence of typhoid among bathers. *Bull. High Inst. Publ. Health Alexandria 12:* 337–351.

Fattal, B., E. Peleg-Olevsky, T. Agurshy, and H.I. Shuval. 1987. The association between seawater pollution as measured by bacterial indicators and morbidity among bathers at Mediterranean beaches of Israel. *Chemosphere 16:* 565–570.

Foulon, G., J. Maurin, N. Quoi, and G. Martin-Boyer. 1983. Relationship between the microbial quality of bathing water and health effects. *Rev. Fr. Sci. Eau 2:* 127–143.

Gantzer, C., A. Maul, J.M. Audic, and L. Schwartzbrod. 1998. Detection of infectious enteroviruses, enterovirus genomes, somatic coliphages, and *Bacteroides fragilis* phages in treated wastewater. *Appl. Environ. Microbiol. 64:* 4307–4312.

Geldenhuys, J.C. and P.D. Pretorius. 1989. The occurrence of enteric viruses in polluted water, correlation to indicator organisms and factors influencing their numbers. *Water Sci. Technol. 21:* 105–109.

Giavenni, R. 1979. Alcuni aspetti zoonosicidelle microbatteriosi di origin ittica. *Riv. Ital. Piscic. Ittiopatol. 14:* 123–126.

Godfree, A., F. Jones, and D. Kay. 1990. Recreational water quality: The management of environmental health risks associated with sewage discharges. *Mar. Pollut. Bull. 21:* 414–422.

Heckert, R.A., S. Elankumaran, A. Milani, and A. Baya. 2001. Detection of a new *Mycobacterium* species in wild striped bass in the Chesapeake Bay. *J. Clin. Microbiol. 39:* 710–715.

Hedrick, R.P., T. McDowell, and J. Groff. 1987. Mycobacteriosis in cultured striped bass from California. *J. Wildl. Dis. 23:* 319–395.

Jiang, S., R. Noble, and W. Chu. 2001. Human adenoviruses and coliphages in urban run-off-impacted coastal waters of southern California. *Appl. Environ. Microbiol. 67:* 179–184.

Kelly, R. 1976. *Mycobacterium marinum* infection from a tropical fish tank: Treatment with trimethroprim and sulphamethoxazole. *Med. J. Aust. 2:* 681–682.

Ko, G., T.L. Cromeans, and M.D. Sobsey. 2003. Detection of infectious adenovirus in cell culture by mRNA reverse transcription-PCR. *Appl. Environ. Microbiol. 69:* 7377–7384.

Langford, G.C., Jr., and P.A. Hansen. 1954. The species of *Erysipelothrix*. Antonie van Leeuwenhoek. *J. Microbiol. Serol. 20:* 87–92.

Le Guyader, F., F.H. Neill, M.K. Estes, S.S. Monroe, T. Ando, and R.L. Atmer. 1996. Detection and analysis of a small round-structured virus strain in oysters implicated in an outbreak of acute gastroenteritis. *Appl. Environ. Microbiol. 62:* 4268–4272.

Linell, F. and A. Norden. 1954. A new acid-fast bacillus occurring in swimming pools and capable of producing skin lesions in humans. *Acta Tuberc. Scand., Suppl. 33:* 1–85.

Metcalf, T.G., J.L. Melnick, and M.K. Estes. 1995. Environmental virology: From detection of virus in sewage and water by isolation to identification by molecular biology, a trip of over 50 years. *Annu. Rev. Microbiol. 49:* 461–487.

Moore, B. 1959. Sewage contamination of coastal bathing waters in England and Wales: A bacteriological and epidemiological study. *J. Hyg. 57:* 435–469.

Mujeriego, R., J.M. Bravo, and M.T. Feliu. 1982. Recreation in coastal waters: Public health implications. *Vier J. Etud. Pollut.,* Cannes, CIESM, pp. 585–594.

Murphy, A.M., G.S. Grohamm, P.J. Christopher, W.A. Lopez, G.R. Davey, and R.H. Millsom. 1979. An Australia-wide outbreak of gastroenteritis from oysters caused by Norwalk virus. *Med. J. Aust. 2:* 329–333.

Nigrelli, R.F. and H. Vogel. 1963. Spontaneous tuberculosis in fishes and in other cold-blooded vertebrates with special reference to *Mycobacterium fortuitum* Cruz from fish and human lesions. *Zoologica (N.Y.) 48:* 131–144.

Nishida, T., H. Kimura, M. Saitoh, M. Shinohara, M. Kato, S. Fukuda, T. Munemura, T. Mikami, A. Kawamoto, M. Akiyama, Y. Kato, K. Nishi, K. Kozawa, and O. Nishio. 2003. Detection, quantitation, and phylogenetic analysis of noroviruses in Japanese oysters. *Appl. Environ. Microbiol. 69:* 5782–5786.

Pina, S., M. Puig, F. Lucena, J. Jofre, and R. Girones. 1998. Viral pollution in the environment and in shellfish: Human adenovirus detection by PCR as an index of human viruses. *Appl. Environ. Microbiol. 64:* 3376–3382.

Price, J.E.L. and W.E.J. Bennett. 1951. The erysipeloid of Rosenbach. *Br. Med. J. 2:* 1060–1063.

Reece, R.J. 1909. 38th annual report to local government board, 1908-9 supplement with report of medical officer for 1909-9, appendix A, No. 6: 90.

Sakanari, J.W., C.A. Reilly, and M. Moser. 1983. Tubercular lesions in Pacific coast populations of striped bass. *Trans. Am. Fish. Soc. 112:* 565–566.

Sheard, K. and H.G. Dicks. 1949. Skin lesions among fishermen at Houtman's Abrolhos, Western Australia, with an account of erysipeloid of Rosenbach. *Med. J. Aust. 2:* 352–354.

Shuval, H.I. 1974. The case for microbial standards for bathing beaches, pp. 95-101. In: A.L.H. Gamesson (ed.), Proceedings of an International Symposium on Discharge of Sewage from Sea Outfalls, London. Pergamon Press, Oxford.

Shuval, H.I. 1986. Thalassogenic diseases. UNEP Regional Seas Reports and Studies No. 79, 40 pp.

Sneath, P.H.A., J.D. Abbott, and A.C. Cunliffe. 1951. The bacteriology of erysipeloid. *Br. Med. J. 2:* 1063–1066.

Snow, D.J.R. 1959. Typhoid and city beaches, Western Australia. Report of the Commissioner of Public Health for 1958, p. 52.

Stevenson, A.H. 1953. Studies of bathing water quality and health. *Am. J. Publ. Health Assoc. 43:* 529–536.

Sugieda, M., K. Nakajima, and S. Nakajima. 1996. Outbreaks of Norwalk-like virus-associated gastroenteritis traced to shellfish: Coexistence of two genogroups in one specimen. *Epidemiol. Infect. 116:* 339–346.

Vogel, H. 1958. Mycobacteria from cold-blooded animals. *Am. Rev. Tuberc. Pulm. Dis. 77:* 823–838.

Wellmann, G. 1950. Pathogenität und Wachstum der auf Fischen vorkommenden Rotlaufbacterien. *Abhl. Fisch. 3:* 489.

Wellmann, G. 1957. Über die Ubiquität des Rotlauferregers (*Erysipelothrix rhusiopathiae*). *Z. Fisch. 6:* 191–193.

Williams, C.S. and D.C. Riordan. 1973. *Mycobacterium marinum* (atypical acid-fast bacillus) infections of the hand. *J. Bone Joint Surg. 55:* 1042–1050.

Wilson, J.W. 1976. Skin infections caused by inoculation of *Mycobacterium marinum* from aquariums. *Drum Croaker 16:* 39–42.

Winslow, C.E.A. and D. Moxon. 1928. Bacterial pollution of bathing beach waters in New Haven Harbor. *Am. J. Hyg. 8:* 299.

Zeligman, I. 1972. *Mycobacterium marinum* granuloma. *Arch. Dermatol. 106:* 26–31.

5 Harmful Algal Blooms in Coastal Waters

INTRODUCTION: ALGAL BLOOMS AND ALGAL TOXICITY

Algal blooms and algal toxicity are natural phenomena. Why, then, should they have a dominant position in a book concerned with coastal pollution? Two good reasons are: (1) augmentation of levels of nitrogen and phosphorous in coastal/estuarine waters — the basis for explosive population growth of planktonic algae — has been shown in many instances to be of human origin, and (2) human transport of toxic algal species to new habitats — with ships' ballast water and by other commercial practices — has been demonstrated and is undoubtedly of common occurrence.

The frequency of occurrence, areal extent, and intensity of algal blooms seems to be increasing on a global scale — a trend that would be expected as a consequence of human contributions of nutrient chemicals to coastal/estuarine waters and of human participation in disseminating alien species. Additionally, the list of types of algal toxins is gradually expanding, as is the geographic extent of reported toxic events and knowledge about the nature of the toxins produced.

Population explosions of planktonic unicellular algae — so-called "algal blooms" or "red tides" (even though many of them are not red) have been observed for centuries and have in some instances caused shellfish in areas such as Puget Sound and northern New England to become temporarily toxic to humans. Paralytic shellfish poisoning (PSP) is the best-known consequence of eating toxic bivalve molluscs, although several other types of poisoning have been described, and new ones are being identified. Not all blooms are toxic, but many that are not may still be harmful in a variety of ways — for example, by reducing light penetration, by reducing dissolved oxygen levels, by forming mucilaginous aggregates, or by interfering with respiration of fish. The overall perception is that many, if not most, algal blooms can be harmful, but not all harmful blooms are toxic.

To make some sense of this, we can create some artificial categories and give a few examples, recognizing that interest in algal blooms is usually stimulated by the danger of toxic effects on humans, usually from eating contaminated shellfish. However, as was pointed out in a recent excellent review (Shumway 1990), fish and shellfish (and other animals) may be affected severely by some of the algal toxins. I have identified, for descriptive purposes, several not–mutually exclusive categories that seem to encompass most but probably not all of the kinds of algal blooms that have been reported. Examples, in the form of vignettes from my long-term field notes, have been inserted occasionally in the following material, to bring some reality to the artificial subdivisions.

ALGAL TOXINS

Some algal species may produce biotoxins that affect humans as well as marine animals. Such toxin-producing organisms, when present in abundance, may be consumed by shellfish, plankton, and plankton-eating fish — sometimes causing fish mortalities and often rendering fish flesh toxic to humans and marine mammals. Five principal types of toxins have been described, based on their effects on humans. One (ciguatera) is found in tropical and subtropical fish flesh and is very toxic to humans; another, neurotoxic fish poisoning (NTP), is found in fish and shellfish; and three are found principally in shellfish: paralytic shellfish poisoning, diarrhetic shellfish poisoning (DSP), and amnesic shellfish poisoning (ASP).

CIGUATERA FISH POISONING

The causative organism of ciguatera fish poisoning (CFP; CTX) is *Gambierdiscus toxicus*, an epibenthic high-light-intolerant species of dinoflagellate associated with macroalgae (Yasumoto et al. 1977). It is common in shallow tropical and subtropical seas between latitudes 28°N and 28°S. Ciguatera is often considered to be the most common form of toxin-based seafood illness in the world (Fleming et al. 1998, 2000). Outstanding manifestations of ciguatera poisoning are neurological, with symptoms persisting for weeks, months, or years, resulting in disabilities and some-times fatalities.

The incidence of ciguatera poisoning appears to be rising in Florida, the Carib-bean, and the Pacific. The *Gambierdiscus* biotoxins pass up the food chain from herbivorous reef fish to larger carnivorous, commercially valuable species. Ciguatera poisoning was traditionally limited to tropical regions, but modern improvements in refrigeration and transport have augmented commercialization of tropical reef fish and increased the frequency of this type of fish poisoning among consumers in temperate regions.

Biomagnification of ciguatoxin at various trophic levels results in high concen-trations in species such as barracuda, groupers, and snappers, associated with coral reefs. Damage to living corals, caused by pollution, dredging, temperature increase, or other human activity, can encourage growth of the macroalgae that *Gambierdiscus* and other toxic microalgae use as substrates — leading eventually to an increase in fish toxicity (*Gambierdiscus* does not form pelagic blooms of motile individuals).

NEUROTOXIC FISH POISONING

Widespread, often seasonal toxic algal blooms can cause extensive fish mortalities, which may occur annually or at least sporadically. Probably the best documented are seasonal outbreaks of neurotoxic poisoning (NTP) of fish and shellfish in the eastern Gulf of Mexico and particularly on the west coast of Florida — beginning early in the 20th century and caused by blooms of the dinoflagellate *Gymnodinium* *breve* (now called *Karenia brevis;* Daugbjerg et al. 2000).* Aided by the Gulf Stream, this neurotoxin-producing organism, traditionally a problem only in Florida, caused closures of major shellfish harvesting areas in North Carolina and South Carolina in 1987 and 1988. Hundreds of Atlantic dolphin died in 1988, probably due to the neurotoxin present in fish consumed by the mammals.

So the neurotoxins may be consumed by shellfish, plankton, and plankton-eating fish, causing mortalities and rendering their flesh toxic to humans and marine mammals — and their toxins may also be *aerosolized* by wave action and onshore winds, to affect humans by still another route — as experienced recently in Florida.

Algal Toxins Make Unwelcome Landfall in Florida

A late afternoon cocktail party on Thanksgiving weekend 2002 at a beach-front house in Sebastian, Florida, came to an abrupt and unpleasant end when onshore winds carrying sea spray increased in intensity. One participant described the episode vividly as follows:

"It was thoroughly disagreeable. Everybody was coughing and sneez-ing. I had an almost instantaneous burning throat, and then began gagging with every breath. We all left the party quickly, but the aftereffects in my case involved more than malaise — I was unable to get out of bed the next day."

Up and down that section of Florida's east coast, many people — especially beachgoers and surfers — suffered from a variety of ocean-related ills — especially irritation of eyes and throat, coughing, and sneezing. The problem persisted for the rest of November and into December 2002, apparently caused by a bloom of <u>Karenia brevis</u> (aka <u>Gymnodinium breve</u>), a well-known neuro-toxic alga most common on the opposite (west) coast of Florida. Reports of oxygen depletion and fish kills were part of the story, with small inshore fish species most commonly involved. A beachfront employee (describing the local situation at Cocoa Beach) stated, "the water feels slimy, and sometimes there are dead fish in it."

From Field Notes of a Pollution Watcher
(C.J. Sindermann, 2003)

A problem that is possibly equal in intensity to effects of algal neurotoxins in southeastern U.S. waters is that of PSP in northeastern and northwestern Canada and United States, caused by blooms of dinoflagellates of the genus *Alexandrium.*

---

* The taxonomy of the microalgae has been in what can be labeled euphemistically as a dynamic state, especially during the past 2 decades. Some proposed changes in specific and generic designations have been accepted quickly, and others more slowly, resulting in a high degree of confusion on the part of the nonspecialist.

Intensive toxin monitoring has resulted in frequent closure of shellfisheries on these North American coasts for more than half a century. The toxin may also cause fish mortalities. Sea herring (*Clupea harengus*) were killed in large numbers in the Bay of Fundy in the late 1970s by feeding on zooplankton (cladocerans and pteropods) that had fed on the toxic algae. In a later experimental study, red sea bream (*Pagrus major*) larvae and juveniles were affected, and many died after feeding on plankton containing the toxins.

Poisoning by *Alexandrium* toxins was also suspected to be a cause of humpback whale mass mortalities (12 to 14 individuals) on Georges Bank off Massachusetts in 1987 and again in 2003 (Geraci et al. 1989, Pearson 2003). The speculation was made that the whales had been feeding on mackerel in which the toxin had been accumulated. (Paralytic shellfish poisoning and other shellfish-borne biotoxins will be explored further, but from a molluscan perspective, in the following section.)

SHELLFISH-BORNE BIOTOXINS

Shellfish have been identified as passive carriers in a number of outbreaks of human illnesses due to toxins of algal origin. The list begins with the well-known and widespread paralytic shellfish poisoning (PSP), which is a consequence of blooms, in temperate coastal waters, of toxic dinoflagellates of the genus *Alexandrium* (formerly *Gonyaulax*), and with neurotoxic shellfish poisoning (NSP), caused by blooms of *Karenia brevis* (formerly *Ptychodiscus brevis* and, before that, *Gymnodinium breve*), responsible also for massive fish kills in the Gulf of Mexico. Next in order of appearance is diarrhetic shellfish poisoning (DSP), caused by other genera of toxic dinoflagellates (*Dinophysis, Prorocentrum*, and others). This form of poisoning was first reported in Europe in 1961 (Korringa & Roskam 1961). It became important in the late 1970s, after outbreaks in Japan and Europe, and persisted as a significant problem in the 1980s (Kat 1987). In December 1987, a "new" toxin, domoic acid, which causes amnesic shellfish poisoning (ASP), was found in mussels from Prince Edward Island in the Gulf of St. Lawrence and was responsible for 129 cases of poisoning and 2 deaths in Canada (Pirquet 1988). Domoic acid can affect the brain and the nervous system of humans. It is concentrated during blooms, in the digestive gland of shellfish, especially mussels, and (during the Canadian outbreak) its origin was reported to be a persistent bloom of the diatom *Pseudo-nitzschia multiseries* in mussel culture areas of Prince Edward Island. Table 5.1 summarizes key information about these common types of shellfish poisoning, as well as ciguatera fish poisoning.

The history of changes in the status of the various fish and shellfish poisonings includes the following:

PSP — Recurrent PSP outbreaks now affect the states of Maine, New Hampshire, Massachusetts, Oregon, Washington, and Alaska. PSP problems constrain the development of a shellfish industry in Alaska. Offshore shellfish on Georges Bank off New England became toxic for the first time in 1989 and have remained toxic since then. Low levels of PSP have been found in Rhode Island, Connecticut, and New York. In 1987, 19 whales are thought to have died from PSP toxin contained in mackerel they had consumed — a significant event, since PSP was previously believed to be a problem principally in shellfish. Resting cysts of PSP-toxin–pro-

TABLE 5.1
Common Types of Poisoning by Biotoxins Acquired by Consuming Fish and Shellfish

Illness	Symptoms	Cause	Description	Useful References
Paralytic shellfish poisoning (PSP)	Numbness of lips, face, and extremities; visual disturbance; staggering gait; difficulty breathing; paralysis	Species of the dinoflagellate *Alexandrium* (formerly *Gonyaulax*)	About 12 forms purine-derived saxitoxins; water soluble; acts by blocking sodium channel needed to transmit nerve impulses	Bricelj & Shumway 1998
Neurotoxic shellfish poisoning (NSP)	Nonfatal but unpleasant neurotoxic symptoms; strong action on cardiovascular system	*Karenia brevis* (formerly *Gymnodinium breve*)	Toxins have unusual polycyclic ether skeletons	Schneider & Rodrick 1995; Fletcher, Hay, & Scott 1998, 2002
Diarrhetic shellfish poisoning (DSP)	Cramps; severe diarrhea; nausea; vomiting; chills; death rare	Species of dinoflagellates *Dinophysis* and *Prorocentrum* in particular	*Dinophysis* toxin and okadaic acid; large, fat-soluble polyethers; can move across cell membranes and make them "leak"	Bricelj et al. 1998; Gayoso, Dover, & Morton 2002
Amnesic shellfish poisoning (domoic acid; ASP)	Nausea; vomiting; muscle weakness; disorientation; loss of short-term memory	Species of the diatom genus *Pseudo-nitzschia*; toxin also found in species of the red alga *Chondria*	Nonessential amino acid; mimics glutamic acid; affects brain and nervous system; found in digestive glands of contaminated shellfish	Wohlgeschaffen et al. 1992; Whyte, Ginther, & Townsend 1995
Ciguatera fish poisoning (CFP)	Paralysis; respiratory failure; death	Dinoflagellate *Gambierdiscus toxicus* (other species possibly linked to ciguatera include *Amphidinium carterae; Coolia monotis; Ostreopsis* spp., *Prorocentrum* spp., and *Thecadinium* spp.)	Neurotoxic microalga associated with tropical/subtropical macroalgae	Yasumoto et al. 1980, Gillespie et al. 1985, Villareal & Morton 2002

Note: In addition to the "common" toxins described above, a number of "new" toxins have been reported. One group causes azaspiracid poisoning (AZP), which causes human illnesses in Europe and has been reported to result from mussel *Mytilus edulis* consumption. Other less clearly defined toxins include yessotoxins (YTX) and pectenotoxin-2 and analogues from Chilean mussels (Lewis 2000).

Source: Adapted in part from Pirquet, K.T. 1988. *Can. Aquac.-lt. 4(2):* 41-43, 46-67, 53.

ducing species from bottom sediments in New England waters were found to be 10 times more toxic than were motile stages. Cysts may be ingested by shellfish, causing toxicity even when blooms are not apparent.

DSP — Diarrhetic shellfish poisoning, with diarrhea as the dominant symptom, was first described in Europe in 1961 and subsequently reported elsewhere in Europe and in Japan. It must be noted, though, that shellfish-associated enteric disorders have a long history, and this type of poisoning may have been present much earlier but not diagnosed correctly. In 1990, the first confirmed outbreak of DSP in North America occurred in Canada. Two more outbreaks occurred in 1992. Scattered, unconfirmed cases of DSP have been positively reported in the United States, and the causative organisms have been positively identified in U.S. waters.

ASP — The toxin of amnesic shellfish poisoning (domoic acid) was detected in Nantucket scallops in 1990 and 1991. Toxic *Pseudo-nitzschia* species have been identified in the Gulf of Mexico, and seabird mortalities in the state of California in 1991 were linked to levels of the toxin found in the flesh of fish that had been consumed. Also in 1991, the ASP toxin occurred in the state of Washington, where contaminated clams and crabs caused human illnesses. The causative organisms have also been identified in U.S. Atlantic coastal waters.

PFIESTERIA — A TOXIC ALGAL PREDATOR

Some algal species may be toxic under certain environmental conditions, but not continuously so. A dinoflagellate, *Pfiesteria piscicida*, and its relatives are examples of such organisms. Examined with great interest since the early 1990s, *Pfiesteria*, with effects in U.S. east coast waters, has been found at times to be toxic to, and even predatory on, estuarine fish (Burkholder et al. 1992, Noga et al. 1993). Furthermore, the organism has been reported to be toxic to humans as a consequence of environmental exposure, causing skin and neural disorders (Burkholder et al. 1995).

The "Microbe from Hell"

It is a hot humid August morning in 1997 in the tiny fishing village of Shelltown, on the banks of the Pocomoke River — an insignificant waterway that forms the extreme southern boundary of Maryland's Eastern Shore. But some of the people on the street are not fishermen or even locals. They look disturbingly like alien invaders, with metallic-appearing full-body protective clothing and with respirators dangling around their necks. They are actually field technicians from the Department of Natural Resources, and they are here to investigate a strange and frightening series of events in this little tributary of Chesapeake Bay. Fish — especially menhaden — are dying in large numbers, and many of the surviving individuals have conspicuous ulcerations on their skins. But beyond this, and of much greater concern, people who had contacted the water near the fish kills have been reporting illnesses — disorientation, loss of recent memory, nausea, and skin lesions.

The fish kills are not that unusual; they have been occurring sporadically for most of the century in varying intensities and locations up and down the Atlantic coast from Long Island to Florida — to the extent that most coastal states have established fish kill reporting offices. But fish kills that involve simultaneous outbreaks of human disease have not been reported before, and this is a matter of great interest to public health officials and politicians (and, of course, to residents).

The culprit seems to be a single-celled aquatic organism discovered in 1988 and subsequently named <u>Pfiesteria piscicida</u>, a member of a mostly planktonic algal group known as dinoflagellates. This is no ordinary member of the group. It is an aggressive fish predator, and it is reported to secrete a toxin (or toxins) that immobilizes the fish, destroys portions of its skin, and kills it by disrupting its nervous system. Individual fish that escape the lethal effects of the toxin often display evidence of the encounter in the form of skin lesions that develop into deep penetrating ulcers.

Fish kills of this type have been common further south in Pamlico Sound, North Carolina, since the early 1980s, and the presumptive cause — <u>Pfiesteria</u> — has been studied there since the early 1990s. Over half of all the fish kills in those waters from 1991 to 1993 have been attributed to the toxic organism, and it has been identified in water and sediment samples from other Atlantic coastal states from Delaware to Alabama — as have high prevalences of skin ulcers in affected fish species.

Events in the Pocomoke River, a scenic two-hour drive from Washington, D.C., became a media focus in late summer of 1997. Scientists were interested because the reality of a dinoflagellate as an aggressive fish predator rather than a passive, occasionally toxic algal form, represented a paradigm shift in their thinking; public health administrators had to be involved because of claims of human illnesses associated with the fish deaths; watermen and seafood dealers had to face substantial consumer resistance and lost income because of unsubstantiated fears of toxin contamination of fish products; and politicians had to weigh their public positions and statements very carefully, because one of the underlying causes of the outbreak was thought to be excessive nutrient contamination from agriculture and from overloaded sewage treatment plants. What a bonanza for an alert news reporter! Portions of the Pocomoke River and two other suspect rivers were closed by the governor of Maryland; public meetings were held at least weekly on the Eastern Shore during that period of uncertainty; and the U.S. Congress rushed to prepare bills authorizing relatively huge sums (well in excess of 10 million dollars) for research and environmental monitoring. Most of the funding will go, of course, to the Centers for Disease Control in Atlanta, because of the human health threat (fish and environment are always "also-rans" in the quest for funds if human diseases are involved).

But then the onset of cool weather in the late autumn of 1997 seemed to reduce the activity of the "microbe from hell" (a newspaper headline writer's invented appellation) and the event was banished to the back pages of the "Washington Post" and the "Baltimore Sun" — with only occasional reawakening of interest when the various committees, commissions and investigative *groups formed at the peak of the excitement made their reports. Most of them pointed to excessive nutrient loading of rivers and sub-estuaries of Chesapeake Bay as a likely reason for the outbreak of <u>Pfiesteria</u>, although evidence was far from robust. All the reports encouraged more research, especially on the human disease implications, the nature of the toxins produced, and the environmental factors responsible for the proliferation of the toxic form of the organism (and possibly its close relatives).*

What we may be seeing in this small tributary of the Chesapeake Bay is a tiny segment of an expanding global problem — an increase in the frequency, intensity, and nature of harmful algal blooms.

From Field Notes of a Pollution Watcher
(C.J. Sindermann, 1998)

Federal funding from several agencies resulted from the intense media attention to the *Pfiesteria* outbreak in 1997. The possibility of human disabilities resulting from environmental exposure to toxins stimulated sizeable research grants from the National Institutes of Health and the Centers for Disease Control. However, as might be expected, blooms of *Pfiesteria* did not recur to any significant extent in the years 1998 to 2003, although the causative organism (and some close relatives) have been identified in estuarine waters of the middle Atlantic states as far south as the Florida border. Toxic episodes with associated fish ulcerations and mortalities have not been reported.

The availability of research funding has led to significant new information about *Pfiesteria piscicida*. The organism has been detected frequently in mid-Atlantic waters, but not in toxic form. A second but nontoxic species, *P. shumwayae*, has been recognized (Vogelbein et al. 2002). Nontoxic forms of the second species, *P. shumwayae*, were reported to attack and wound fish such as young menhaden, providing possible entry points for fungal spores, with subsequent development of ulcers. In other reports, the ulcers thought to be caused by exposure to the toxins of *P. piscicida* were considered to be caused (at least in part) by invasion of the fungus *Aphanomyces invadans*.

ALGAL BLOOMS AND AQUACULTURE

Algal blooms, toxic and nontoxic, in the vicinity of aquaculture production operations may cause fish mortalities. Farmed fish, especially Atlantic salmon, have been killed, often in large numbers, by such blooms. Mortalities of farmed salmon in Scotland in 1979 and 1982 were caused by blooms of species of *Olisthodiscus* or *Chattonella*. These algae had been identified earlier in connection with unusual blooms, but occurrences were rare and geographically restricted. Blooms of the naked dinoflagellate *Gyrodinium aureolum* were observed to cause mortalities of marine organisms in England in 1978, in Scottish salmon cages in 1980, and in wild as well as captive fish on the coast of Norway in 1981 and 1982. A large bloom of

Chrysochromulina polylepsis caused extensive mortalities in sea cages (Bruno, Dear, & Seaton 1989). Widespread blooms of this toxin-producing species occurred in the waters adjacent to a number of North European countries during much of the 1980s, seriously affecting salmon aquaculture (Vagn-Hansen 1989).

Some species of diatoms and spiny armored dinoflagellates may bloom in the vicinity of sea cages, where they can cause fish mortalities from excess mucus production and resulting suffocation. Mortalities of farmed salmon in the U.S. Pacific Northwest due to blooms of the diatom *Chaetoceros* and the chloromonad *Heterosigma* have been a serious impediment to the development of this industry.

Major destructive blooms have also occurred in many other parts of the world, with severe effects on aquaculture. Notable in this respect have been extensive and recurrent blooms (especially in the 1970s and 1980s) in parts of the Seto Inland Sea of Japan, which have affected yellowtail and sea bream production (Imai, Itakura, & Ito 1991). Other sporadic outbreaks have had impacts on mussel culture in Spain and Canada, and on bay scallop production in Long Island Sound (NY) waters.

MUCILAGINOUS ALGAE

Some kinds of algal blooms may be accompanied by extensive mucous aggregations that foul beaches and fishing nets and may cause bottom water anoxia, with accompanying mortalities of benthic animals. Such mucilaginous blooms have occurred in recent decades in the North Sea and the Adriatic Sea, caused respectively by the diatoms *Phaeocystis pouchetti* and *Skeletonema costatum*. Mucilaginous *Phaeocystis* blooms in North Sea coastal waters have caused so-called "foam banks" on the beaches of Germany, France, and the Netherlands. Five acute *Skeletonema* episodes have occurred in the northern Adriatic Sea (Croatian coast) since 1988. Each event was characterized by areas of loose gelatinous mucous aggregates consisting largely of extracellular polysaccharides. Other diatoms (for example, *Chaetocerus affinis*) and types of phytoplankton other than diatoms may produce polysaccharide exudates under the proper environmental conditions (Heil, Maranda, & Shimizu 1993).

Some of the chemistry of mucilaginous blooms has been elucidated by research done with *Skeletonema* in the Adriatic Sea (Thornton, Santillo, & Thake 1999). These investigators theorize that drought conditions result in lower river flow and limitation of nutrients and calcium in coastal waters. Mucilage is produced by the phytoplankton under those conditions and is stabilized to form a gel by contact and intermixing with higher calcium levels offshore — leading to formation of extensive aggregates, some of which move onto beaches or into shallow water, where fish and bottom-dwelling animals may be killed.

COASTAL/ESTUARINE AND OFFSHORE
ALGAL BLOOMS

Algal blooms of varying dimensions, persistence, and toxicity occur in ever-increasing frequency in coastal/estuarine waters, where they may cause hypoxia or anoxia, with fish and shellfish mortality, or may cause decline in submerged attached vegetation as a consequence of decreased light penetration. Here is a good example.

"Brown Tide" in Long Island Waters

The New York Air National Guard plane made its final crisis response flight for the month over the abnormally brown waters of Great South Bay on the outer coast of Long Island. It was late summer 1986 — the second year that bays on the Island had been discolored and choked by the massive growth of a planktonic microalgal species just recently identified as Aureococcus anophag- efferens, *an organism not previously known to cause blooms in that area of the coast. The findings from the day's survey were grim: current abundance of the toxic organism in the bay was 1,000,000,000 algal cells per liter, similar to what it had been all summer, resulting in severe reduction of light penetration of the water. This had caused significant reduction in eel grass abundance and distribution in Long Island bays, and profound disturbances in other components of the shallow water ecosystem.*

One of the animal species most affected by the algal bloom was the bay scallop Argopecten irradians, *the base for an important commercial shellfishery. In the summer of 1985, when the bloom began, most of the scallop larvae had died, resulting in a massive recruitment failure. New York scallop landings in that year were only 58% of the average for the preceding four years. Natural restocking was precluded by recurrence of the bloom in the summer of 1986, and its reappearance in some previously affected areas in 1987. The concurrent loss of critical eel grass habitat may serve to further inhibit reestablishment of the Long Island bay scallop fishery.*

The problem was not confined to Long Island waters. The same alga, Aureococcus anophagefferens, *bloomed from Narragansett Bay, Rhode Island, southward to Barnegat Bay, New Jersey, in 1985. It caused mortalities of mussels* Mytilus edulis *in excess of 95% in Narragansett Bay, and significant growth suppression in hard clams* Mercenaria mercenaria *in Long Island bays. Mats of dead eel grass littered the shores in New York and New Jersey.*

The problem did not disappear with the passage of time, either. A 1991 "NOAA News Bulletin" reported that an Aureococcus *bloom had reoccurred in eastern Long Island Sound in June of that year, with cell densities eight times that known to harm marine animals. Bay scallops larvae were again assumed to have been killed by the early stages of the bloom.*

Major brown tides occurred along the south Texas coast from 1990 through 1992, and on the New Jersey coast in 1999 and 2000.

A relationship of recurrent algal blooms such as these to modifications by humans of coastal/estuarine waters is often suggested, and some evidence exists. Nutrient enrichment from agricultural runoff, sewerage outfalls, and some industrial effluents may be involved, as may be the transport of toxic algae to new locations in ships' ballast water or with introduced marine animals. What-

ever the causation, there seems to be a real increase in the frequency, intensity, and geographic areas affected by algal blooms on a worldwide basis, and shellfish populations are among the impacted groups.

From Field Notes of a Pollution Watcher
(C.J. Sindermann, 1994)

Algal blooms such as these brown tides in estuaries usually get great attention from scientists and from the news media, because they are close at hand and present such a drastic change in the appearance of inshore waters. It is important to note, though, that massive offshore toxic or nontoxic blooms may occur and may be transported inshore by ocean currents. Decline of such blooms may result in extensive areas of hypoxia or anoxia in continental shelf waters, often with accompanying mass mortalities of shellfish and some fish species. An excellent example of this entire process can be found (logically) in the next chapter, on anoxia.

BLOOMS OF CYANOBACTERIA (BLUE-GREEN ALGAE) IN COASTAL WATERS

This discussion of harmful algal blooms in coastal waters would not be complete without some attention to the cyanobacteria or blue-green algae — best known as causes of toxicity and other problems in freshwater lakes and impoundments, but represented in marine waters as well (Falconer 1993). A few genera, such as *Nodularia*, *Trichodesmium*, *Ocillatoria*, *Schizothrix*, and *Aphanizomenon*, can occur in tropical and subtropical oceanic waters (of the Caribbean, for example; Hawser and Codd 1992), or in low-salinity coastal waters (Codd 1994, Kononen & Sellner 1995, Sellner 1997). Members of this taxonomic group (the Cyanobacteriales) produce two types of toxins that may affect humans: alkaloid neurotoxins resembling saxitoxin and neosaxitoxin (PSPs), and protein or peptide hepatotoxins (Sivonen et al. 1989). In severe exposures, the toxins can cause respiratory distress or liver failure within minutes or hours. Brief contact (by bathers, for example) can cause transient skin and eye irritations, allergic reactions, and gastroenteritis.

As with other phytoplankton species, the frequency and extent of occurrence of cyanobacterial blooms in brackish coastal waters, such as parts of the Baltic Sea, have increased dramatically in recent decades, probably as another consequence of eutrophication resulting from nutrient loading of human origin. One study, using satellite imagery, demonstrated a marked increase in surface coverage by cyanobacterial blooms (principally *Nodularia spumigena*) in the Baltic from 1982 to 1993, reaching over 62,000 km^2 in 1992 (Kahru, Horstmann, & Rud 1994).

The planktonic cyanobacteria and the occurrence of toxic blooms have a global distribution, with temperature optima for bloom formation from 15 to 30°C (Skulberg, Codd, & Carmichael 1984). Members of the group are of increasing concern as environmental risks to aquaculture operations, since fish mortalities — possibly toxin-related — have been reported frequently in coastal waters during blue-green algal blooms.

CONCLUSIONS

Examining the robust literature on harmful algal blooms leads us easily to a number of conclusions, especially to the overriding one — that *human activities*, such as nutrient enrichment of coastal/estuarine waters by agricultural runoff, sewage discharges, and industrial effluents, undoubtedly *enhance the likelihood of algal bloom formation and persistence*. For some of the outbreaks, the dominant algal species had been unknown in the affected area, or had been reported only rarely. This leads to the postulations that significant nutrient chemical changes had occurred, or that the toxic organisms may have been imported in ships' ballast water or attached to introduced shellfish species. This, of course, is a form of biological pollution, and some limited evidence for both methods of transfer has been reported from studies in Ireland and Australia (O'Mahony 1993; Hallegraeff & Bolch 1991, 1992).

Survival or death in the presence of toxic algal blooms has undoubtedly been part of the evolutionary history of coastal/estuarine animal species, and survival mechanisms have had to be developed on three levels: (1) the individual, which adapts or dies, such as the clam that closes its valves tightly when it first senses a toxin; (2) the population, which, after generations of exposure to toxins, consists mostly of individuals that have developed defense mechanisms and have modified reproductive strategies to counter the effects of toxins; and (3) the community, which is modified in its species composition and dominance by the selective pressures of the toxins (and by many other influences). What is different in the recent past is the frequency, severity, and widespread distribution of toxic blooms in coastal waters — combined with the likelihood that at least some are caused by nonindigenous organisms introduced by humans as consequences of commercial practices.

From an ecosystem perspective, it is possible to envision major shifts in the dominance of certain algal species, in which those of different sizes and with different nutrient requirements may supplant previous species assemblages. Such changes could affect feeding relationships of higher links in food webs, eventually even affecting the abundance of fish and shellfish. A good example of this process at work was seen during the 1985–1987 blooms of *Aureococcus anophagefferens* in Long Island bays. Hard-shell clams (*Mercenaria mercenaria*), which are filter feeders, showed evidence of starvation in the midst of plenty, probably because the bloom organisms were too small to be accepted as food, and because shell valves remained closed when toxins were present. Similarly, bay scallops (*Argopecten irradians*) showed a 76% reduction in adductor muscle weight compared with the year preceding the bloom (Cosper et al. 1987, Shumway 1988).

Beginning in the late 1980s, a new perception of mechanisms of algal toxicity has emerged, with the description of so-called "predatory algae" with a life cycle characterized by encysted forms that can be induced to bloom and produce toxins very quickly if fish are present in the vicinity. The toxins are rapidly lethal to fish, as determined by experimental exposures, and some of the life cycle stages of the algae (dinoflagellates) may actually attack the fish, even when they are in a nontoxic state (Smith, Noga, & Bullis 1988; Burkholder et al. 1992, 1993; Vogelbein et al. 2002). At least some (and possibly many) previously unexplained fish kills may have resulted from predation by such aggressive microorganisms.

Until the late 1980s, any discussion of human illness caused by environmental exposure to toxins and industrial toxicants would have been severely circumscribed — pretty much limited to anecdotal accounts of algal toxins in sea spray during blooms, causing bronchitis or eye irritation in local residents. But the story is changing. Laboratory exposures to toxin(s) from cultures of the dinoflagellate *Pfiesteria* in the late 1980s led to association of a sequence of human ills — asthmatic bronchitis, skin lesions, eye irritations, short-term memory loss — following unprotected contact with some cultured life history stages of the organism. Then, in 1997, the effects noted in laboratory workers were observed among fishermen, field technicians, and even a water skier who were in the vicinity of a *Pfiesteria* outbreak with accompanying fish mortalities, and who had contact with the water in a tiny tidal creek on Maryland's Eastern Shore. The toxic organism, along with its relatives, exists in other mid-Atlantic estuaries, especially in Pamlico Sound, North Carolina — but its most noteworthy effect to date is to cause sudden fish kills and skin lesions in survivors. There is some indication that a number of related species are present in those waters and those further south, as far as Florida.

These and other kinds of toxic or nontoxic algal blooms — if they continue to increase in scale, frequency, and diversity — may severely compromise or even eliminate some coastal aquaculture ventures. Concern has also been expressed about instances of greater duration of toxicity in natural populations of bivalve molluscs and the increasing costs of monitoring toxicity levels for public health purposes.

The paths to ecological distress and disturbance are becoming apparent: humans change nutrient balances in coastal/estuarine waters, transfer marine organisms promiscuously, and still expect to achieve continuing commercial harvests from severely stressed habitats and populations. The human hand writes large in coastal events that involve the very important primary producers — the planktonic microalgae that have been the subjects of this chapter — but the resulting script is flawed and incomplete, and the acquisition of understanding is too slow.

The role of anthropogenic nutrient loading of coastal waters in increasing the risks of biotoxin-induced human illnesses may be greater than present data will support, although there is some suggestive information available. At least three possible environmental situations exist in which nutrient loading could have an indirect effect on human health:

- Proliferation of known or unknown toxin-producing microalgae — including forms, such as *Pfiesteria piscicida*, that have neurotoxic capabilities — may be encouraged. A cause-and-effect relationship of proliferation of such forms with nutrient loading from agricultural sources has been proposed but needs further substantiation.
- It is possible that some microalgal species not known as toxin producers may become toxic if environmental nutrient concentrations are augmented from human sources (Smayda and Fosonoff 1989, Burkholder 1998).

- Proliferation of salinity-tolerant or salinity-requiring potentially pathogenic bacterial populations (*Vibrio* and *Aeromonas* in particular) may
occur in brackish-water habitats in which nutrient concentrations have
been increased from anthropogenic sources (Cabelli 1978).

REFERENCES

Note: The existing literature on harmful algal blooms is almost overwhelming.
Advances in understanding have been aided by a long and continuing series of
international conferences, beginning in the early 1970s (LoCicero 1975), each with
a published volume of technical papers. International organizations, such as the
Intergovernmental Oceanographic Commission (IOC) of UNESCO and others, have
recognized the global significance of harmful blooms, as have a plethora of regional
groups, such as the International Council for the Exploration of the Sea.

Toxic blooms have become important enough as a global problem to result in
the establishment by the United Nations of a newsletter titled "Harmful Algae News."
Almost predictably, a new scientific publication, *Journal of Natural Toxins*, has also
been established (in 1993), and a new international scientific organization was
formed (International Society for the Study of Harmful Algae) in 1999.

When all of this organizational attention is augmented by other edited books
and countless journal reports and reviews, the total literature becomes truly massive.
Some references relative to the discussion in this chapter follow.

Bricelj, V.M. and S.E. Shumway. 1998. An overview of the kinetics of PSP toxin transfer in
    bivalve molluscs, pp. 431–436. In: B. Reguera, J. Blanco, M.L. Ferandez, and T.
    Wyatt (eds.), *Harmful Algae*. Xunta de Galacia and Intergovernmental Oceanographic
    Commission of UNESCO, Vigo.
Bricelj, V.M., J.E. Ward, A.D. Cembella, and B.A. MacDonald. 1998. Monitoring toxic
    phytoplankton and paralytic shellfish poisoning (PSP) in mollusks in the St.
    Lawrence: 1989–1994. *J. Shellfish Res. 17*(1): 319–320.
Bruno, D.W., G. Dear, and D.D. Seaton. 1989. Mortality associated with phytoplankton
    blooms among farmed Atlantic salmon, *Salmo salar* L., in Scotland. *Aquaculture 78:*
    217–222.
Burkholder, J.M. 1998. Implications of harmful microalgae and heterotrophic dinoflagellates
    in management of sustainable marine fisheries. *Ecol. Appl. 8 (Suppl.):* S37–S62.
Burkholder, J.M., E.J. Noga, C.H. Hobbs, and H.B. Glasgow, Jr. 1992. New "phantom"
    dinoflagellate is the causative agent of major estuarine fish kills. *Nature 348:* 407–410.
Burkholder, J.M., H.B. Glasgow, E.J. Noga, and C.W. Hobbs. 1993. The role of a new toxic
    dinoflagellate in finfish and shellfish kills in the Neuse and Pamlico estuaries. Rep.
    No. 93-08, Albemarle-Pamlico Estuarine Study, North Carolina of Environment,
    Health & Natural Resources and U.S. Environmental Protection – National Estuary
    Program. 58 pp.
Burkholder, J.M., H.B. Glasgow, Jr., and K.A. Steidinger. 1995. Stage transformations in the
    complex life cycle of an ichthyotoxic "ambush-predator" dinoflagellate, pp. 567–572.
    In: P. Lassus, G. Arzul, E. Erard-Le Denn, P. Gentien, and C. Marcaillou-Le Baut
    (eds.), *Harmful Marine Algal Blooms*. Lavoisier Publishing, Paris.

Cabelli, V.J. 1978. Swimming associated disease outbreaks. *J. Water Pollut. Control Fed. 50:* 1374–1377.

Codd, G.A. 1994. Cyanobacterial (blue-green algal) toxins in marine and estuarine waters. *Scott. Assoc. Mar. Sci. Newsl. 9:* 7.

Cosper, E.M., W.C. Dennison, E.J. Carpenter, V.M. Bricelj, J.G. Mitchell, S.H. Kuenstner, D. Colflesh, and M. Dewey. 1987. Recurrent and persistent brown tide blooms perturb coastal marine ecosystem. *Estuaries 10:* 284–290.

Daugbjerg, N., G. Hansen, J. Larsen, and Ø. Moestrup. 2000. Phylogeny of some of the major genera of dinoflagellates based on ultrastructure and partial LSU rDNA sequence data, including the erection of three new genera of unarmoured dinoflagellates. *Phycologia 39:* 302–317.

Falconer, I.R., Editor. 1993. *Algal Toxins in Seafood and Drinking Water.* Academic Press, New York.

Fleming, L.E., D.G. Baden, J.A. Bean, R. Weisman, and D.G. Blythe. 1998. Seafood toxin diseases: Issues in epidemiology and community outreach, pp. 245–248. In: B. Reguera, J. Blanco, M.L. Ferandez, and T. Wyatt (eds.), *Harmful Algae.* Xunta de Galicia and Intergovernmental Oceanographic Commission of UNESCO, Vigo.

Fleming, L.E., D. Katz, J.A. Bean, and R. Hammond. 2000. Epidemiology of seafood poisoning, pp. 287–310. In: Y.H. Hui, D. Kitts, and P.S. Stanfield (eds.), *Seafood and Environmental Toxins. Foodborne Disease Handbook,* Vol. 4. Marcel Dekker, Monticello, NY.

Fletcher, G.C., B.E. Hay, and M.G. Scott. 1998. Detoxifying Pacific oysters (*Crassostrea gigas*) from the neurotoxic shellfish poison (NSP) produced by *Gynmnodinium breve. J. Shellfish Res. 17:* 1637.

Fletcher, G., B. Hay, and M. Scott. 2002. Reducing neurotoxic shellfish poison (NSP) in Pacific oysters (*Crassostrea gigas*) to levels below 20 mouse units · 100 G^{-1}. *J. Shellfish Res. 21:* 465–469.

Gayoso, A.M., S. Dover, and S. Morton. 2002. Diarrhetic shellfish poisoning associated with *Prorocentrum lima* (Dinophyceae) in Patagonian gulfs (Argentina). *J. Shellfish Res. 21:* 461–463.

Geraci, J.R., D.M. Anderson, R.J. Timperi, D.J. St. Aubin, G.A. Early, J.H. Prescott, and C.A. Mayo. 1989. Humpback whales (*Megaptera novaeangliae*) fatally poisoned by dinoflagellate toxin. *Can. J. Fish. Aquat. Sci. 46:* 1895–1898.

Gillespie, N.C., M.J. Holmes, J.B. Burke, and J. Daley. 1985. Distribution and periodicity of *Gambierdiscus toxicus* in Queensland, Australia, pp. 183–188. In: D.A. Anderson A. White, and D. Baden (eds.), *Toxic Dinoflagellates.* Elsevier, New York.

Hallegraeff, G.M. and C.J. Bolch. 1991. Transport of toxic dinoflagellate cysts via ships' ballast water. *Mar. Pollut. Bull. 22:* 27–30.

Hallegraeff, G.M. and C.J. Bolch. 1992. Transport of diatom and dinoflagellate resting spores in ships' ballast water: Implications for plankton biogeography and aquaculture. *J. Plankton Res. 14:* 1067–1084.

Hawser, S.P. and G.A. Codd. 1992. The toxicity of *Trichodesmium* blooms from Caribbean waters, pp. 319–330. In: E.J. Carpenter, D.G. Capone, and J.G. Reuter (eds.), *Marine Pelagic Cyanobacteria:* Trichodesmium *and Other Diazotrophs.* Kluwer, Dordrecht.

Heil, C.A., L. Maranda, and Y. Shimizu. 1993. Mucus-associated dinoflagellates: Large scale culturing and estimation of growth rate, pp. 501–506. In: T.J. Smayda and Y. Shimizu (eds.), *Toxic Phytoplankton Blooms in the Sea.* Elsevier, Amsterdam.

Imai, I., S. Itakura, and K. Itoh. 1991. Life cycle strategies of the red tide causing flagellates *Chattonella* (Raphidophyceae) in the Seto Inland Sea, pp. 165–170. In: *Environmental Management and Appropriate Use of Enclosed Coastal Seas — EMECS '90,* Kobe, Japan.

Kahru, M., U. Horstmann, and O. Rud. 1994. Satellite detection of increased cyanobacteria blooms in the Baltic Sea: Natural fluctuations or ecosystem change? *Ambio 23:* 469–472.

Kat, M. 1987. Diarrhetic mussel poisoning. Measures and consequences in the Netherlands. *Rapp. P.-V. Reun. Cons. Int. Explor. Mer 187:* 83–88.

Kononen, K. and K.G. Sellner. 1995. Toxic cyanobacteria blooms in marine, estuarine and coastal ecosystems, pp. 858–860. In: P. Lassus, G. Arzul, E. Erard, and C. Marcaillou (eds.), *Harmful Algal Blooms.* Lavoisier Intercept Ltd., London.

Korringa, P. and R.T. Roskam. 1961. An unusual case of mussel poisoning. *Int. Counc. Explor. Sea Doc. C.M.* 1961/41, 2 pp.

Lewis, R.J. 2000. Marine toxins. *Harmful Algae News* No. 20 (May), p. 5.

LoCicero, V.R., Editor. 1975. Proceedings of the First International Conference on Toxic Dinoflagellate Blooms. Massachusetts Science Technical Foundation, Wakefield, MA. 541 pp.

Noga, E.J., S.A. Smith, J.M. Burkholder, C. Hobbs, and R.A. Bullis. 1993. A new ichthyotoxic dinoflagellate: Cause of acute mortality in aquarium fishes. *Vet. Rec. 133:* 48–49.

O'Mahony, J.H.T. 1993. Phytoplankton species associated with imports of the Pacific oyster *Crassostrea gigas* from France to Ireland. *Int. Counc. Explor. Sea Doc. C.M.* 1992/F:26, 8 pp.

Pearson, H. 2003. Toxic algae suspected in whale deaths. *Nature Science Update* (October 16), 2 pp.

Pirquet, K.T. 1988. Poisonous secrets: Shellfish testing in Canada. *Can. Aquacult. 4*(2): 41–43, 46–67, 53.

Schneider, K.R. and G.E. Rodrick. 1995. The use of ozone to degrade *Gymnodinium breve* toxins, pp. 277–289. In: R. Poggi and J.-Y. Le Gall (eds.), *Shellfish Depuration.* Second International Conference on Shellfish Depuration, Rennes. IFREMER, Plouzan.

Sellner, K.G. 1997. Physiology, ecology, and toxic properties of marine cyanobacteria blooms. *Limnol. Oceanogr. 42:* 1089–1104.

Shumway, S.E., Editor. 1988. Toxic algal blooms: Hazards to shellfish industry. *J. Shellfish Res. 7:* 587–705.

Shumway, S.E. 1990. A review of the effects of algal blooms on shellfish and aquaculture. *J. World Aquacult. Soc. 21:* 65–104.

Sivonen, K., K. Kononen, W.W. Carmichael, A.M. Dahlem, K.L. Rinehart, J. Kiviranta, and S.I. Niemelä. 1989. Occurrence of the hepatotoxic cyanobacterium *Nodularia spumigena* in the Baltic Sea and structure of the toxin. *Appl. Environ. Microbiol. 55:* 1990–1995.

Skulberg, O.M., G.A. Codd, and W.W. Carmichael. 1984. Toxic blue green algal blooms in Europe: A growing problem. *Ambio 13:* 244–247.

Smayda, T.J. 1990. Novel and nuisance phytoplankton blooms in the sea: Evidence for a global epidemic, pp. 29–40. In: E. Granéli, B. Sundström, L. Edler, and D.M. Anderson (eds.), *Toxic Marine Phytoplankton.* Elsevier, New York.

Smayda, T.J. 1992. Global epidemic of noxious phytoplankton blooms and food chain consequences in large ecosystems, pp. 275–307. In: K. Sherman, L.M. Alexander, and B.D. Gold (eds.), *Food Chains, Models and Management of Large Marine Ecosystems.* Westview Press, San Francisco.

Smayda, T.J. and P. Fosonoff. 1989. An extraordinary, noxious brown-tide in Narragansett Bay. II. Inimical effects, pp. 133–136. In: T. Okaichi, D.M. Anderson, and T. Nemoto (eds.), *Red Tides: Biology, Environmental Science, and Toxicology.* Elsevier, New York.

Smith, S.A., E.J. Noga, and R.A. Bullis. 1988. Mortality in *Tilapia aurea* due to a toxic dinoflagellate bloom, pp. 167–168. In: *Proceedings of 3rd International Colloquium on Pathology in Marine Aquaculture*. European Association of Fish Pathologists, Aberdeen, Scotland.

Thornton, D.C.O., D. Santillo, and B. Thake. 1999. Prediction of sporadic mucilaginous algal blooms in the northern Adriatic Sea. *Mar. Pollut. Bull. 38:* 891–898.

Vagn-Hansen, K. 1989. Toxic algae in Danish waters, pp. 21–29. In: Report of the ICES Working Group on Harmful Effects of Algal Blooms on Mariculture and Marine Fisheries. *Int. Counc. Explor. Sea Doc. C.M.* 1989/F:18, 80 pp.

Villareal, T.A. and S.L. Morton. 2002. Use of cell-specific PAM-fluorometry to characterize host shading in the epiphytic dinoflagellate *Gambierdiscus toxicus. Mar. Ecol. 23:* 127–140.

Vogelbein, W.K., V.J. Lovko, J.D. Shields, K.S. Reece, P.L. Mason, L.W. Haas, and C.C. Walker. 2002. *Pfiesteria shumwayae* kills fish by micropredation not exotoxin secretion. *Nature 418:* 967–970.

Whyte, J.N.C., N.G. Ginther, and L.D. Townsend. 1995. Accumulation and depuration of domoic acid by the mussel, *Mytilus californianus*, pp. 531–537. In: P. Lassus, G. Arzul, E. Erard-LeDenn, P. Gentien, and C. Marcaillou-Le Baut (eds.), *Harmful Marine Algal Blooms*. Proceedings of the Sixth International Conference on Toxic Marine Phytoplankton, Nantes, France. Lavoiser, London.

Wohlgeschaffen, G.D., K.H. Mann, D.V. Subba Rao, and R. Pocklington. 1992. Dynamics of the phycotoxin domoic acid: Accumulation and excretion in two commercially important bivalves. *J. Appl. Phycol. 4:* 297–310.

Yasumoto, T., I. Nakajima, E. Chunque, and R. Adachi. 1977. Finding of a dinoflagellate as a likely culprit of ciguatera. *Bull. Jpn. Soc. Sci. Fish. 43:* 1021–1026.

Yasumoto, T., A. Inoue, T. Ochi, Y. Oshima, Y. Fukuyo, R. Adachi, R. Bagnis, and K. Fujimoto. 1980. Environmental studies on a toxic dinoflagellate (*Gambierdiscus toxicus*) responsible for ciguatera. *Bull. Jpn. Soc. Sci. Fish. 46:* 1397–1404.

6 Anoxia in Coastal Waters

The Day of the Tall Ships

The afternoon sun was bright on the masts and riggings of an aggregation of major sailing ships that will probably never again be duplicated. It was July 3, 1976, the eve of the bicentennial celebration of the Declaration of Independence of the United States of America, and tall ships from all the nations of the world with seafaring traditions had assembled in Sandy Hook Bay, within sight of the skyline of New York City — awaiting the grand procession past the Statue of Liberty scheduled for the next morning.

Some of those magnificent vessels were anchored for the night within a stone's throw of the Sandy Hook Laboratory, a unit of the National Marine Fisheries Service, with a large program in ocean pollution research. It was the weekend, and most of the laboratory staff members and their families had gathered on the lawns and the broad porches of the early 20th century buildings occupied by the agency, to picnic, drink beer, and watch this once-in-a-lifetime spectacle. Research problems were far away, and for once the almost inevitable "shop talk" indulged in by scientists was muted. The afternoon passed pleasantly, and the significance of the celebration was duly noted by the good federal employees of the lab. Early the next morning, sails were unfurled and the ships sailed away on the outgoing tide to their rendezvous in the lower Hudson River, where the country's President and an array of dignitaries waited to greet them.

As if to reinforce these human events of epic proportions that were taking place in the harbor, a massive oceanic event was developing simultaneously in continental shelf waters off the Middle Atlantic States. A gigantic bloom of algae (the armored dinoflagellate Ceratium tripos; see Figure 6.1), extending at its minimum along more than 100 miles of coastline from Long Island to Delaware, and as far as 50 miles seaward, was beginning to affect and to kill marine life. First reports of mass mortalities of fish and shellfish were made by fishermen and scuba divers on the Fourth of July, just as the speeches were being given, and the numbers and intensity of such reports increased during the following weeks. Observers described sea bottoms littered with dead lobsters, crabs, and fish, and complete destruction of clam beds — all occurring in an extensive abnormally turbid water mass that at times contained no oxygen at all. Surveys by laboratory vessels were greatly intensified, and for all of the rest of that bicentennial summer of 1976 the efforts of the Sandy Hook research group were focused almost exclusively on documenting and trying to interpret the ongoing large scale oceanographic phenomenon — a huge algal bloom with resulting severe oxygen depletion, high sea water temperatures, abnormally diminished

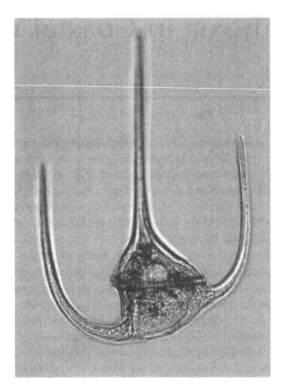

FIGURE 6.1 *Ceratium tripos,* an armored dinoflagellate that produced an extensive and destructive bloom in the New York Bight in 1976 (original magnification 160×). (Photograph courtesy of P. Falkowski and W. Marin, Brookhaven National Laboratory.)

current patterns, and nutrient loading from terrestrial (human) sources — that had probably contributed jointly to the formation of a persistent and widespread zone of death for marine animals.

From our more or less continuous surveys, we estimated that at least 60% of the surf clam Spisula solidissima population off the central New Jersey coast — some 147,000 tons — had been destroyed, with significant but lesser mortalities of ocean quahogs and sea scallops. Lobster catches were reduced by almost 50% during the period (Swanson and Sindermann 1979).

Mortalities of reef-dwelling fish were observed; estuarine species did not perform their usual offshore migrations, and coastal migratory species such as bluefish made a major seaward diversion around the anoxic zone.

By October, the algal bloom had dissipated, the mortalities ceased, the water column was reoxygenated, and denuded coastal waters were being repopulated by immigrants from unaffected areas or by larvae spawned outside the killing zone. The new set of surf clams (one of the principal commercial species most severely affected by mortalities) was excellent, and stocks rebuilt rapidly in the ensuing several years. The tall ships went on to exotic destinations elsewhere in the world; the bicentennial year passed into history; the Sandy Hook scientists went on with their surveillance, but without the earlier crisis motivation; and the perturbed environment gradually moved toward a state resembling — but not identical to — its preexisting one.

From Field Notes of a Pollution Watcher
(C.J. Sindermann, 1980)

This vignette provides an excellent example of the role that periodic or occasional oxygen depletion can play in coastal/estuarine environments. Dissolved oxygen, critical to animal life in the sea, is in a dynamic state, subject to daily, seasonal, and longer term fluctuations, and influenced by temperature, depth, nutrient levels, currents, and abundance of microbial populations. Oxygen saturation levels in coastal/estuarine waters can reach ±10 to 12 mg/l, but this amount varies widely, being dependent on the variables listed above, as well as others such as amounts of organic material present, intensity of photosynthetic activity, and rate of bacterial decomposition.

Oxygen depletion can result in hypoxia (a term generally indicating any amount of oxygen below saturation level, but often limited to levels <50% oxygen saturation) or anoxia (defined as absence of oxygen). Resource species (fish and shellfish) vary widely in responses to hypoxia, but in general they exhibit stress responses when subjected to oxygen levels <50% saturation (about 4 mg/l) and begin to die at levels of 2 mg/l (24% saturation at 20°C) or less. At oxygen concentrations <50%, "some fish species begin to show avoidance, reduced growth, and other signs of physiological stress" (Breitburg 2002). Exposure to anoxic water is rapidly fatal to resource and other species, as is the accompanying phenomenon of sulfide production in anoxic zones.

Anoxia can be a natural condition, especially in some coastal/estuarine waters with limited circulation (fjords, enclosed or semienclosed seas, estuaries with restricted tidal currents). Some bottom waters (for example, parts of the Black Sea and Baltic Sea) are permanently anoxic, whereas others (for example, the deeper parts of the Chesapeake Bay and parts of western Long Island Sound) are seasonally or sporadically anoxic.

Important to the present discussion of pollution effects in the observation that hypoxic/anoxic zones in the world's oceans are increasing in occurrence, extent, duration, and severity. The seasonal extent of hypoxic/anoxic bottom water in coastal waters (estuaries and semienclosed seas) is increasing, presumably as a consequence of ever-greater inputs of organic material of anthropogenic origin, or of increased levels of nitrate and phosphate from agricultural and other industrial sources. A good example of such changes can be seen in the deep channel of the Chesapeake Bay, where hypoxia has been observed to increase in area with the passing decades since 1950. Hypoxic zones have been characterized as "one of the most widespread and accelerating human-induced deleterious impacts on the world's marine environments" (Rabalais et al. 1999). It has been further observed that "hypoxia related to human activities currently threatens many of the major coastal embayments and estuaries in the world (Diaz & Rosenberg 1995).

Hypoxic/anoxic conditions in the sea have a certain morbid fascination for ocean scientists — probably because anoxia represents an extreme environmental change that can be rapidly lethal to resource species and to all other marine animals, and partly because its causation can be described satisfactorily using present ecological concepts. Continuing scientific examination of a number of anoxic zones has resulted in an expanding body of literature — some of it spanning decades. Favorite long-term study sites have included the anoxic zones in the Baltic Sea, with their seasonal and annual fluctuations, influenced significantly by North Sea currents, and the permanently anoxic zones of the Black Sea. Recent favorite sites in U.S. coastal waters include the seasonally and annually variable anoxic zones of the Chesapeake Bay, and an at times remarkably extensive area of hypoxia in the northern Gulf of Mexico, extending westward from Louisiana to Texas.

Since each of these anoxic zones has unique features as well as commonalities, and since coastal pollution can be identified as a partial contributor to increasing severity of the phenomenon, we should examine each of the selected problem areas to some modest extent at this point.

BALTIC SEA ANOXIA

The Baltic Sea can be described as a semienclosed brackish sea, connected to, and with periodic inflow from, the saltier North Sea. It is permanently stratified, with the saline North Sea water forming a deep poorly oxygenated layer. Before the 1940s, the Baltic was a nutrient-poor body of water, with oxygen depletion confined to sporadic episodes in the deepest troughs (such as the Gotland Deep). Since 1940, increasing eutrophication has resulted in episodic severe oxygen depletion (Jansson & Dahlberg 1999). Nutrients, especially nitrogen and phosphorous, have been added to the Baltic from excessive chemical fertilization of crops, burning fossil fuels, and urban or industrial effluents. Recent levels of nitrogen have been estimated to be three times the 1940 quantities, and phosphorous about five times their 1940 levels (Wulff, Stigebrandt, & Rahm 1990).

Severe oxygen depletion, sometimes accompanied by anoxia and sulfide formation, has occurred during recent decades and has extended intermittently over 70,000 km^2 (one-third) of the total bottom area of the Baltic proper.

BLACK SEA ANOXIA

A floating international symposium on "The Black Sea in Crisis" took place on a ship, the *El Venezelos* (circumnavigating that dreadful body of water), late in 1997. More than 400 participants, many of them scientists, viewed and reviewed the disastrous environmental conditions surrounding them and suggested remedies. Natural and man-made horrors of this much-abused sea include the following:

- Approximately 90 percent of the volume of the Black Sea is permanently anoxic — a result of depth, strong salinity stratification, and extremely limited exchange of bottom waters with more highly oxygenated systems (Breitburg 2002). The time period for water and pollutant exchange through the connecting Bosporus Strait is an amazing *140 to 160 yr* (compared with — for example — 1 to 4 yr for the North Sea exchange rate).

- The Black Sea is and has been the recipient of riverborne wastes (fertilizers, raw sewage, heavy metals, petrochemicals, and radioactive byproducts) derived from *22 countries* (six of which border on the sea; Anonymous 1997).
- Twenty of the 26 commercial fish species have become commercially extinct in recent decades.
- Eutrophication has compressed the oxygenated zone of the sea, reducing habitat for remaining fish and invertebrates.
- Populations of an introduced comb jelly, the ctenophore *Mnemiopsis leidyi*, increased enormously beginning in 1982, reaching an estimated 95% of the Black Sea biomass during the late 1980s and early 1990s (Platt 1995).

Proposed remedies for some of the man-made damage — mostly beyond the capabilities of countries contributing to the problems — included the standard ones of reducing toxic industrial contamination and reducing nutrient loading from sewage and agriculture (Zaitsev 1991, 1993). Some of the more flagrant abuses — continuing industrial pollution, oil discharges from shipping and spills, overfishing of remaining stocks, mineral exploration, and alteration of river discharges — seem amenable to international abatement efforts, but the future of the Black Sea seems, well, very black.

CHESAPEAKE BAY ANOXIA

The deep channel of the Chesapeake Bay was observed to become anoxic in summer as long ago as 1936. Its extent has varied annually, but the trend has been an increase in area and duration, especially since 1950, with a shift from sporadic to consistent anoxia in summer by 1970 (Officer et al. 1984). Increase in oxygen demand for benthic respiration has been the primary causative factor. The effects of expansion of hypoxic and anoxic zones on resource species can be profound, sometimes resulting in total mortalities of benthic forms in anoxic areas, and in 50% or more mortalities when oxygen declines to <2.0 mg/l. Abundance of bottom-feeding fish, such as croaker (*Micropagonias undulatus*) and spot (*Leiostomus xanthurus*), may be affected negatively, whereas plankton-feeding species such as menhaden (*Brevoortia tyrannus*) may be less severely affected. Extensive summer oxygen depletion may, therefore, be an important factor in reducing the ability of the bay to support resource species (Breitburg 1990).

Associated with areal expansion of hypoxic/anoxic deeper water in Chesapeake Bay is sporadic intrusion of that oxygen-depleted bottom water into nearshore habitats under the influence of wind forcing and tidal cycles. Mobile species such as most fish and blue crabs can move away from the disturbance, but less mobile or immobile benthic species may be stressed or killed (Breitburg 1992). With recent expansion of hypoxic/anoxic bottom waters of the bay, partly as a consequence of increased anthropogenic nutrient loading, the frequency and severity of such nearshore intrusions will undoubtedly increase correspondingly (Smith, Leffler, & Mackiernan 1992).

HYPOXIA AND ANOXIA IN THE NORTHERN GULF OF MEXICO

The largest known zone of oxygen-depleted waters in the western Atlantic occurs in summer along the inner shelf of the northern Gulf of Mexico from Louisiana to Texas. First noted in the early 1970s, the extent of hypoxic/anoxic bottom water has increased in recent years, from 8000 to 9000 km^2 in 1985–1992, to 16,000 to 18,000 km^2 in 1993–1997 (Rabalais et al. 1991, 1996, 1998, 1999; Rabalais, Turner, & Scavia 2002). Patterns and extent of depletion are correlated with river discharges and nutrient fluxes (Justic et al. 1993; Justic, Rabalais, & Turner 2002). Nitrogen (principally from agricultural practices) is the principal nutrient in excess, and its export by the Mississippi River system has increased between twofold and sevenfold during the past century.

Possible effects of the hypoxic/anoxic zone on Gulf of Mexico fisheries include the following:

- Reduced food for fish and shrimp in the affected waters (Rabalais et al. 1995)
- Reduced abundance of fish and shrimp in the affected waters (Pavela, Ross, & Chittenden 1983)
- Decline in shrimp catches since the affected area has expanded (Zimmerman, Nance, & Williams 1997)
- Loss of potential production due to interference with migration of juvenile shrimp to offshore waters (Nance, Martinez, & Klima 1994; Zimmerman & Nance 2001)

Modeling studies indicate that hypoxia/anoxia in the northern Gulf of Mexico could be alleviated by reduction of nutrient loading — presumably by better management (nutrient abatement) of the entire Mississippi drainage area. This will be a herculean task requiring decades, but one that is feasible if adequately funded. A recent study, using three numerical models of varying complexity, indicated that a 40 to 45% reduction (0.64 to 0.72 × 10⁶ tons) in annual nitrogen loads would be needed to limit the hypoxic area to <5000 km^2 (Scavia, Justic, & Bierman 2004). This was a goal set by a task force formed to reduce, mitigate, and control hypoxia in the northern Gulf of Mexico (CENR 2000).

CONCLUSIONS

What is emerging, as some control over the dumping of toxic industrial wastes into coastal/estuarine waters is being achieved, is an appreciation for the destructive effects of nutrient loading (especially of nitrogen and phosphorus), resulting in the occurrence of massive algal blooms — some of them toxic — and subsequent oxygen depletion, with formation of anoxic zones as blooms decay. The nutrients may come from terrestrial application of manure and chemical fertilizers, sewage plant effluents, certain industrial wastes, and runoff from urban areas. They supply elements essential for growth of algal populations which, when they die and decompose, use oxygen and form hypoxic or anoxic zones at or near the bottom.

Some consequences of hypoxia/anoxia to living resources in coastal/estuarine waters include the following:

- *Exclusion*, temporary or permanent, of sedentary species of shellfish and fish
- *Change* (usually temporary) in fish migration patterns
- *Mortalities* of many sessile benthic invertebrates during acute anoxic events, often accompanied by fish mortalities
- *Reduced species diversity* in zones subject to repeated anoxic events
- *Habitat loss* and avoidance of hypoxic/anoxic waters by motile invertebrates
- *Negative impacts* on growth, development, and reproduction of some fish and shellfish
- *Decreased recruitment* of some fish species: Baltic cod (*Gadus morhua;* Bagge 1993; MacKenzie, St. John, & Wieland 1996), smelt (*Osmerus eperlanus*) in the Elbe estuary (Thiel et al. 1995), and American shad (*Alosa sapidissima*) in Delaware Bay (Chittenden 1974, Weisberg et al. 1996)

Baltic cod eggs may sink into hypoxic/anoxic bottom waters, and negative correlations have been seen in annual vertical extent of waters with <2 mg/l oxygen and subsequent year class survival. High mortalities of smelt larvae in hypoxic zones of the Elbe are thought to result in periodic recruitment failures (Thiel et al. 1995). Spawning migrations and seaward returns of anadromous species such as shad are thought to have been blocked by hypoxic water in Delaware Bay, leading to long-term decline in the stocks (Weisberg et al. 1996).

International and regional conventions, commissions, plans, and programs, such as the Bucharest Convention on the Protection of the Black Sea from Pollution, the Helsinki Commission for the Baltic Sea, the Baltic Action Plan, and the Chesapeake Bay Program, can be useful in focusing on the unacceptable consequences — such as increasing anoxic zones in the sea — of coastal/estuarine eutrophication. However, the attention spans of humans and their parliaments are notoriously short. Solutions can come only from long-term commitments from countries and societies that, unfortunately, are already faced with many competing crises.

A final admonition was expressed 3 decades ago by F.A. Richards (1971):

[A]noxic systems [are] those in which any part of the water column becomes completely devoid of oxygen. The condition may be permanent or intermittent. *The intermittently anoxic systems and systems that have very small oxygen contents may be of particular interest in the context of the impingement of man on the oceans, because they may be in a critical state of balance and therefore particularly sensitive to man's activities* [emphasis is mine]. (p. 214)

REFERENCES

Anonymous. 1997. The Black Sea crisis. *Environ. Health Persp. 105:* 1288–1289.

Bagge, O. 1993. Possible effects on fish reproduction due to the changed oceanographic conditions in the Baltic proper. *Int. Counc. Explor. Sea, Coll. Pap. J31:* 1–7.

Breitburg, D.L. 1990. Near-shore hypoxia in the Chesapeake Bay: Patterns and relationships among physical factors. *Estuarine Coastal Shelf Sci. 30:* 593–609.

Breitburg, D.L. 1992. Episodic hypoxia in Chesapeake Bay: Interacting effects of recruitment, behavior, and physical disturbance. *Ecol. Monogr. 62:* 525–546.

Breitburg, D.L. 2002. Effects of hypoxia and the balance between hypoxia and enrichment on coastal fishes and fisheries. *Estuaries 25:* 767–781.

CENR (Committee on Environment and Natural Resources). 2000. Integrated assessment of hypoxia in the northern Gulf of Mexico. National Science and Technology Council, Washington, DC.

Chittenden, M.E., Jr. 1974. Trends in the abundance of American shad, *Alosa sapidissima,* in the Delaware River basin. *Chesapeake Sci. 15:* 96–103.

Diaz, R.J. and R. Rosenberg. 1995. Marine benthic hypoxia: A review of its ecological effects and the behavioural responses of benthic macrofauna. *Oceanogr. Mar. Biol. Annu. Rev. 33:* 245–303.

Jansson, B.-O. and K. Dahlberg. 1999. The environmental status of the Baltic Sea in the 1940s, today, and in the future. *Ambio 28:* 312–319.

Justic, D., N.N. Rabalais, R.E. Turner, and W.J. Wiseman, Jr. 1993. Seasonal coupling between riverborne nutrients, net productivity and hypoxia. *Mar. Pollut. Bull. 26:* 184–189.

Justic, D., N.N. Rabalais, and R.E. Turner. 2002. Modeling the impacts of decadal changes in riverine nutrient fluxes on coastal eutrophication near the Mississippi River Delta. *Ecol. Model. 152:* 33–46.

MacKenzie, B., M. St. John, and K. Wieland. 1996. Eastern Baltic cod: Perspectives from existing data on processes affecting growth and survival of eggs and larvae. *Mar. Ecol. Prog. Ser. 134:* 265–281.

Nance, J.M., E.X. Martinez, and E.F. Klima. 1994. Feasibility of improving the economic return from the Gulf of Mexico brown shrimp fishery. *N. Am. J. Fish. Manage. 14:* 522–536.

Officer, C.B., R.B. Biggs, J.L. Taft, L.E. Cronin, M.A. Tyler, and W.R. Boynton. 1984. Chesapeake Bay anoxia: Origin, development and significance. *Science 223:* 22–27.

Pavela, J.S., J.L. Ross, and M.E. Chittenden, Jr. 1983. Sharp reductions in abundance of fishes and benthic macroinvertebrates in the Gulf of Mexico off Texas associated with hypoxia. *Northeast Gulf Sci. 6:* 167–173.

Platt, A.E. 1995. Dying seas. *World Watch* (January–February), pp. 10–19.

Rabalais, N.N., R.E. Turner, W.J. Wiseman, Jr., and D.F. Boesch. 1991. A brief summary of hypoxia on the northern Gulf of Mexico continental shelf: 1985–1988, pp. 35–46. In: R.V. Tyson and T.H. Pearson (eds.), *Modern and Ancient Continental Shelf Anoxia.* Geological Society Special Publ. No. 58. The Geological Society, London.

Rabalais, N.N., L.E. Smith, D.E. Harper, Jr., and D. Justic. 1995. The effects of bottom water hypoxia on benthic communities of the southeastern Louisiana continental shelf. OCS Study MMS 94-0054, U.S. Dept. Interior, Minerals Management Service, Gulf of Mexico OCS Region, New Orleans, LA. 105 pp.

Rabalais, N.N., R.E. Turner, D. Justic, Q. Dortch, W.J. Wiseman, Jr., and B.K. Sen Gupta. 1996. Nutrient changes in the Mississippi River and system responses on the adjacent continental shelf. *Estuaries 19:* 386–407.

Rabalais, N.N., R.E. Turner, W.J. Wiseman, Jr., and Q. Dortch. 1998. Consequences of the 1993 Mississippi River flood in the Gulf of Mexico. *Regulated Rivers: Research & Management 14:* 161–177.

Rabalais, N.N., R.E. Turner, D. Justic, Q. Dortch, and W.J. Wiseman, Jr. 1999. Characterization of hypoxia: Topic 1 report for the integrated assessment of hypoxia in the Gulf of Mexico. NOAA Coastal Ocean Program Decision Analysis Series No. 15, U.S. Dept. Commerce, National Oceanic and Atmospheric Administration, Silver Spring, MD. 167 pp.

Rabalais, N.N., R.E. Turner, and D. Scavia. 2002. Beyond science into policy: Gulf of Mexico hypoxia and the Mississippi River. *BioScience 52:* 129–142.

Richards, F.A. 1971. Anoxic versus toxic environments, pp. 201–217. In: D.W. Hood (ed.), *Impingement of Man on the Oceans.* Wiley Interscience, New York.

Scavia, D., D. Justic, and V.J. Bierman, Jr. 2004. Reducing hypoxia in the Gulf of Mexico: Advice from three models. *Estuaries 27:* 419–425.

Smith, D.E., M. Leffler, and G. Mackiernan (eds.). 1992. Oxygen dynamics in the Chesapeake Bay: A synthesis of recent research. Sea Grant Program, Univ. Maryland, College Park. 234 pp.

Swanson, R.L. and C.J. Sindermann, Editors. 1979. Oxygen depletion and associated benthic mortalities in the New York Bight. NOAA Prof. Pap. No. 11, U.S. Dept. Commerce, National Oceanic and Atmospheric Administration, Silver Spring, MD. 345 pp.

Thiel, R., A. Sepulveda, R. Kafemann, and W. Nellen. 1995. Environmental factors as forces structuring the fish community of the Elbe estuary. *J. Fish Biol. 46:* 47–69.

Weisberg, S.B., H.T. Wilson, P. Himchak, T. Baum, and R. Allen. 1996. Temporal trends in abundance of fish in the tidal Delaware River. *Estuaries 19:* 723–729.

Wulff, F., A. Stigebrandt, and L. Rahm. 1990. Nutrient dynamics of the Baltic Sea. *Ambio 19:* 126–133.

Zaitsev, Y.P. 1991. Cultural eutrophication of the Black Sea and other European seas. *La Mer 19:* 1–7.

Zaitsev, Y.P. 1993. Impacts of eutrophication on the Black Sea fauna, pp. 64–86. In: *Fisheries and Environment Studies in the Black Sea System.* U.N. Food and Agriculture Organization, FAO General Fisheries Council for the Mediterranean, Rome.

Zimmerman, R.J. and J. Nance. 2001. Effects of hypoxia on the shrimp fishery of Louisiana and Texas, pp. 293–310. In: N.N. Rabalais and R.E. Turner (eds.), *Coastal Hypoxia: Consequences for Living Resources and Ecosystems. Coastal and Estuarine Studies 58,* American Geophysical Union, Washington, DC.

Zimmerman, R., J. Nance, and J. Williams. 1997. Trends in shrimp catch in the hypoxic area of the northern Gulf of Mexico, Mississippi, pp. 64–75. In: J.D. Giattina (ed.), *Proceedings of the First Gulf of Mexico Hypoxia Management Conference.* EPA-55-R-97-001, U.S. Environmental Protection Agency, Gulf of Mexico Program Office, Stennis Space Center, MS.

7 "Black Tides": Petroleum in Coastal Waters

INTRODUCTION

Major spills in coastal waters have been an episodic problem for the massive global production and transport system of the petroleum industry. Tanker accidents, with resulting environmental contamination by released oil, attract intense but transient news media attention, as "black tides" overwhelm local waters and as oiled birds and sea mammals die. Fisheries and tourism are immediately impacted, and an atmosphere of crisis is created. Predictions of disaster abound, and flocks of lawyers descend. Public protest meetings are held, and industry representatives deny or minimize culpability.

These events (and subsequent research and monitoring) have been repeated often enough, especially during the past 3 decades, to have provided at least partial answers to some persistent questions, such as: "What were the actual quantitative effects on fish, shellfish, bird, and mammal populations?" "What has been the record of fish and shellfish landings since the event?" "How long was any evidence of oil contamination observable?" "What long-term changes were observed in impacted ecosystems?" and "To whom were damages paid?" Only when answers to hard questions like these have been evaluated objectively can we decide whether oil spills are truly "catastrophes" as described by news media at the time of their occurrence.

To provide a little background for responses to these questions, I propose to skip lightly through several of the more recent (1979 to the present) major oil contamination events, trying to select those that varied in latitude, method of release, type of oil released, duration, type of environment impacted, and findings from research and monitoring. My choices are: the IXTOC 1 oil spill in the Gulf of Mexico (1979), the *Exxon Valdez* oil spill in Alaska (1989), and the recent *Prestige* oil spill off the coast of Spain (2002). This is a minimal sample. Even when these events are augmented by records of other major spills (see Table 7.1), they do not include the hundreds of yearly intermediate and small spills that result from the daily activities of an oil-based global economy. Although the focus here is on incidents that are objects of greatest public concern, some attention should also be paid to the numerous lesser events that receive little recognition from the national news media. One well-known oil spill consultant has pointed out that during his 14-yr involvement (1978 to 1992), in addition to major spills that received international attention, there were 57 spills in excess of 10×10^6 gal each (Hayes 1999).

TABLE 7.1
Some of the Major Oil Spills of the Past 3 Decades, with
Estimated Amounts of Oil Released to the Environment

Name Used for Reference	Location	Year	Estimated Amount of Oil Spilled (in millions of gallons)
Metula	Chile	1974	13.9
Urquiola	Spain	1976	28.4
Amoco Cadiz	France	1978	68.7
IXTOC-1	Gulf of Mexico	1979	140
Nowruz Field	Arabian Gulf	1983	80
Exxon Valdez	Alaska	1989	10.8
Gulf War (Kuwait)	Arabian Gulf	1991	240
Nakhodka	Sea of Japan	1997	1.5
Erika	France	1999	1.2
Prestige	Spain	2002	2.0

But even these estimates of overwhelming aggregate quantities of oil spilled do not disclose the total impacts of such events. Important variables that must be considered include the following:

- Type, weight, amount, and relative toxicity of the oil
- Sea temperatures at the time of and subsequent to the spill
- Location of the event (estuary, marsh, open coast, etc.)
- Season
- Prevailing winds and storms
- Possible economic impacts on fisheries, tourism, and other coastal industries
- Marine mammals and birds in the impacted area
- Esthetic values of coastal communities

Understandably, oil spills have received increasing research attention, beginning in the 1970s, and a growing body of literature has resulted. Concepts of oil spill behavior have been developed and distilled by scientists, as an aid to damage assessments. In 1978, a so-called "oil spill vulnerability index" was proposed (Grundlach et al. 1978) to enable estimation of damage to be expected as a consequence of oil impacting on a particular type of shoreline. According to the index, vulnerability (and expected damage) increases following a scale of 1 to 10 (see Table 7.2).

Subsequent to the proposal of a "vulnerability index" for estimating damage, one of its coauthors, M.O. Hayes, published a memoir (Hayes 1999) containing a series of so-called "rules of oil spill behavior," based on his experience, some of which are listed here:

- "The ultimate impact of any oil spill will never be as bad as that projected by the media."

TABLE 7.2
Oil Spill Vulnerability Index

1	Exposed rocky cliffs — Under high wave energy, oil spill cleanup is usually unnecessary.
2	Exposed rocky platforms — Wave action causes a rapid dissipation of oil, generally within weeks. In most cases, cleanup is not necessary.
3	Flat fine sand beaches — Due to close packing of the sediment, oil penetration is restricted. Oil usually forms a thin surface layer that can be efficiently scraped off. Cleanup should concentrate on the high tide mark; lower beach levels are rapidly cleared of oil by wave action.
4	Medium- to coarse-grained beaches — Oil forms thick oil-sediment layers and mixes down to 1 m deep with the sediment. Cleanup damages the beach and should concentrate on the high water level.
5	Exposed tidal flats — Oil does not penetrate in compacted sediment surface, but biological damage results. Clean up only if oil contamination is heavy.
6	Mixed sand and gravel beaches — Oil penetration and burial occur rapidly; oil persists and has a long-term impact.
7	Gravel beaches — Oil penetrates deeply and is buried. Removal of oiled gravel is likely to cause future erosion of the beach.
8	Sheltered rocky coast — The lack of wave activity enables oil to adhere to rock surfaces and tidal pools. Severe biological damage results. Cleanup operations may cause more damage than if the oil is left untreated.
9	Sheltered tidal flats — Long-term biological damage results. Removal of the oil is nearly impossible without causing further damage. Clean up only if the tidal flat is very heavily oiled.
10	Salt marshes and mangroves — Long-term deleterious effects result. Oil may continue to exist for 10 or more years.

Source: From Gerlach, S.A. 1981. *Marine Pollution: Diagnosis and Therapy.* Springer-Verlag, Berlin.

- "Salt marshes are susceptible to long-term impacts from oil spills."
- "On exposed high-energy rocky coasts, the cleanup of spilled oil by natural processes is rapid (hours to days)."
- "The depth of penetration and burial on beaches increases with increasing nodimont grain size." (pp. 117, 134, 168)

Hayes has also developed the concept of sensitivity mapping of shorelines as an important approach to impact assessments.

A recent recruit to the procession of books about preventing and responding to oil spills is *Oil Spills First Principles: Prevention and Best Response* (Ornitz & Champ 2002). Described as both a handbook and a textbook that provides a history of oil spills, recommends responses, and reviews environmental impacts (Pearce 2003), the volume is deemed to be an important addition to the literature on oil in the sea.

Field and laboratory observations and experiments have provided a good but not perfect body of information about oil in the environment and its effects on marine organisms. Answers have been found to seemingly simple questions like "Does oil sink or float?" (answer: both), or more complex ones like "How does a clam metabolize oil it ingests with food?" (answer: with great difficulty). The information from many investigations is voluminous, but unfortunately a sizeable portion of it is sequestered as "proprietary" in the files of petroleum companies and their lawyers, and much of the rest of it is difficult to obtain from the records of a multiplicity of commissions, committees, task forces, conferences, and other temporary entities formed and then terminated as investigations of spills proceeded and arguments between regulatory agencies and industry representatives escalated.

OIL IN THE GULF OF MEXICO, 1979

The Great IXTOC-1 Oil Spill, 1979

With the expansion of offshore petroleum production in the Gulf of Mexico, residents of the southeast coastal zone have become inured to the minor spills and leaks that make oil contamination an everyday reality there. Even these pollution-adapted citizens had to be impressed, though, by the magnitude of an oil spill that began on June 3, 1979. IXTOC-1, a Mexican exploratory well located 58 miles off the Gulf of Campeche coast of the Yucatan Peninsula, had blown on that date, and was initially gushing an estimated million gallons of oil per day in the Gulf. Prevailing winds and currents had caused a gigantic oil slick to reach the exquisite white sand beaches of Padre Island off the southern Texas coast by late August of that year.

Totally by chance, arrival of the oil coincided with an annual meeting of the World Aquaculture Society, a technical association with a membership of over 3000 aquatic scientists and entrepreneurs, being held in Corpus Christi, not far from those by then despoiled beaches of Padre Island. The meeting organizing committee quickly laid on field trips to the impacted zone, so that participants could witness at close hand the reality of what is always a concern of aquaculturists — spilled oil polluting productive waters.

The scene on the beaches was truly awesome. A black rim of oil stretched along the shore in both directions to the furthest horizon; oil fumes and an oil spray overpowered the senses; great floating globs of brown chocolate mousse-like material (called "mousse") sloshed in the surf and coated the sand; front-end loaders charged up and down the beach in a futile effort to scoop up the oil-soaked sand; and NOAA crisis response teams in helicopters and small boats assessed the condition and movement of the oil — all combining to create an unreal science fiction-like stage set. At risk, in many peoples' minds as they looked at the devastation, was the major Gulf shrimp fishery. Predictions of disaster for the shrimp industry from mass mortalities and tainted catches were universal, and the future well-being of these estuarine-dependent resource species in a contaminated environment was thought to be in jeopardy.

The oil flow was reduced by late summer, but the well was not finally capped until March 1980, after releasing an estimated 140 million gallons of oil into the Gulf (some estimates were much higher). The aquaculturists flew home from their meeting in Corpus Christi with many "oh my" photographs of the damaged beaches and floating oil; a few reports of oil-tainted shrimp catches were highly publicized by the Texas news media; the oil slick dissipated rapidly, and it soon became difficult to locate any bottom deposits of oil, except in the immediate vicinity of the damaged wellhead. The event faded into history almost without a trace. The anticipated impacts of the spill on subsequent year classes of shrimp and other resource species of the Gulf were examined by fisheries scientists, with equivocal results, because they were unable to separate effects of the oil from those of many other environmental factors that could influence population abundance (it should be noted, though, that annual Gulf shrimp landings since 1979 have fluctuated around 100,000 tons, with no significant declines).

So, from the perspective of the human intruder, standing on an oil-covered shore, the event seemed catastrophic, but from that of the resource species and the entire Gulf ecosystem, it probably represented a relatively mild perturbation, quite likely of lesser overall significance than the passage of an average-strength hurricane.

From Field Notes of a Pollution Watcher
(C.J. Sindermann, 1990)

I was fortunate to have on the scene of this major oil spill a long-term environmental observer with impeccable credentials — my brother Bob — then a faculty member at San Antonio College in Texas. His unpublished report (Sindermann 1980; excerpted here with his permission) contains a summary of events during that exciting time when the giant IXTOC oil slick was approaching its predicted landfall on the south Texas coast:

Mexico "delivered" millions of gallons of badly-needed crude oil to Texas free, albeit in unusable form. The oil slick from Ixtoc took ten weeks to reach the Texas coast, just as scientists familiar with the Gulf currents predicted that it would. The currents in this western part of the Gulf flow north in the spring and summer, reversing to a southerly direction in fall and winter.

With massive slicks and patches of oil looming offshore, it was deemed impossible by the Coast Guard, leader of the Federal Response Team, to attempt to contain it offshore with booms and skim it from the surface. The oil had "weathered" and become less toxic during its 500-mile journey from Campeche, and had been transformed into surface sheen, brown "mousse" emulsion, and tar balls. Containment booms whose plastic skirts extended a scant two feet above and below the surface could hardly be a match for submerged oil (divers reported finding tar balls as much as forty feet below the surface), so a strategy was devised using the natural barrier islands along the Texas coast: Padre, Mustang, San Jose, etc., as a *first* line of defense. The booms, skimmers, and vacuums would be thrown into the fray primarily to keep the oil from entering the five passes or openings between these barrier islands, to protect the sensitive bays and estuaries, nurseries for sea life.

The "black tide" finally did wash ashore in mid-August, first lightly and sporadically; then heavily until the Texas coast was oiled from the mouth of the Rio Grande to beyond Aransas Pass. Oil impacting on vacation beaches was collected manually or with machinery, while unused beaches were left untouched. Mother Nature took its course there, by burying the oil under eight inches of sand, or depositing it in the intertidal "trough", neither of which seems desirable environmentally.

The adequacy of the response in terms of manpower, equipment, and tactics or strategy has been called into serious question, and perhaps rightly so. *Could* the oil have been contained offshore with booms and skimmed from the surface? No one knows, but the attempt was not made. Some 56,000 feet of containment booms were available from cooperatives along the Gulf Coast, but this resource went untapped. ... [D]isaster was averted, with relatively little oil apparently entering the sensitive bays and estuaries beneath the booms. ... But Texas and the U.S. were *lucky*, I suggest, that ocean currents *reversed* themselves *before* there was a massive, unstoppable onslaught of black or brown waves on Aransas Pass, Cedar Bayou, or Pass Cavallo, which would have proven as indefensible with present technology and equipment as would Galveston Bay. *Only* the reversal of ocean currents and winds saved the Texas Gulf Coast from outright *disaster* last fall, if the *truth* were known (pp. 17–19).

By March 1980, the well had been capped, the oil slick had been diverted by southerly winds and currents, and local tourist bureaus were promoting a "Coast Is Clear" program to entice return of tourists. Tar balls were a persistent nuisance on the beaches, but local officials pointed out that they could be from natural seepage, and, anyway, were known to have occurred for centuries along the Texas and Mexican coasts, even being used in the past by Indians to build fires.

A comprehensive 3-yr environmental study, proposed by a federal interagency team at the time of the blowout, was not funded, so the opportunity to acquire good data on the dangers of a massive spill was lost. Effects of the spill on commercial fisheries were considered to be slight. The event faded quickly from public attention, even though persistent residues of the spill are still present in small tidal backwaters.

The IXTOC-1 blowout, estimated at 140×10^6 *gal,* was the second largest oil spill on record, surpassed only by the deliberate contamination of the Persian Gulf by Saddam Hussein and his cohorts during the Gulf War in Kuwait (1991). Dubbed "the mother of all spills," that awful episode resulted in an estimated oil pollution of 240×10^6 gal, and its effects are still occasionally visible on the shores of the Gulf (Hayes 1999). Hayes's resurvey of part of the affected coastline of Saudi Arabia near Jubial in 1997 — 6 yr after the Gulf War — found almost no change since earlier surveys in 1991 and 1993. Liquid oil remained in crab burrows, some marshes were still oiled, and an asphalt-like pavement persisted in some intertidal zones. Some of this damage was thought to have resulted from oil spilled in 1983 during the earlier (1980 to 1988) Iran/Iraq War (the *Nowruz* oil spill, of an estimated 80×10^6 gal — the third largest oil spill on record.

EXXON VALDEZ OIL SPILL IN ALASKA, 1989

The grounding of the supertanker *Exxon Valdez*, with its subsequent massive oil spill in an environmentally sensitive area of Alaskan coastline (Prince William Sound), stimulated an outpouring — forced or otherwise — of funds ($2 billion from Exxon) to assess damage, clean up contaminated shorelines, and reimburse victims (and inevitably to support a virtual army of lawyers). Some money was also made available for research and monitoring that would provide information about effects of oil on animals and ecosystems in Subarctic regions. Government and academic scientists (as well as private consultants) were immersed in a time-sensitive fact-finding frenzy that peaked early after the event but has continued at a reduced level, and mostly in a monitoring mode, to the present time.

Life in the Wake of the *Exxon Valdez*

The sheer size of the vessel defied adequate description, except with such superlatives as "stupendous" or "gigantic." Nine hundred feet long and loaded with 550,000 tons of Prudhoe Bay crude oil, it was in the uncertain hands of an alcoholic captain on a dark and stormy night in March 1989. The ship was outward bound in the hazardous seaward passage through Prince William Sound in southeastern Alaska, but it stopped abruptly, well off course, on a rocky outcropping, and one-fifth (10.8 million gallons) of its cargo flooded into the sea.

Birds and sea mammals were oiled and killed (1000 sea otters, 36,000 birds, including 150 bald eagles), but other populations were affected too, and the tale is far from finished even now — over a decade later — especially the chronicle of fisheries' biological events that followed. The spill occurred just before the Pacific herring (<u>Clupea pallasi</u>) moved close to shore to spawn, although according to reports by observers being paid by the Exxon Corporation, only a small part of the spawning area was actually oiled (Pearson, Moksness, & Skaiskl 1995; Elston et al. 1997). Catches of adult fish in 1990, 1991, and 1992 were good, suggesting that stocks in existence at the time of the spill had not been severely affected, but in 1993 — the year when herring born during the oil spill year would have entered significantly into the spawning stocks — population biomass decreased by more than 70%, and the decline continued in 1993 and 1995, causing closure of the fishery.

To complicate the story further, an epizootic of viral hemorrhagic septicemia (VHS), a lethal viral disease, was reported in the spawning stocks in 1993 and 1994 (Meyers et al. 1994). Furthermore, later studies of samples collected in 1994 disclosed epizootic levels of a lethal systemic fungal disease caused by <u>Ichthyophonus hoferi</u>, 18% prevalence in one study by Elston et al. (1997) and 29% in a separate study by Marty et al. (1995).

Although it is tempting to make some association of the oil spill, the population decline four years later, and the simultaneous presence at the time of the decline of two lethal diseases at epizootic levels (an unusual event by itself), a conclusion about direct relationships does not seem warranted. Spawning in the year of the oil spill could have been a partial failure, with high egg and larval mortality resulting in poor contribution of the 1989 year class to the 1993 and 1994 spawning stocks. Alternatively (or additionally), the short- and long-term stress effects of oil in the environment could result in immunosuppression, as could effects of other environmental factors such as food scarcity or low temperatures. Some field and experimental data have disclosed depressed immune responses in other species of fish with detectable levels of polycyclic aromatic hydrocarbons (PAHs) (Arkoosh et al. 1991; Arkoosh, Stein, & Casillas 1994). These findings suggest, but certainly do not confirm, a complex relationship of oil contamination, epizootic disease, and population decline.

Unfortunately, the still-emerging Prince William Sound oil spill story has been clouded and is still being clouded by persistent differences in conclusions reached by state government researchers and others under contract with the Exxon Corporation, making it difficult for interested observers to discern where reality lies, in the results of what could be a classic large-scale environmental experiment. The <u>Exxon Valdez</u> spill and its aftermath dramatize how conclusions reached by scientists can be polarized and become subjects of public and legal confrontations — about the size of the spill, its impact in various locations, and its effects on environment and resources — when private science consultants hired by the oil industry compete with and often disagree with scientists from government and universities. Admittedly, the problem is complex — almost too complex — from a living resource perspective at least:

- *an oil spill at spawning time of a major resource species;*
- *an outbreak three years later of two lethal diseases; and*
- *failure of recruitment four years after the spill.*

I wish, sometimes, that science could be simpler — or at least not quite so vulnerable to societal and economic pressures.

From Field Notes of a Pollution Watcher
(C.J. Sindermann, 2000)

Some valuable and relevant information was gained through research and monitoring during and after the *Exxon Valdez* spill — an interesting time of action with funding from Exxon for support. Here are a few examples:

- Most petroleum floats. It has a specific gravity less than seawater, but a popular misconception is that it sinks to the bottom after a spill. The reality is that some oil does sink — but not on its own. It sinks if it adsorbs to suspended particles in the water column — on particulates such as clay, sand, powdered rock, or organic detritus.
- The fate of *Exxon Valdez* oil 1 month after the spill was determined from multiple sources:
  - Evaporation of aromatic fractions was high (20%).
  - A significant part was dispersed in the water column (20 to 25%).
  - Some was beached within the sound (40 to 45%).

- Some was carried offshore out of the sound (25%); bits of the slick reached land 500 mi away, and tar balls and mousse came ashore on the Kenai Peninsula and on Kodiak Island (Wheelwright 1994).
- Long-term persistence of oil residues in Prince William Sound and along the shores of the Gulf of Alaska have been monitored sporadically since the spill. In 1995 — 6 yr after the event — mean hydrocarbon concentration was more than twice background levels in sediments and in mussels (*Mytilus trossulus*) at most sampling sites (Carls et al. 2001). Those authors predicted that it would take 30 yr for most mussel beds to return to background levels of total polynuclear aromatic hydrocarbons.
- Petroleum and its components have been found in a number of studies to reduce reproductive rates in fish populations (National Research Council 1985). Effects include alterations in levels of reproductive hormones, inhibition of gonad development, and reduced egg and larval viability (Carls, Rice, & Hose 1999; Sol et al. 2000). Several coastal fish species were subjected to high levels of oil exposure during the *Exxon Valdez* spill, and 1 yr later negative correlations were observed between exposure indicators and the reproductive parameters cited above. In subsequent years, however, those parameters tended to return to the normal range (Brown et al. 1994).

Meanwhile, the *Exxon Valdez*, presumably fully recovered from its encounter with the rocky coast of Prince William Sound, still rides the world's oceans, though now with a new and almost poetic name — the *Sea River Mediterranean*. Watch for this notorious vessel on your next cruise.

SINKING OF THE TANKER *PRESTIGE* OFF THE COAST OF SPAIN, 2002

The Galician coast of extreme northwestern Spain is a dangerous place for all shipping. Known as the "Coast of Death," this stormy sector of the North Atlantic has been the site of three major oil spills in a single decade: the *Agean Sea* wreck in 1992, the *Erika* spill in 1999, and the *Prestige* sinking in 2002 — together constituting an unconscionable degree of ecological punishment for one coastal area.

The *Prestige* oil spill was somewhat unique in that an estimated 80% of its cargo of 20×10^6 gal of heavy fuel oil went with the two halves of the ship to the bottom, 150 mi off the coast at a depth of 2 mi. Release of that oil and its appearance at the surface has been gradual and sporadic, amounting to an estimated 21,000 gal/d. By January 2003, the oil had contaminated Spain's Galician coast — an important fishing ground — and the oyster culture areas of the adjacent Aquitaine region of the French coast, as far north as the famed oyster-producing Bay of Arcachon.

Described in *Scientific American* as "Spain's worst ecological disaster ever" (Ariza 2003), the long-term fate of the oil still in the ship is uncertain and is subject to continuing debate. Bottom temperatures around the vessel sections are 2.75°C, but the rates of cooling and solidification of the heavy oil are determined by a number of variables. Pressure chamber studies by the French Research Institute for Exploitation of the Sea suggest that the fuel oil will *never* solidify, even at below-zero temperatures. Concern has been expressed, therefore, about possible decades-long slow releases becoming a "permanent source of pollution," affecting beaches and coastal fishing. Evidence supporting this grim prediction exists from an earlier spill in another part of the world. The tanker *Nakhodka* sank in water 1.5 mi deep off Japan in 1997 and *still* releases relatively small quantities of oil. Many sunken ships in shallower water are known to leak oil decades after their demise. Deterioration of the tanker hull and its compartments is also a cause for concern about pollution from long-term releases.

EFFECTS OF PETROLEUM ON FISH AND SHELLFISH

Oil spills in coastal waters can affect fish and shellfish in certain specific ways. Fish eggs and larvae that are pelagic at or near the surface may be killed or chemically damaged by aromatic fractions of oil slicks. Eggs that are deposited intertidally or subtidally on the bottom or on fixed macroalgae (herring, for example) may also be affected by oil at the surface or in the water column. Adult fish are not usually killed by spills, but spawning females of certain bottom-dwelling species (flounders and sole, for example) may accumulate tissue hydrocarbon concentrations that can affect egg survival and normal development of embryos (Black, Phelps, & Lapan 1988; Johnson et al. 1988, 1995; Idler et al. 1995).

Molluscan shellfish may be killed or retarded in growth by exposure to initial and residual levels of oil in sediments. A few examples from the literature should offer insights into the nature and extent of the problem that oil exposure presents to fish and shellfish.

Effects of Petroleum on Fish Eggs and Larvae

As part of an examination of the tanker *Argo Merchant* oil spill near Nantucket Island, Massachusetts, in December 1976, cod embryos and larvae were exposed experimentally to various components of the resulting oil slick (Kühnhold & Lef-court 1977). High mortality was evident in early embryos exposed to 500 ppb of total extractable hydrocarbons (water soluble fraction), and survivors exhibited severe abnormalities (abnormal cell division, delayed development, and irregular and reduced heart rate). Abnormalities were reduced at lesser concentrations of oil extracts. A later study (Tilseth, Solberg, & Westrheim 1984) extended the list of sublethal effects of water-soluble fractions of *Ekofisk* (North Sea) crude oil on early larval stages of cod.

Another study, this time on winter flounder (spawning females and developing eggs and larvae) provided more definitive evidence that petroleum contamination can affect survival of larvae (Kühnhold et al. 1978). Experimental exposures of mature female winter flounders and their developing eggs and larvae to low concentrations of No. 2 fuel oil produced results that should be useful in population analyses. Exposure to 100 ppb throughout the gonad maturation of parents and during fertilization and embryogenesis resulted in a 3- to 9-d delay in hatching, a 19% reduction in viable hatch, and a 4% prevalence of spinal defects in hatched larvae. Larvae produced from gametes contaminated during parental gonadal maturation but then reared in clean water had a mortality coefficient of 0.130, considered by the authors to be much higher than the calculated mortality coefficient for untreated, laboratory-reared winter flounder larvae of 0.036 to 0.059. Growth of larvae hatched from gametes from oil-exposed spawners was also slower.

Despite these and other reports (see, for example, Longwell et al. 1983, Thomas & Budiantara 1995) of damage to developing fish eggs and larvae from exposure to oil, a cautionary note is necessary. Pronounced species differences have been observed in effects of oil on eggs and larvae. In one Norwegian study (Føyn and Serigstad 1989), growth and survival of cod eggs were very susceptible to experimental oil contamination, whereas herring eggs and larvae seemed unaffected at similar dosage levels. Potential damage to populations of fish from an oil spill depends on the species involved, the developmental stage present at the time, and the intensity of exposure. Eggs, larvae, and postlarvae in contact with critical concentrations of water-soluble fractions of oil offer the best opportunity for population impacts. Reductions in recruitment of a year class of fish can reach 30 to 45%, depending on the distribution of the population, the developmental stages present at the time of spill, and the extent of contamination (Føyn and Serigstad 1989). (It might be noted here that the topic of acute and chronic oil pollution in the sea was examined in detail from a fishery perspective by McIntyre [1982]. He concluded a detailed review by stating that "no long-term adverse effects on fish stocks can be attributed to oil, but local impacts can be extremely damaging in the short term" [p. 411].)

The wreck of the *Exxon Valdez* in southeastern Alaska waters in March 1989 coincided with the spawning of Pacific herring *Clupea pallasi* in that area, providing the scientific world with an epic environmental experiment to determine effects of spilled oil on eggs and larvae of the 1989 year class and subsequent year classes. Several research groups — from the University of Alaska, the Alaska Department of Fish and Game, and the University of California, as well as assorted private consultants — charged in for a piece of the action. Their studies resulted in an outpouring of technical papers published during the 1990s, containing significant new information about the effects of oil on early life stages of herring. Results included.

- Larvae from oiled sites were shorter, had ingested less food, had slower growth, had ascites, and had higher percentages of cytogenetic abnormalities" (Marty et al. 1996a).
- Mean egg-larval mortality in oiled areas was twice that in unoiled areas, and larval growth rates in oiled areas were about half those measured in populations from other areas of the North Pacific" (McGurk & Brown 1996).
- Many larvae from oiled sites exhibited abnormalities associated with oil exposure in laboratory experiments and in other oil spills: morphological malformations, genetic damage, and small size. (Growth between May and June 1989 was the lowest ever reported for field-caught herring; Norcross et al. 1996.)

- Histopathology in larvae from oiled sites included ascites, hepatocellular vacuolar change, and necrosis of skeletal muscle, retinal, and brain cells (Marty et al. 1996b).
- In 1990 and 1991, oil-related developmental and genetic effects were undetectable (Hose et al. 1996).

Findings clearly supported the hypothesis of short-term damage to herring embryos and larvae resulting from exposure to oil, even though conclusions about population impacts are still controversial.

EFFECTS OF PETROLEUM ON MOLLUSCAN SHELLFISH

The effects of spilled oil have been examined repeatedly in molluscan shellfish, using a combination of field observations and field or laboratory experiments, since the combined impacts of reduction in growth rates and increase in mortality rates of molluscs are common phenomena in polluted habitats. Several studies have demonstrated these effects. As an example, the growth rates of soft clams (*Mya arenaria*) declined by 65% when clams were transplanted from a clean site to an oil-polluted site on the Maine coast (Dow 1975). Survival of the transplants at the polluted site was only 12%, compared with 78% at a reference site.

Effects of experimental transplants of clams to and from oiled sites were also examined in connection with research following the *Exxon Valdez* oil spill. A study of contamination levels 5 yr after the event in Pacific little-neck clams (*Protothaca staminea*) indicated that residual oil still affected survival and growth rates (Fukuyama, Shigenaka, & Hoff 2000). Effects of long-term exposure on mortality and growth rates, as well as high tissue levels of hydrocarbons, of transplanted clams were persistent over the 2-yr study period (a finding that parallels the East Coast finding of Dow [1975]). Based on their results, the authors of the West Coast study warn against premature clam restoration attempts after an oil spill, before environmental levels of oil are sufficiently reduced — with the possibility of compromising the success of the restoration effort.

Arguably the most thorough examination of the effects of spilled oil on molluscan shellfish followed the wreck of the tanker *Amoco Cadiz* on the Brittany coast of France in 1978, spilling 68.7×10^6 gal of crude oil. Effects on the commercially important oyster populations of the area have been examined repeatedly since that accident, through a combination of field surveys and experimental introductions of clean oysters (*Ostrea edulis* and *Crassostrea gigas;* Berthou et al. 1987). Some of the findings relevant to quantitative effects of oil pollution are these:

1. Immediate mortalities of periwinkles and limpets occurred on and near the oiled coastline, and 2 weeks later mass mortalities of razor clams occurred at a site (St. Efflan) 95 km from the wreck — possibly from toxic oil components in the water column or in sediments (Chasse 1978).
2. Oyster mortalities of 20 to 50% were reported in the most heavily polluted sites during the first 3 months after the spill.

3. Lesions, principally necrosis or atrophy of the digestive tract epithelium, were observed frequently during 1978 and in several succeeding years, gradually decreasing to background levels by 1983.
4. No evidence was found for a relationship of oil pollution with increased prevalences or effects of two lethal protozoan pathogens — *Marteilia refringens* and *Bonamia ostreae* — in *O. edulis*.
5. Severe lesions of the reproductive system were seen in 1978 in the form of destruction of the duct epithelium, intense inflammatory reaction, and final atrophy. Total suppression of spawning of *O. edulis* was postulated for 1978 (Neff & Haensly 1982).
6. Neoplasms have been reported previously in *O. edulis*, but prevalences did not increase after the oil spill. Neoplasms with prevalences up to 40% were, however, observed in cockles (*Cerastoderma edule*) from the impacted area in 1983, but a relationship with oil pollution was not established (Poder & Auffret 1986).
7. Hydrocarbon levels residual in oysters in 1985 — 7 yr after the spill — were still two to five times the values found in an unpolluted reference site, but pathological signs, seen earlier in the form of digestive gland and gonadal lesions, had disappeared. The so-called "indifference" of oysters to chronic petroleum pollution, as measured by the results of pathological examination, was suggested. This, however, is not consistent with findings from experimental exposures of another species of oyster, *Crassostrea virginica*, to chronic low levels of crude oil, in which severe alterations in the digestive tract and the gonads were seen (Barszcz et al. 1978).

One factor that has been of great importance in evaluating impacts of the *Amoco Cadiz* oil spill on molluscan shellfish populations is that the University of Paris has operated a marine laboratory on the affected coastline for more than a century, and scientists have unusually complete information about the populations *before* the oil spill as well as after. Due in part to this favorable coincidence, investigations of the effects of this single massive oil pollution event on the French coast on molluscan species have been useful in providing information about immediate as well as long term damage.

CONCLUSIONS

So many oil spill events have risen to fleeting prominence in the world news — to cause dismay and revulsion in abused coastal areas, to captivate the news media, and then to fade quickly from the headlines. Stories of panic-driven actions, financial distress, and personal trauma are enfolded in each event, but these tend to disappear into sterile technical and legal descriptions of amounts of coastline despoiled, numbers of birds oiled and killed, losses of tourist dollars, and attempts by perpetrators to deny culpability. A few books have tried to capture the realities — especially the horror and disbelief — that accompany major spills (Wheelwright 1994, Hayes 1999), and more attempts to go beyond the statistics should be made. Few of man's many atrocities committed against coastal environments are more graphic or more disturbing to the innocent observer than an ongoing major oil spill.

The published literature about some of man's failed attempts to keep oil and water from mixing leads on one hand to dismay at the repetitions of regional environmental insults from petroleum spills, and on another hand to some satisfaction with the accretion of information about the behavior of spilled oil in the sea and about its effects on living organisms and their habitats. Examination of portions of this literature leads to a number of conclusions:

- At the time of a major spill, and during several weeks following, the event is viewed by local inhabitants and the news media as catastrophic for the marine environment. Shorebirds and marine mammals are killed, and shorelines are heavily but temporarily impacted by the oil. The rate of clearance of the contamination is proportional to many variables such as temperature, density and amount of oil, winds, tides, storms, type of shoreline, and coastal currents. Dynamic coastlines in lower latitudes usually have shorter recovery periods.

- Residual evidence of oil contamination may persist in sediments for decades, especially on low-energy shorelines, but the history of major spills, such as IXTOC-1 in 1979 and the Persian Gulf in 1991, indicates rapid disappearance of major effects on resource species.

- Scientific studies are always necessary to develop estimates of the extent of resource and ecosystem damage. Effects on fish and shellfish populations, and other economic impacts, are assessed for the short and long term by teams sharply divided into two groups: (1) government scientists and their contractors, and (2) consultants (university or private) supported by the petroleum industry. Findings and conclusions are often equivocal or conflicting, even though some may be based on the same data sets.

- News media squeeze every possible bit of information (and misinformation) from the event, keeping the story alive as long as a trace of oil can be found. They are notably fickle, though, prone to moving on to other breaking stories without hesitation. Furthermore, they delight in exploiting any controversies that develop among visiting experts or regulatory agencies.

- Fishermen have a genuine and legitimate concern about impacts of spilled oil on fished stocks and fish habitats. They are often chagrined by the relative inconclusiveness of scientific findings and conclusions, but they can be an effective lobbying group for legal and governmental actions (such as awards of damages to injured parties and proper assignment of disaster relief funds).

- The petroleum industry and its segments that are immediately involved in the event typically try to minimize public perceptions of the effects of spilled oil on marine resources and habitats. The industry tends to fund research by a cadre of academicians or other scientific consultants who have developed reputations for espousing conclusions and opinions favorable to their clients. The petroleum industry is also willing to commit huge amounts to dispute claims against it, and it is prepared for protracted legal maneuvering sometimes extending over decades.

- Government environmental and resource protection agencies, and federal and state courts, attempt to fix blame and to develop opinions and judgments that will result in award of financial compensation to those individuals and entities (fishermen, tourist industries, fish processing plants, etc.) who have legitimate claims against polluters.

- Claims for damages as consequences of oil spills can take years to reach settlement. The *Exxon Valdez* spill in Alaska is an excellent example: more than 10 yr after the event (1989), the company (now Exxon-Mobil) was still embroiled in a legal battle with commercial fishermen, native Alaskans, and others about damages resulting from the spill. Claims reached $15 billion!

In closing this chapter on oil spills, I would like to point out another perspective that may easily be overlooked. Many of the lesser oil spills that are an everyday reality in the world are not publicized because they produce *no detectable effects*. Negative findings about the effects of some of those smaller spills are rarely reported — leading to misconceptions about overall impacts of oil spills on the environment. A recent editorial in the *Marine Pollution Bulletin* by Edgar, Snell, and Lougheed (2003) makes several points on this subject:

> Oil spill studies revealing no detectable impacts are generally more difficult to publish than studies identifying impacts, hence the oil spill literature is probably dominated by outcomes of the most massive spills where clear effects are indicated. In turn, this can arguably lead to general misconceptions about the overall impacts of moderate- and small-size spills. Nevertheless, it is also possible that, because of a lack of baseline monitoring data, subtle and long-term effects generally go unrecognised. (p. 275)

REFERENCES

Ariza, L.M. 2003. Oiling up Spain. *Scientific American Newscum* (April), pp. 23–24.

Arkoosh, M.R., E. Casillas, E. Clemons, B. McCain, and U. Varanasi. 1991. Suppression of immunological memory in juvenile chinook salmon (*Oncorhynchus tshawytscha*) from an urban estuary. *Fish Shellfish Immunol. 1:* 261–277.

Arkoosh, M.R., J.E. Stein, and E. Casillas. 1994. Immunotoxicology of an anadromous fish: Field and laboratory studies of b-cell mediated immunity, pp. 33–48. In: J.S. Stolen and T.C. Fletcher (eds.), *Modulators of Fish Immune Responses,* Vol. 1. Fair Haven, NJ.

Barszcz, C.A., P.P. Yevich, L.R. Brown, J.D. Yarbrough, and C.D. Minchew. 1978. Chronic effects of three crude oils on oysters suspended in estuarine ponds. *J. Environ. Pathol. Toxicol. 1:* 879–895.

Berthou, F., G. Balouet, G. Bodennec, and M. Marchand. 1987. The occurrence of hydrocarbons and histopathological abnormalities in oysters for seven years following the wreck of the *Amoco Cadiz* in Brittany (France). *Mar. Environ. Res. 23:* 103–133.

Black, D.E., D.K. Phelps, and R.L. Lapan. 1988. The effect of inherited contamination on egg and larval winter flounder, *Pseudopleuronectes americanus. Mar. Environ. Res. 25:* 45–62.

Brown, E.D., T.T. Baker, F. Funk, J.E. Hose, R.M. Kocan, G.D. Marty, M.D. McGurk, B.L. Norcross, and J.W. Short. 1994. The *Exxon Valdez* oil spill and Pacific herring in Prince William Sound, Alaska: A summary of injury from 1989–1994. In: E.D. Brown and T.T. Baker (eds.), *Injury to Prince William Sound Following the* Exxon Valdez *Oil Spill: Final Report for Fish/Shellfish Study No. 11.* Alaska Dept. Fish & Game, Div. Commercial Fisheries Management & Development, Anchorage, AK.

Carls, M.G., S.D. Rice, and J.E. Hose. 1999. Sensitivity of fish embryos to weathered crude oil: Part I. Low-level exposure during incubation causes malformations, genetic damage, and mortality in larval Pacific herring (*Clupea pallasi*). *Environ. Toxicol. Chem. 18*(3): 481–493.

Carls, M.G., M.M. Babcock, P.M. Harris, G.Y. Irvine, J.A. Cusick, and S.D. Rice. 2001. Persistence of oiling in mussel beds after the *Exxon Valdez* oil spill. *Mar. Environ. Res. 51:* 167–190.

Chasse, C. 1978. The ecological impact on and near shores by the *Amoco Cadiz* oil spill. *Mar. Pollut. Bull. 9:* 298–301.

Dow, R.L. 1975. Reduced growth and survival of clams transplanted to an oil spill site. *Mar. Pollut. Bull. 6:* 124–125.

Edgar, G.J., H.L. Snell, and L.W. Lougheed. 2003. Impacts of the *Jessica* oil spill: An introduction. *Mar. Pollut. Bull. 47:* 273–275.

Elston, R.A., A.S. Drum, W.H. Pearson, and K. Parker. 1997. Health and condition of Pacific herring, *Clupea pallasi,* from Prince William Sound, Alaska, 1994. *Dis. Aquat. Org. 31:* 109–126.

Føyn, L. and B. Serigstad. 1989. How can a potential oil pollution event affect the recruitment to fish stocks? Int. Counc. Explor. Sea (ICES) Doc. C.M.1989/No. 5, 23 pp.

Fukuyama, A.K., G. Shigenaka, and R.Z. Hoff. 2000. Effects of residual *Exxon Valdez* oil on intertidal *Protothaca staminea:* Mortality, growth, and bioaccumulation of hydrocarbons in transplanted clams. *Mar. Pollut. Bull. 40:* 1042–1050.

Gerlach, S.A. 1981. *Marine Pollution: Diagnosis and Therapy.* Springer-Verlag, Berlin. 218 pp.

Grundlach, E.R., M.O. Hayes, C.H. Ruby, L.G. Ward, A.E. Blount, I.A. Fisher, and R.J. Stein. 1978. Some guidelines for oil-spill control in coastal environments, based on field studies of four oil spills. ASTM Spec. Tech. Publ. 659. 98–110.

Hayes, M.O. 1999. *Black Tides.* Univ. Texas Press, Austin, TX. 287 pp.

Hose, J.E., M.D. McGurk, G.D. Marty, D.E. Hinton, E.D. Brown, and T.T. Baker. 1996. Sublethal effects of the *Exxon Valdez* oil spill on herring embryos and larvae: Morphologic, cytogenetic, and histopathological assessments, 1989–1991. *Can. J. Fish. Aquat. Sci. 53:* 2355–2365.

Idler, D.R., Y.P. So, G.L. Fletcher, and J.F. Payne. 1995. Depression of blood levels of reproductive steroids and glucuronides in male winter flounder (*Pleuronectes americanus*) exposed to small quantities of Hibernia crude, used crankcase oil, oily drilling mud and harbour sediments in the 4 months prior to spawning, p. 187. In: G.W. Goetz and P. Thomas (eds.), *Proceedings of the Fifth International Symposium on the Reproductive Physiology of Fish,* Austin, TX.

Johnson, L., E. Casillas, T. Collier, B. McCain, and U. Varanasi. 1988. Contaminant effects on ovarian maturation in English sole (*Parophrys vetulus*) from Puget Sound, Washington. *Can. J. Fish. Aquat. Sci. 45:* 2133–2146.

Johnson, L.L., J.E. Stein, T. Hom, S. Sol, T.K. Collier, and U. Varanasi. 1995. Effects of exposure to Prudhoe Bay crude oil on reproductive function in gravid female flatfish. *Environ. Sci. 3:* 67–81.

Kühnhold, W. and P. Lefcourt. 1977. Effects of oil on developing embryos, pp. 107–108. In: P.L. Gross and J.S. Mattson (eds.), *The* Argo Merchant *oil spill: A preliminary scientific report.* U.S. Dept. Commerce, NOAA. 133 pp.

Kühnhold, W.W., D. Everich. J.J. Stegeman, J. Lake, and R.E. Wolke. 1978. Effects of low levels of hydrocarbons on embryonic, larval and adult winter flounder (*Pseudopleuronectes americanus*), pp. 677–711. In: *Proceedings of the Conference on Assessment of Ecological Impacts of Oil Spills, Keystone, Colorado.* American Institute of Biological Science, Washington, DC.

Longwell, A.C., D. Perry, J.B. Hughes, and A. Herbert. 1983. Frequencies of micronuclei in mature and immature erythrocytes of fish as an estimate of chromosome mutation rates — results of field surveys on windowpane flounder, winter flounder and Atlantic mackerel. Int. Counc. Explor. Sea, Doc. C.M.1983/E:55.

Marty, G.D., E.F. Freiberg, T.R. Meyers, J.A. Wilcock, C.R. Davis, T.B. Farver, and D.E. Hinton. 1995. *Ichthyophonus hoferi*, viral hemorrhagic septicemia virus, and other causes of morbidity in Pacific herring spawning in Prince William Sound in 1994. Project 94320S, *Exxon Valdez* Oil Spill Restoration Annual Report.

Marty, G.D., C.R. Davis, E.F. Freiberg, D.E. Hinton, T.R. Meyers, and J. Wilcock. 1996a. Causes of morbidity in Pacific herring from Sitka Sound and Prince William Sound, Alaska, in spring 1995. Project 95320S, *Exxon Valdez* Oil Spill Restoration Project Annual Report.

Marty, G.D., J.E. Hose, M.D. McGurk, E.D. Brown, and D.W. Hinton. 1996b. Histopathology and cytogenetic evaluation of Pacific herring larvae exposed to petroleum hydrocarbons in the laboratory or in Prince William Sound, Alaska, after the *Exxon Valdez* oil spill. *Can. J. Fish. Aquat. Sci. 54:* 1846–1 857.

McGurk, M.D. and E.D. Brown. 1996. Egg-larva mortality of Pacific herring in Prince William Sound, Alaska, after the *Exxon Valdez* oil spill. *Can. J. Fish. Aquat. Sci. 53:* 2343–2354.

McIntyre, A.D. 1982. Oil pollution and fisheries, pp. 401–411. In: R.B. Clark (ed.), *The Long Term Effects of Oil Pollution on Marine Populations, Communities, and Ecosystems.* Phil. Trans. R. Soc. Lond. B297.

Meyers, T.R., S. Short, K. Lipson, W.N. Batts, J.R. Winton, J. Wilcock, and E. Brown. 1994. Association of viral hemorrhagic septicemia virus with epizootic hemorrhages of the skin in Pacific herring *Clupea harengus pallasi* from Prince William Sound and Kodiak Island, Alaska, USA. *Dis. Aquat. Org. 19:* 27–37.

National Research Council. 1985. *Oil in the Sea: Inputs, Fates, and Effects.* National Academic Press, Washington, DC.

Neff, J.M. and W.E. Haensly. 1982. Long-term impact of the *Amoco Cadiz* oil spill on oysters *Crassostrea gigas* and plaice *Pleuronectes platessa* from Aber-Benoit and Aber-Wrach, Brittany, France, pp. 269–328. In: *Ecological Study of the* Amoco Cadiz *Oil Spill.* NOAA-CNEXO Report, U.S. Dept. Commerce, Washington, DC.

Norcross, B.L., J.E. Hose, M. Frandsen, and E.D. Brown. 1996. Distribution, abundance, morphological condition and cytogenetic abnormalities of larval herring in Prince William Sound, Alaska, following the *Exxon Valdez* oil spill. *Can. J. Fish. Aquat. Sci. 53:* 2376–2387.

Ornitz, B.E. and M.A. Champ. 2002. *Oil Spills First Principles: Prevention and Best Responses.* Elsevier, Oxford, UK.

Pearce, J.B. 2003. Book review: Ornitz, B.E. and M.A. Champ, 2002. *Oil Spills First Principles: Prevention and Best Responses. Mar. Pollut. Bull. 46:* 1630.

Pearson, W.R., E. Moksness, and J.R. Skaiskl. 1995. A field and laboratory assessment of oil spill effects on survival and reproduction of Pacific herring following the *Exxon Valdez* spill, pp. 626–661. In: P.G. Wells, J.N. Butler, and J.S. Hughes (eds.), Exxon Valdez *Oil Spill: Fate and Effects in Alaska Waters.* STP 1219, ASTM, Philadelphia, PA.

Poder, M. and M. Auffret. 1986. Sarcomatous lesions in the cockle *Cerastoderma edule. Aquaculture 56:* 1–8.

Sindermann, R.P. 1980. *The Untold Story of the Mexican Oil Spill.* Unpublished manuscript, San Antonio College, Dept. Government, San Antonio, TX.

Sol, S.Y., L.L. Johnson, B.H. Horness, and T.K. Collier. 2000. Relationship between oil exposure and reproductive parameters in fish collected following the *Exxon Valdez* oil spill. *Mar. Pollut. Bull. 40:* 1139–1147.

Thomas, P. and L. Budiantara. 1995. Reproductive life history stages sensitive to oil and naphthalene in Atlantic croaker. *Mar. Environ. Res. 39:* 147–150.

Tilseth, S., T.S. Solberg, and K. Westrheim. 1984. Sublethal effects of the water-soluble fraction of Ekofisk crude oil on the early larval stages of cod (*Gadus morhua* L.). *Mar. Environ. Res. 11:* 1–16.

Wheelwright, J. 1994. *Degrees of Disaster.* Simon and Schuster, New York. 348 p.

8 Biological Pollution: Invasions by Alien Species

An Alien Pathogen of Oysters in American Waters

In early summer of 1960, I came for the first time to the Oxford Biological Laboratory on Maryland's Eastern Shore. As a new shellfish research laboratory of the National Marine Fisheries Service, it had been confronted in the year before my arrival with high mortalities of oysters due apparently to a disease of unknown origin. Oysters of the Middle Atlantic region had died in great numbers beginning in 1957 in Delaware Bay and in 1959 in Chesapeake Bay, and production had declined precipitously. The laboratory's programs were being reoriented to develop information about the cause. During the 1960s and 1970s, staff members worked full-time on the problem, as did research scientists at several state and university laboratories in the Middle Atlantic area. The pathogen (a protozoan named <u>Haplosporidium nelsoni</u>) and its life cycle were described, as were some controlling environmental factors (especially salinity and temperature), but the origin of the disease agent and its method of natural transmission remained a mystery for many years thereafter.

Impact of the disease varied annually, with periods of severe mortalities and intervals of decreased prevalence and less severe effects. Today the pathogen is still present and is still killing oysters. Research has continued, but at reduced intensity, except at the Haskin Laboratory of Rutgers University and at the Virginia Institute of Marine Science, where major programs have continued to the present — emphasizing attempts to develop disease-resistant stocks.

Knowledge about the possible origin of H. nelsoni was finally achieved during the late 1990s (Burreson, Stokes, & Friedman 2000), using newly developed genetic techniques. Its genetic makeup was found to be almost completely comparable to that of a rare parasitic organism found in Japanese oysters (which had been grown on the Pacific coast of North America since early in the 20th century — and had been, sporadically over many decades, imported in small quantities by individual oystermen for "testing" in oyster-producing areas of the Atlantic coast). The Japanese oysters are apparently resistant to the pathogen, but carry it in the tissues as a benign parasite wherever the species is transferred — capable of infecting susceptible species such as the eastern oyster.

So the major and continuing economic losses to the oyster industry of the U.S. Atlantic coast seem to have been caused by the presence of an alien pathogen, introduced with its host, the Japanese oyster, first to the Pacific coast

of North America and eventually to the U.S. east coast, where it has invaded
susceptible populations of eastern oysters. The consequences have been almost
half a century of variable mortalities of oysters, which in epizootic years have
destroyed up to 90% of existing populations.

From Field Notes of a Pollution Watcher
(C.J. Sindermann, 2003)

This tale of an itinerant oyster pathogen (*H. nelsoni*) and its success in a susceptible host population illustrates part of the problem associated with introductions of nonindigenous marine species — the probability of importing disease organisms that can become epizootic in native species — but other problems also lie ahead for us in this chapter.

The geographic distribution of marine organisms is rarely static; population boundaries in the sea are flexible in the extreme, being affected by a host of factors, including ocean currents, temperature, salinity, food supply, predator abundance, presence of toxins, and so forth. Human activities have increasingly contributed to expansion of population boundaries and to introductions of nonindigenous species into new coastal habitats. Some of the introductions have been accidental (for example, animals or plants carried in ships' ballast water; Carlton & Geller 1993), and some have been intentional (principally species such as salmon, shrimp, and oysters introduced for aquaculture purposes). Some introductions have had favorable economic consequences (such as the Japanese oyster on the U.S. Pacific Northwest coast), but many (such as the green crab on that same coast) have had deleterious effects on native species.

An extreme position on introductions, accidental or deliberate, might be that they should be considered (at least initially) as potential forms of biological pollution (defining *pollution* as human activities causing negative effects on marine life, human health, resources, or amenities). This chapter is built around the perception that any introduction, accidental or not, can create three principal concerns:

Those related to possible *ecological changes* (especially changes in habitats, competition, or predation)
Those related to possible *genetic influences* on native species
Those related to the *introduction of pathogens* not endemic in the receiving area

Each of these concerns will be explored in the following sections.

ECOLOGICAL CHANGES RESULTING FROM INVASIONS BY ALIEN SPECIES

Alien species may cause drastic changes in receiving ecosystems, or they may fit into existing ecosystems without any or with only minor perturbations, especially where niches exist or are occupied inefficiently (an example would be displacement of the West Coast Olympia oyster, *Ostrea lurida*, during the past century by the introduced Pacific oyster, *Crassostrea gigas*). Explorations of new environments by alien species are usually natural phenomena, but the frequency of occurrence of probes and the geographic distances covered by "advance parties" have been enormously enhanced by human actions. Furthermore, successful invasions (in the sense of establishing extensive populations) are rare unless the process is artificially supported by aquaculture techniques (protection from predators, induced spawning, provision of food, etc.).

Humans have had major roles — deliberate or accidental — in a great number of past invasions of coastal/estuarine waters by alien species. Some of those invasions have produced drastic and often long-term modifications in existing ecosystems. In this section, we will consider two good examples: the invasion of the Mediterranean Sea by the alga *Caulerpa taxifolia* and the invasion of the Black Sea by the ctenophore *Mnemiopsis leidyi*. Other choices, such as the ongoing invasion of the Pacific Northwest coast by the green crab (*Carcinus maenas*), originally from Europe, or the earlier invasion of those same coastal areas by the Japanese littleneck clam (*Venerupis philippinarum*), or the invasion of San Francisco Bay, California, by the Asian clam (*Potamocorbula amurensis*; Carlton et al. 1990), can be found easily but will not be exploited here.

AN AGGRESSIVE INTRODUCED MACROALGA: *CAULERPA TAXIFOLIA*

Thus far in this book, we have considered (in Chapter 5) only the introduction of unattached planktonic or benthic microalgae as possible forms of biological pollutants, but it is important to point out that blooms of macroalgae, often introduced from other habitats, have occurred and are occurring. Effects on some coastal ecosystems have been profound; one such invasion is ongoing in the Mediterranean, as sketched briefly in the following vignette.

Spread of the Introduced "Killer Alga" *Caulerpa taxifolia* in the Mediterranean Sea and Beyond

This tale of alien invasion began innocently enough — in the display aquaria of the prestigious Oceanographic Museum of Monaco, whose director at the time (1984) was the famed marine publicist Jacques-Yves Cousteau. Those aquaria were routinely planted with a bright green alga received in 1982 from a public facility in Stuttgart, Germany, and propagated because of its ability to prosper in the cool winter temperatures of the Mediterranean. The organism was <u>Caulerpa taxifoli</u>, an unusual algal species normally found only in tropical waters.

<u>But it had escaped from the aquarium</u>, and was found growing in open waters in front of the Oceanographic Museum! Fortunately, its existence in the coastal waters was noted in 1984 by an algologist from the University of Nice, located just down the coast from Monaco. Serendipitously, this scientist, Dr. Alexandre Meinesz, was a specialist in the algal group, the siphonous green algae, to which the alien species belonged. He quickly recognized the potential

invasiveness of a cold-adapted strain of C. taxifolia, and immediately proposed steps toward complete eradication — the first move in what was to be a ten-year career-changing episode in his life.

His suggestions were resisted by local, regional, and national environmental bureaucracies, including the directorate of the Oceanographic Museum (since its aquarium was the likely source of the initial contamination). Years passed without effective control, and the alga spread up and down the coast, into French and Italian coastal waters out to depths of 50 meters. It competed successfully with the native flora, forming vast green meadows covering the sea bottom, and it secreted toxins that repelled fish and other marine animals. By the year 2003, Caulerpa taxifolia had invaded an estimated 40,000 acres of sea bottom in French coastal waters; it was spreading rapidly on the Spanish and Italian coasts; and it had even reached Tunisian shores (covering almost 1000 acres off the port of Sousse) (Langar et al. 2000). The alga has also been found at several locations near San Diego and Los Angeles, California, and on the Australian coast south of Brisbane.

Aroused national and international bureaucracies have finally responded with recommendations, workshops, committees, surveys, and conferences, and even a Caulerpa hotline (www.caulerpa.org). At some point early in this saga of invasion, a news reporter coined the inflammatory term "killer alga," maybe because it so completely displaces native plants and animals (but does not actually kill them, insofar as we know, even though toxins have been identified).

Dr. Meinesz and his associates conducted extensive research (see references from 1991 to 1999 at the end of this chapter) and fought the good fight for control measures, but adequate government responses did not materialize early enough — so he compensated by writing a good book called "Killer Algae" (Meinesz 1999).

From Field Notes of a Pollution Watcher
(C.J. Sindermann, 2004)

This vignette provides a reasonably coherent introduction to a more technical discussion of the invader, *Caulerpa taxifolia*, especially about unique features of the organism and its global spread.

It is a member of the Bryopsidales (Vroom & Smith 2001), an order characterized by a unique vegetative growth in which the cytoplasm flows from old growth to new, in a process of perpetual replacement. (Another noted invasive alga, *Codium fragile*, that caused problems on the New England coast beginning a few decades ago, and later on the New Zealand coast, is a close relative.)

The alga has a peculiar structure consisting of a single giant cell — multinucleate but not divided by cell walls. Its form is fern-like, with rhizoids for attachment, and it reproduces by cloning (fragmentation).

Caulerpa taxifolia may elaborate as many as nine toxins, including caulerpenyne and caulerpicine, some of which probably function as repellents for herbivorous animals or as inhibitors of attachment of other algae or microorganisms. There are as yet no reports of toxicity being transmitted by herbivores (sea urchins) feeding on the alga, and no toxicity risk to humans has been demonstrated.

The alga is an aggressive colonizer, easily displacing native Mediterranean flora, including its indigenous relative *C. prolifera*, as well as red and brown algae and gorgonians. Biodiversity is thus drastically curtailed in invaded territories.

Rates of invasion have been simulated, with the finding that one cutting of *C. taxifolia* can produce invasion coverage of 25 acres in about 6 yr (Meinesz 1999).

No sexual reproduction has been observed in the invasive strain of the species; only male gametes are formed (possibly because the clone from which that Mediterranean strain was derived was from a male plant).

Spread of this aggressive temperature-tolerant strain of *C. taxifolia* (now labeled the Mediterranean strain) may well have been aided by releases from private and public aquaria — especially since it is still advertised for sale on the Internet by aquarium suppliers.

The invasive Mediterranean strain of *C. taxifolia*, unlike its tropical counterparts of the same species, is adapted to temperate waters and can even survive for 3 months at 10°C. However, at temperatures below 13°C the fronds begin to regress, and growth does not resume until the water temperature exceeds 18°C (Meinesz 1999).

Biological control is often an option of choice (or of last resort) for control of invasive introduced species (Center, Frank, & Dray 1997). The search for an herbivore that might selectively feed on *C. taxifolia* was conducted by Meinesz (1999) for several frustrating years, and his best candidates were two species of small tropical sea slugs (saccoglossans), *Elysia subornata* and *Oxynoe azuropunctata*, that were known to feed on tropical *Caulerpa* in the Caribbean. These slugs turn out to be fascinating in their habits:

- They suck out the algal cytoplasm, and the frond eventually dies (since *Caulerpa* is a single giant cell, unlike a majority of other native algae, which are multicellular).
- They ingest chloroplasts of *Caulerpa* and store them under the skin, so they can survive in the presence of light for several weeks without food, as pseudo-plants.
- They ingest *Caulerpa* toxins, modify them, and become poisonous to predators.

Unfortunately, these slugs, being tropical forms, only survive in the Mediterranean summer; and, furthermore, the French Ministry of the Environment did not fund and did not authorize field testing that was required for their use as biological control agents.

Genetic studies of *Caulerpa taxifolia* have provided definitive information about possible routes of colonization by the so-called Mediterranean strain (Jousson et al. 1998). Sequences of the gene ITS rDNA in samples from Monaco and elsewhere in the Mediterranean are identical and are remarkably similar to samples from Australia (Meusnier et al. 2001, Wiedenmann et al. 2001), suggesting a possible origin there, since the species has been reported to have existed on the southeast coast (Moreton Bay) for half a century (Meinesz 1999).

Caulerpa taxifolia was discovered in several locations in California in the year 2000 — first in lagoons near San Diego and then farther north, at Port Huntington near Los Angeles (Dalton 2000, Kaiser 2000). The samples from California closely resembled the invasive Mediterranean strain, morphologically and physiologically, and were genetically identical, as determined by sequencing of the gene ITS rDNA (Jousson et al. 2000).

The invasive nature of *C. taxifolia* has been recognized officially but belatedly. In 1995, a recommendation was adopted by the European Council, to control the spread of the alga in the Mediterranean — calling for collaboration, mapping, and eradication. A similar recommendation was made in that year in the U.N. Action Plan for the Mediterranean (Meinesz 1999). Nothing happened.

Invasion of coastal areas by *Caulerpa taxifolia* results in significant ecosystem alterations. Indigenous algae are overgrown and displaced; the sea bottom and all its features are covered with undulating bright green masses of algal fronds; marine fish, and other animals become scarce; and each year brings greater spatial expansion of the invader. The hour is late in the Mediterranean for human intervention; natural ecosystem dynamics will assume control of the future course of events there.

THE INTRODUCED CTENOPHORE *MNEMIOPSIS LEIDYI* IN THE BLACK SEA

Mnemiopsis leidyi, an alien predatory comb jelly (phylum Ctenophora) was first observed in the Black Sea in 1982, presumably introduced with ships' ballast water from North America. It increased in numbers dramatically in ensuing years, reaching the astounding level of 10^9 tons of biomass (wet weight) in 1989 for the entire Black Sea. Its abundance declined in 1991–1992 but increased again in 1994 to an estimated 10^7 tons.

Population Explosion of the Introduced Comb Jelly *Mnemiopsis leidyi* in the Black Sea

As a first-year graduate student in 1950, I and 80 others from many universities took the Invertebrate Zoology course at the Marine Biological Laboratory in Woods Hole, Massachusetts. The experience was superb; we took field trips with the famed Libby H. Hyman, and perused trematode life cycles with the equally famed Horace W. Stunkard — among other excellent instructors. Some of the animals we studied were dredged from the tidal "Eel Pond" located down the street from the classroom building. One of the most interesting members of the fauna was a gelatinous walnut-shaped translucent blob with long rows of vibrating hairs, moving ever-so-slowly and gracefully through the water column. The beast was the ctenophore <u>Mnemiopsis leidyi</u> which existed in the saltwater pond in seemingly countless numbers.

My next contact with the organism occurred five years later. With a fresh new Ph.D., I had begun teaching the course in Invertebrate Zoology at Brandeis University. One of the labs in the course was on so-called "minor phyla" — which included the Ctenophora as well as other little-known invertebrate groups. There *in a large jug were preserved <u>Mnemiopsis leidyi</u> — getting little respect, as usual, since they were not the kind of animals that lent themselves to complex dissection by undergraduates. After a few years of university teaching, I moved on to marine disease research, and completely lost track of <u>Mnemiopsis</u> and its world.*

But the organism was to gain my attention again, and the world's respect, beginning in 1987, when it appeared in overwhelming numbers in the Black Sea — apparently introduced in the early 1980s in ballast water of ships from North America! Estimates of its total biomass (10^9 tons in 1989, for example) were absolutely mind-boggling, and the entire ecosystem of that sea was severely disturbed. Part of the disruption was that, as plankton feeders, the ctenophore hordes virtually swept the waters clean, leaving little food for plankton-eating fish, and consuming their eggs and larvae, thus participating as a causal agent in a drastic decline in fish production. The massive outbreak of <u>Mnemiopsis</u> subsided in the early 1990s, but then increased in abundance again in 1994, continuing its impact on the beleaguered Black Sea, already suffering from the human abuses of overfishing, industrial pollution, and eutrophication — all imposed on a body of water that is, in part, permanently anoxic. What evil we inflict, even in little-known parts of the planet!

From Field Notes of a Pollution Watcher
(C.J. Sindermann, 1998)

This brief history of my fleeting contacts with *Mnemiopsis leidyi* is of course meant to be an attention-getting device. But examination of the sudden abundance of an introduced predator in the Black Sea and its multiple impacts on the already disturbed ecosystem that it has invaded can be instructive in a much broader ecological sense. The event can also serve as a warning of a bleak future to those humans who degrade and mismanage enclosed or partially enclosed seas.

Since much of my career in science has been concerned with marine fisheries, I tend to think about fish as principal actors in ecosystem dramas, especially those that involve disturbed coastal/estuarine environments. A major role for an introduced ctenophore as a dominant predator is disconcerting, and even a tad frightening. Somehow, jellyfish and comb jellies should belong in bit parts, and not as stars of the show — which is the situation we have at hand. I am uneasy enough to feel the need to explore this outbreak in more detail here, to squeeze out some ecological sense from what is happening. More is going on than a population explosion of an introduced species, although that event is important in itself.

I have been preceded in this exploration by the many good scientists of GESAMP (Group of Experts on the Scientific Aspects of Marine Pollution). That group made a very thorough review of the *Mnemiopsis* outbreak and published a detailed report in 1997. I have referred to it liberally in the summary discussion that follows, as well as to the published literature from many countries, especially those bordering on the Black Sea.

That body of water, almost landlocked except for a narrow strait (the Bosphorus), was treated abominably during the years of Soviet domination (Mee 1992).

Several major polluted rivers (the Danube, Dnieper, and Dniester, in particular) empty into it, carrying industrial and agricultural pollutants and nutrient loads from a large part of central Europe. Massive water diversion and extraction projects reduced the flow of those rivers and raised estuarine salinities; industrialization of the sea coast was extensive; toxic wastes were pumped directly into the sea; and fishing intensified on disappearing stocks — first on predators and then on forage species (sprat, anchovies, and horse mackerel). All these factors helped to create what ecologists call a "disturbed ecosystem," an environment conducive to success of invasive alien species.

Mnemiopsis leidyi appeared in 1982 and multiplied profusely to almost obscene numbers by 1987. Three principal impacts of the burgeoning ctenophore population were identified by GESAMP (1997) as:

- Predation on fish eggs and larvae (Tsikhon-Lukanina, Reznichenko, & Lukasheva 1992, 1993)
- Feeding on adult and larval zooplankton (Zaika 1993)
- Accelerating ongoing eutrophication and extension of anoxic zones through death and decay of an enormous biomass

Collapse of the coastal pelagic fisheries in the Black Sea in 1989 was attributed by GESAMP to combined effects of overfishing, industrial pollution, eutrophication, and the *Mnemiopsis* invasion. Biodiversity had been reduced by the early 1990s to the point where pelagic plankton feeders (the sprat, anchovies, and horse mackerel) composed 95% of catches, but those catches had been reduced from 350,000 tons in the mid-1980s to only 30,000 tons, principally by the *Mnemiopsis* outbreak (Caddy & Griffiths 1990, Caddy 1993).

The role played by the ctenophore was placed in proper context in the following summary by GESAMP (1997):

> The *Mnemiopsis* invasion had a grave impact on the Black Sea fishery (FAO 1993). After 1988, the remaining commercially important pelagic fish stocks drastically decreased. In addition, the recruitment of two yearclass generations of pelagic fish (anchovies and horse mackerel) was reduced to negligible levels. These events were a consequence, at least in part, of the direct grazing impact of the new predatory intruder, which competed with these fishes for food and ate their eggs and larvae. The drop in fish catch by the CIS countries was also related to the decrease in fishing efforts on this species, aggravated by the deterioration of the socio-economic situation in CIS countries. Thus, a general decrease in the volume of fish landings after 1988–1990 resulted not only from the invasion of the basin by *Mnemiopsis*, but also by the socio-economic crisis in the former communist Black Sea countries. (p. 79)

It is interesting to note that outbreaks of the jellyfish *Aurelia aurita* had occurred in the Black Sea in the late 1970s (Zaitsev & Polischuk 1984), as had increases in biomass of small pelagic fish (Ivanov & Beverton 1985), probably as consequences of increasing eutrophication (Zaitsev 1992a, 1992b). Decline in benthic animals and demersal fish stocks were also caused in part by the same processes of nutrient loading and increasing anoxic water. With this as background, Zaitsev (1993a,

1993b) summarized his interpretation of events leading to the *Mnemiopsis* outbreak as follows:

> [T]he invasion of *Mnemiopsis leidyi* of the pelagic biome in the mid- to late 1980s encountered a high standing stock of pelagic zooplankton, whereas a heavily exploited small pelagic fish population, with annual exploitation rates of up to 50 per cent or more, competed with *Aurelia* for the role of dominant primary carnivore. Such high exploitation rates of traditionally important zooplanktivores (anchovy, horse mackerel, kilka, etc.) may have acted jointly, perhaps synergistically, with eutrophication to open up an ecological niche for the comb jelly predator in the pelagic ecosystem of the Black Sea. This role was first filled by *Aurelia* and then by *Mnemiopsis*. (Zaitsev 1993b, p. 84)

Strategies for control of *Mnemiopsis* were of course important discussion points in the GESAMP deliberations, but the melancholy conclusion was reached that only ecological control by predator introduction or enhancement had a realistic chance of success (Harbison & Volovik 1993, 1994). That august group of experts also pointed out that "it is essential that the predators that inhabit the home range of the pest be investigated, since predators that have evolved in the same habitat as the pest will be best equipped to prey on it" (GESAMP 1997, p. 81).

It turns out that gelatinous zooplankton constitute a major dietary component for more than 100 fish species worldwide. Of these, GESAMP (1997) selected three possible candidates: Baltic cod (*Gadus morhua callarias*), chum salmon (*Oncorhynchus keta*), and the Atlantic butterfish (*Peprilus triacanthus*). Of this short list, and despite some disadvantages, the butterfish may be the predator of choice, because it is a highly specific feeder on gelatinous zooplankton and is known to feed on both *Aurelia aurita* and *Mnemiopsis leidyi*. Furthermore, it has been identified as possibly controlling *M. leidyi* populations in parts of the North American coastline (Oviatt & Kremer 1977).*

In view of the escalating number of introductions of alien species in coastal/estuarine waters of the planet and the increasing environmental impacts of successful invasions — such as that of *Mnemiopsis leidyi* in the Black Sea — a global mechanism (maybe a GESAMP with fangs) is needed to accumulate and disseminate information about new problems, to examine post-introduction histories, and to develop and implement control methods for aggressive invasive species.

GENETIC INFLUENCES OF ALIEN SPECIES ON NATIVE SPECIES

In approaching a topic that combines studies of population genetics with introductions of genetic material from other gene pools, it is tempting to rush off in many directions. Should we first consider the possibility of actually changing the genetic

---

* But the tale of the *Mnemiopsis* invasion has an additional chapter. Another accidentally introduced ctenophore, *Beroe ovata*, a voracious predator on other members of the phylum, began to increase in numbers and to effect a reduction in *Mnemiopsis* populations in 1999 (Kideys et al. 2005). The interaction resulted in positive changes in the Black Sea ecosystem, including increases in numbers of zooplankton, fish eggs, and fish larvae (Shiganova et al. 2004).

to genetically alter native populations, reduce local adaptation, and negatively affect population viability and character" (p. 965; i; italics mine).

- "Transferred fish, many of which have been kept in culture, will differ genetically from native populations. ... Adaptive differences are also likely to be increased by genetic changes incurred by transferred fish while in culture" (p. 970).
- *"If adaptive differences exist between transferred and native salmon, and transferred fish survive to breed, then interbreeding will lead to an overall reduction in survival and recruitment in the native population"* (p. 970; italics mine).

These generalizations have been supported by other investigations in several countries, since, from the earliest commercial sea pen rearing efforts with Atlantic salmon off the unforgiving coasts of Norway, Scotland, and North America, a great concern was the possibility that escapees from farmed populations might produce adverse changes, through interbreeding, in the genetic composition of native populations (Hindar, Ryman, & Utter 1991). This seemed entirely probable, because homing in the species is precise, resulting in small geographically and genetically restricted subpopulations whose adaptations to local conditions can be overwhelmed by gene flow from outside. Small-scale escapes from sea pens were and are common, and storms have caused and still cause large-scale escapes of tens to hundreds of thousands of cage-adapted, hatchery-reared fish (Gausen & Moen 1991).

Salmon brood stock used to produce sea pen juveniles are genetically distinct from native salmon; they may be derived from geographically distant parent stocks and may have been selected for growth and size attributes, as well as through inadvertent and other directional selection (Skaala et al. 1990). (The sheer overwhelming size of the sea pen operation in Europe (especially in Norway) can be easily seen in the numbers: European farmed Atlantic salmon production exceeded 400,000 tons annually by 1996 — far beyond production from wild stocks.)

A comparative study in Ireland of genetic changes resulting from escaped farmed Atlantic salmon breeding with wild populations was reported by McGinnity et al. (1997). An experimental study was designed to compare freshwater performance of the progeny of wild, farmed, and hybrid spawners. Juveniles were assigned to group parentage by DNA profiling based on composite genotypes at seven loci. Findings were:

- Survival of progeny of farmed salmon to the smolt stage was significantly lower than that of wild salmon, with increased mortality being greatest in the period from eyed egg to the first summer.
- Progeny of farmed salmon (probably because of earlier selection for favorable aquaculture traits) grew fastest and displaced the smaller native fish downstream.
- Growth and performance of hybrids were generally either intermediate or not significantly different from wild fish.

The authors concluded that escaped farmed salmon can produce long-term genetic changes in natural populations, affecting single-locus or quantitative traits (such as growth and sea age at maturity). Such changes are likely to reduce population productivity.

This, then, is my example of the kinds of genetic concerns that emerge when populations are manipulated by humans for commercial reasons. One advantage that can be seen in the results of investigations cited here is that with the advent of molecular approaches, such as DNA profiling, genetic changes resulting from human interventions can be determined with precision and followed in long-term studies designed to assess population impacts.

But beyond understanding some of the genetic effects of transfers for aquaculture purposes, what about inserting genes for favorable traits into an aquaculture species (disease resistance or rapid growth, for example)? Beginning in the 1980s, with the development of recombinant DNA technology, it became possible to introduce genetic material from one animal into another (Ferguson 1990). The Atlantic salmon was again an early participant; in 1988 the gene for so-called "antifreeze protein" was introduced into salmon from winter flounder (*Pseudopleuronectes americanus;* Fletcher et al. 1988). The procedure involved injecting genetic material from the flounder into the developing egg. The foreign gene is incorporated into the genome of the recipient and codes for some permanent desirable change in its morphology or physiology (in this case for tolerance to low temperature).

Experimental studies with transgenic fish have been carried out since that time with a number of species, but particularly with coho salmon (*Oncorhynchus kisutch;* Devlin et al. 1994, 1995; Hill, Kiessling, & Devlin 2000). Dramatic increases in growth rates were achieved in transgenic salmon carrying a growth hormone gene construct — although with a tradeoff of slower swimming speed (Lee, Devlin, & Farrell 2003). Growth-enhanced transgenic coho salmon in one study were, at age one, more than twice as long and more than 10-fold heavier than nontransgenic controls (Farrell, Bennett, & Devlin 1997). Although the point can be argued, such transgenic fish can logically be considered an "introduced" species.

INTRODUCTION OF PATHOGENS NOT ENDEMIC IN THE RECEIVING AREA

Much recent attention has been properly directed to the third principal concern about introduced species: disease and its implications. This is entirely logical, since disease may have profound effects on populations of native species, especially those of economic value, and since some marine disease problems have proved to be remarkably intractable for extended periods.

The possible introduction of severe pathogens with movements of fish and shellfish leads directly to an important concept — that *all introductions of pathogens are "accidental" even though the introduction of host animals can be deliberate.* This fact is often overlooked in discussions of "intentional" vs. "unintentional" introductions, and even in drafting legislation concerning introduced species. Emphasis must therefore be on control and exclusion of diseased animals and infected resistant carriers — through programs similar to those that have evolved to control disease dissemination in terrestrial animals.

I propose to consider here five pathogen groups that will help make the case for the importance of introduced disease agents. They are (1) oyster pathogens, (2) shrimp viruses, (3) an eel nematode, (4) a protozoan parasite of bay scallops, and (5) a pilchard herpesvirus. Such an emphasis may seem like overkill, but introduced diseases have had and are having demonstrable and severe effects on marine populations. We can begin with the oyster pathogens, which have been subjects of research in many countries for over half a century.

OYSTER DISEASES

Oysters have long been known to be subject to mass mortalities, but the past 4 decades have been especially troublesome. Major and widespread mortalities have occurred and in some cases are still occurring in the United States, Europe, and Japan. Specific pathogens, often viruses or protozoans, have been identified as causative organisms in most instances.

Oysters have also been moved from place to place probably more frequently than any other marine animals group except for shrimp, providing excellent vehicles for dissemination of pathogens and parasites. Transfer networks comparable to those known for shrimp have been created — especially for the Pacific oyster (*Crassostrea gigas;* Chew 1990), and important pathogens have been transported, sometimes by complex pathways (Elston, Farley, & Kent 1986).

One of the largest experiments in the introduction of marine species was the importation to the coast of France of the Pacific oyster (*C. gigas*) during 1966–1977. Introduced as seed and as adults from Japan and British Columbia, *C. gigas* replaced declining populations of the so-called "Portuguese oyster," *C. angulata*, which had been the industry mainstay for decades. The Pacific oyster prospered; reproduction was successful in some parts of the French coast, and production now exceeds 150,000 tons (Grizel & Héral 1991).

What have we learned from this massive French experiment from a disease perspective? A remarkable series of epizootics in the two native species of oysters occurred simultaneously with the Pacific oyster introductions. Beginning in 1966, the native oyster *C. angulata* died in large numbers from a viral gill disease (Comps & Duthoit 1976, Comps et al. 1976). By 1973, when the epizootic subsided, populations of *C. angulata* had been largely destroyed, and the introduced Pacific oyster (*C. gigas*) had replaced them in most growout areas. During the same period, populations of the European flat oyster (*Ostrea edulis*) were also affected severely by epizootic disease (Comps 1970, Comps, Tige, & Grizel 1980). Beginning in 1968, the protistan parasite *Marteilia refringens* affected oyster growing areas. As that epizootic waned in the mid-1970s, *Bonamia ostreae*, another protistan parasite, increased to epizootic proportions and further reduced populations of *O. edulis*. This epizootic continues at present, and, as a result, *O. edulis* culture is at a standstill. There is some evidence that the pathogen *Bonamia* was introduced to France with imports of *O. edulis* seed from a California hatchery that had been rearing offspring of a stock that originated decades earlier in the

Netherlands and was first introduced into Connecticut in the 1950s (Elston, Farley, & Kent 1986).

The occurrence within a single decade of three major oyster epizootics, each with accompanying mass mortalities, is unique in the long and well-documented history of oyster culture. Also unique is the scale of importation of a replacement species (*C. gigas*) during this same period. Despite extensive research by European pathologists, the role that this massive introduction of a nonindigenous species may have played in disease outbreaks in the native species is unknown, and a direct relationship has not been demonstrated, although suggestions have been made and some associations proposed — like the one describing the history of *Bonamia*.

The worldwide experience with introduced *C. gigas* has indicated that at least three categories of disease risks exist.

1. One risk is from the known pathogens of the introduced species, which may or may not be transferred to the native species. For *C. gigas*, five pathogens are known:
   a. An iridovirus that causes "larval velar disease" (and which is similar to the virus that killed *C. angulata* in France)
   b. A bacterium, *Nocardia* sp., which causes fatal inflammatory bacteremia
   c. An ascetosporan protozoan parasite, *Marteilioides chungmuensis*, which affects ova
   d. A presumptive ascetosporan protozoan, *Mikrocytos mackini*, which caused "Denman Island disease" in *C. gigas*, introduced on the west coast of North America, but yet unreported from native stocks in Japan (Farley, Wolf, & Elston 1988)
   e. A parasitic copepod, *Mytilicola orientalis*, found in the gut

2. Another risk is from other microorganisms with unknown pathogenicity to *C. gigas*, but possibly pathogenic to other species of oysters (an example of this category would be the reported occurrence of a haplosporidan parasite, similar to *Haplosporidium nelsoni*, a severe pathogen of American oysters, in *C. gigas* from Korea, but not known to be pathogenic to *C. gigas* (Genetic studies of this haplosporidan, isolated from introduced *C. gigas* from California, indicated that it was identical to *H. nelsoni* from eastern oysters.)

3. A third risk arises from still other microorganisms, rare or unrecognized in populations of *C. gigas*, that may be pathogenic to related species of oysters (examples suggested, without supporting evidence, are *Marteilia* and *Bonamia* in flat oysters, *O. edulis*).

The uncertainties present in this listing are reflections of the relative youth of marine pathology as a scientific discipline. Many diseases of oysters and other species are unknown or poorly understood, so they would not form part of normal inspection protocols.

Closer to home, the native East Coast American oyster *C. virginica* has been hard hit in the Middle Atlantic states by epizootics of two protozoan pathogens that began in the late 1950s and still persist. Oysters are frequently shipped from state to state, and documentation is good for the transfer of the diseases, caused by the protozoa *Haplosporidium nelsoni* and *Perkinsus marinus* to states where they had been previously unknown.

The transfer network for some cultured species can be extraordinarily complex, and some pathways can be quickly obscured or never revealed. An excellent example is seen in the attempt by Elston, Farley, & Kent (1986) to trace the origin of the flat oyster pathogen, *Bonamia ostreae*, that has destroyed most European production of that species beginning in 1979. Export of infected seed oysters to France from a hatchery on the California coast was proposed as the immediate source of the disease. That hatchery had received brood stock (presumably infected) from the Milford Laboratory in Connecticut, which, in turn, had received the original introductions from the Netherlands in the 1950s. Somewhere in the early part of this chain of events, possibly at Milford, *Bonamia* infected the introduced species. The severe consequences of bonamiasis for European culture of flat oysters, *Ostrea edulis*, have been described in detail by Balouet, Poder, & Cahour (1983); Figueras (1991); Hudson & Hill (1991); McArdle et al. (1991); Stewart (1991), and Van Banning (1991).

This and other examples illustrate the reality that *unless early scientific attention is directed to the occurrence of a new parasite or microbial pathogen, the history of its introduction and subsequent dissemination may be quickly lost*. However, use of new genetic techniques provides a partial window to the past, in comparing gene sequences in pathogens from geographic subpopulations, as was done with *H. nelsoni*.

But my *Bonamia* story is not finished yet. In 2005, nearly half a century after our first contact with what we now accept as an alien oyster pathogen (*H. nelsoni*) — introduced with its Japanese host, *Crassostrea gigas* — several Middle Atlantic states are deliberating about introducing another alien oyster, the Asian or suminoe oyster, *Crassostrea ariakensis*, originally from China. This species seems resistant to the parasite (*H. nelsoni*) that has been so devastating to native East Coast oysters, but it is apparently host to a *Bonamia* species, which may be the pathogen that has already caused major disease problems in European flat oysters (*Ostrea edulis*) beginning in the early 1980s and continuing to the present. Worse still, an outbreak of bonamiasis, with mortalities, has been reported to have occurred in an experimental population of *C. ariakensis* being held in North Carolina in 2003!

So what is the best approach to reviving a dying East Coast oyster fishery? Developing strains of resistant native oysters has been the method of choice for decades, but resistance has been slow to appear and to be disseminated in surviving stocks. The alternative of introducing the Asian oyster is now (in 2005) being seriously considered by the affected states, despite the known presence in that species of at least one lethal oyster pathogen, and possibly others as yet unknown.

VIRAL DISEASES OF SHRIMP

Worldwide interest in penaeid shrimp culture has led to a frenzy of transfers and introductions of species with desirable culture characteristics. During all this commercial activity, it has become painfully clear that viral diseases, transported with their hosts, are serious impediments to successful culture, wherever the operation may be located. Among the 20 or so viruses known to infect shrimp, four have been recognized recently as especially severe problems for industry development (Lightner & Redman 1998). These four lethal pathogens are known as (1) infectious hypodermal and hematopoietic necrosis virus, (2) Taura syndrome virus, (3) yellowhead virus, and (4) white spot syndrome virus. Since these four viruses can have profound effects on culture production of shrimp, and since they all have been disseminated widely, they deserve a little attention here:

Infectious Hypodermal and Hematopoietic Necrosis Virus (IHHNV) — IHHNV disease was first recognized in 1981 in Hawaiian shrimp culture facilities rearing *Penaeus stylirostris* introduced from Panama (Lightner et al. 1983). Soon after, the role of *Penaeus vannamei* as a carrier of the virus was recognized. This highly lethal pathogen poses a serious threat to shrimp culture (Lightner, Redman, & Bell 1983). Found in cultured *P. stylirostris, P. monodon, P. japonicus,* and *P. vannamei,* and in wild-caught *P. monodon* brood stock in Southeast Asia (Lightner, Bell, & Redman 1989), it has been introduced into aquaculture facilities in Florida, Texas, Tahiti, the Philippines, Guam, and elsewhere. *Penaeus stylirostris* is especially susceptible, and extensive mortalities from IHHNV infections have been reported. *Penaeus vannamei* harbors subacute infections that may cause stunting (Kalagayan et al. 1991).

Taura Syndrome Virus (TSV) — TSV emerged in Ecuador in 1991–1993 and quickly spread to all Latin American shrimp production areas; by 1996 it had spread to all shrimp farming regions of the Americas, including Texas, Florida, and South Carolina. At the time of the virus' appearance, shrimp culture in the Americas had come to depend on rearing *Penaeus vannamei* almost exclusively (for example, in 1992 *P. vannamei* production made up over 90% of the total shrimp aquaculture output of the Americas; Lightner & Redman 1998). TSV is especially lethal to this shrimp species, and major production losses resulted.

Yellowhead Virus (YHV) — YHV is a lethal pathogen of *Penaeus monodon* in Asia. It was first recognized in 1991 in Thailand and soon became widespread in Southeast Asia and Indo-Pacific shrimp farming regions, where *P. monodon* was the species of choice for aquaculture at the time. Major mortalities occurred in those regions in the early to mid-1990s, and the pathogens have been exported with their frozen hosts to the Americas and to other shrimp-consuming countries.

White Spot Syndrome Virus (WSSV) — WSSV was first recognized during a 1992 epizootic in three penaeid species — *P. monodon, P. japonicus,* and *P. penicillatus* — in Taiwan. In 1993 it caused mortalities in shrimp farms in Japan, Korea, and China; in 1994 it appeared in Thailand and India; and in 1995 infections with the virus were found in a shrimp farm in Texas. By 1996 most shrimp farming areas of eastern and southern Asia as well as Indonesia had been affected, and WSSV had been isolated on six occasions from penaeid shrimp on the east and gulf coasts of the United States (Lightner & Redman 1998).

These four viral pathogens — IHHNV, TSV, YHV, and WSSV — are the present-day heavy hitters in penaeid shrimp culture pathology. The first two (IHHNV and TSV) have had their principal effects on shrimp culture in the Americas; the last two (YHV and WSSV) have had major impacts in Asia and the Indo-Pacific region. All have been spread widely by commercial practices.

A major concern in shrimp disease investigations has always been that one or more of the lethal viruses found in aquaculture facilities and transported worldwide will escape into native shrimp populations — and indeed this seems to have happened in at least two documented instances:

1. An example of an introduced virus infecting native stocks and producing an epizootic, accompanied by population reduction, was reported by Morales-Covarrubias et al. (1999). IHHNV, not present in Mexico before 1987, was diagnosed in that year in *Penaeus vannamei* postlarvae imported for aquaculture. By 1990 the virus had been diagnosed in all coastal hatcheries and farms raising *P. stylirostris* (Lightner et al. 1992a, 1992b), and was reported in wild populations of that species in the Gulf of California (Pantoja 1993; Pantoja, Lightner, & Holtschmit 1999) — accompanied by a decline of about 50% in *P. stylirostris* fishery landings. An epizootic of IHHNV was suggested by Morales-Covarrubias et al. (1999) to have been a factor in the decreased landings, although other factors could have contributed. Sampling wild *P. stylirostris* stocks in the same area of the Mexican coast in 1996 disclosed high levels of IHHNV infections (56 to 89%) in adult shrimp, but the fishery was beginning to recover.
2. WSSV has been isolated "on at least six occasions from apparently normal or from diseased wild, captive wild, and cultured stocks of freshwater crayfish and marine penaeid shrimp from sites on the eastern, southeastern, and Gulf of Mexico coasts of United States" (Nunan, Poulos, & Lightner 1998).

Establishment of the introduced viral pathogens in wild stocks could have serious but as yet unknown implications, including major epizootics. Stay tuned!

Some appreciation for the extent of the viral problem in shrimp can be seen by examining the movement of stocks during the past 2 decades from centers of research and commercial development in Hawaii, Tahiti, Panama, Japan, and elsewhere. Such extensive movements create a transfer network, often operating without adequate disease inspection. IHHNV and other viruses have already entered this network and are being introduced into farms far from their original geographic range (see Figure 8.1).

Some of these introductions have resulted in huge economic losses to shrimp farms, and even — in two documented instances — transfer of the pathogens to wild stocks. At present, shrimp culture is still largely dependent on capture of wild shrimp for brood stock — a situation that permits maximum vulnerability of a giant industry to disease. Development of specific pathogen-free (SPF) or specific pathogen-resistant (SPR) stocks is expected to alleviate the problem eventually, but such stocks are not yet at the stage of commercial utility. Disease remains an overriding problem in shrimp culture, and introduced viruses constitute a large part of that problem.

The concept of a transfer network is extremely important. Humans are busy creating these complex transfer networks with many cultured marine species, but especially with shrimp, oysters, salmon, and eels. Severe pathogens are entering these networks and moving along them, killing substantial numbers of introduced and native stocks. As pointed out by Lightner (1990),

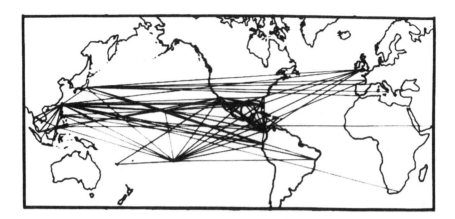

FIGURE 8.1 A "shrimp transfer network." (Prepared by and courtesy of Dr. D.V. Lightner, University of Arizona, 1990, from published and unpublished reports of shrimp stock movements.)

It is apparent that if an unrecognized pathogen entered any facility in the transfer network, the mechanism exists for it to be rapidly transferred to several other facilities. Further, if the pathogen remained undetected, it could be easily introduced into all of the facilities in the transfer network. (p. 5)

INVASION OF EUROPEAN EELS BY ALIEN NEMATODES

A recent and severe disease problem in Europe is the spread of nematode worms, *Anguillicola crassus*, in native European eel, *Anguilla anguilla*, populations. The worms were introduced with shipments of live Japanese eels, *A. japonica*, from Asia. They are large bloodsucking organisms that occlude the swim bladder (see Figure 8.2), cause emaciation, result in mortality in holding pens, and interfere with spawning migrations (Koie 1991).

Infection of native eels was first noticed in Germany in 1982 (Neumann 1985), probably as a consequence of release of infected eels shipped from Taiwan in 1980 (Koops & Hartmann 1989). The parasites now occur in most of the countries of Europe (the Netherlands, Denmark, Poland, England, Spain, Greece, and others), often with high prevalences and intensities (Peters & Hartmann 1986). Infections can be acquired as early as the elver stage, in which an acute inflammatory reaction occurs in the swim bladder; small crustaceans serve as intermediate hosts.

Population expansion of the worm following introduction has been rapid. In one river in England (where the worm was first reported in 1987), prevalence levels of 100% and average intensities of 6.7 worms were attained in 1 yr (Kennedy & Fitch 1990). An active network of eel transfers in Europe has undoubtedly favored the rapid expansion of nematode populations throughout the subcontinent. Eel farms have been seriously affected, with reduction in growth rates, emaciation, and mortalities of up to 65% in captive populations. Like many introduced parasites, *Anguillicola* will continue to spread by natural movements of hosts, but principally by

FIGURE 8.2 Nematodes (*Anguillicola crassus*) occluding swim bladder of European eel, *Anguilla anguilla*. (Photograph courtesy of Dr. P. van Banning, Netherlands Institute for Fisheries Research.)

human transport for stocking aquaculture ponds and for market within and across national boundaries.

AN IMPORTED PROTOZOAN DISEASE OF BAY SCALLOPS IN CANADA

The previous sections on oyster and shrimp diseases illustrate some of the complexities and uncertainties of disease transmission through transfer networks. One disturbing fact not yet pointed out, however, is that pathological examination of candidate species for introduction may not always disclose the presence of known disease agents and is unlikely to identify unknown agents except by chance. Known disease agents may have cryptic stages or may be extremely rare in the samples examined, and they may elude recognition. Pathological effects of undescribed pathogens may be labeled "idiopathic lesions" or "nonspecific granulomas" and thus escape deserved attention.

An excellent case history of just such an event can be found in the introduction into Canadian waters of a parasitic protozoan — a species of *Perkinsus*, with its host, the bay scallop, *Argopecten irradians*. Histopathological examination of the original imports from United States in 1979–1980 disclosed only chlamydia-like and rickettsial infections and nonspecific granulomas. The scallops were reared in quarantine for three generations before field release. Then, in 1989, after open-water culture had begun, hatchery brood stocks were examined, and a *Perkinsus* agent was recognized in the formerly described "granulomas," subsequently described as *P. karlssoni* by McGladdery, Cawthorn, & Bradford (1991). A similar or identical organism had been seen much earlier in bay scallops from U.S. waters (Ray & Chandler 1955), and more recently by Karlsson (1990). This episode illustrates very well the principle that *disease risks from introductions are never zero, even when adequate regulatory and inspection systems exist*. Canada has been diligent in developing regulations and a regionally based infrastructure to control introductions of nonindigenous species, yet, despite this concern and action, the pathogen *P. karlssoni* was introduced into Canadian waters and is now present in Canadian bay scallop populations. The pathogenicity of *P. karlssoni* to bay scallops has yet to be clearly established, however, and interspecies transmission to other molluscs has not been observed (McGladdery, Cawthorn, & Bradford 1991).

AN IMPORTED HERPESVIRUS IN AUSTRALASIAN FISH

A more recent example of possible effects of an introduced viral pathogen was seen in two extensive mortalities of Australasian pilchards (*Sardinops sagax neopilchardus*) in Australia and New Zealand, in 1995 and again in 1998 (Whittington et al. 1997, Ward et al. 2001). A herpesvirus was considered the causative agent, and an estimated 70% of South Australian adult pilchard stocks died in each outbreak. The prevailing hypothesis pointed to introduction of the viral pathogen with untreated frozen imported forage species (including *Sardinops sagax*) used as a food source for an expanding sea pen aquaculture of southern bluefin tuna, *Thunnus maccoyii*, in South Australian coastal waters. Large quantities (over 10,000 t/yr) of sardines (*Sardinops sagax*) and other clupeoids were imported from California, Peru, Chile, and Japan, beginning in 1993–1994, and fed to tuna in sea cages offshore (Whittington et al. 1997). Mortalities began on the central South Australian coast at Anxious Bay and spread westward into Western Australian waters and eastward as far as Queensland, encompassing about 5000 km of coastline.

A continuing investigation of population impacts of the mortalities (Ward et al. 2001) disclosed that the estimated spawning biomass of pilchards in South Australian waters fell from 165,000 t in early 1995 to only 37,000 t following the epizootic and then a fairly rapidly spawning biomass reached an unlimited 147,000 t in early 1998, only to fall again as a consequence of the 1998 mortalities to only 36,000 t. Furthermore, juveniles died in large numbers in South Australia during the 1998 outbreak, whereas the 1995 mortalities involved only adult fish.

Mortalities also occurred in 1995 in New Zealand, but these were less extensive than those in South Australia, being confined largely to 500 km of the North Island coast. Pathology was consistent across the entire range of mortalities in Australia and New Zealand, suggesting a common causation. Rapid spread of the disease was postulated to be aided by fish-eating birds, especially by gannets (*Morus serrator*).

These mortalities of pilchards were the largest ever reported in Australia for a single fish species, and future impacts of the disease are uncertain, since the two epizootics occurred with only a 4-yr interval between them. If the pathogen was really introduced with frozen forage species (as is suspected), an excellent opportunity will be provided for long-term investigations of the effects of a new disease agent on a naive marine fish population.

EMERGING CONCEPTS AND GENERALIZATIONS ABOUT INTRODUCED PATHOGENS AND THE DISEASES THEY CAUSE

The preceding discussion of selected case histories helps to support a sequence of emerging concepts:

1. The development and functioning of global transfer networks of aquaculture species along which pathogens can and do move
2. The reality that most introductions of aquatic pathogens have been accidental, even though introductions of host animals may have been intentional
3. The reality that aquaculture populations may act as sources or reservoirs of infection pressure on wild stocks of the same or other species (for example, salmon copepod parasites)
4. The possibility (demonstrated in shrimp) of transfer of introduced pathogens from aquaculture populations to native wild stocks
5. The need for restrictions of *in vivo* experimental studies of fish pathogens to facilities within the zone where the organism is enzootic, unless failsafe containment facilities exist elsewhere
6. The need for early scientific attention to the appearance of a new parasite or pathogen
7. The reality that disease risks from introductions are never zero
8. The demonstration that once an introduced pathogen is established in a marine population, it is difficult if not impossible to control its further spread
9. The observation that the colonization potential of many parasites in new environments is predictable, on theoretical grounds, but not the relative importance of different methods of dissemination (especially human-assisted movements of infected hosts)

Examination of the foregoing and other examples of the relationship of diseases to importation of nonnative species supports a number of generalizations:

1. Although evidence is less than robust and associations must often be made by deduction or inference, it is likely that many of the recent outbreaks of disease in marine populations of commercial importance — especially shellfish — are results of introductions of pathogens from other geographic areas. This may not be the case for salmonids, however, in which the most devestating outbreaks of disease are often due to agents already enzootic, which either are amplified in acquaculture situations or which can infect an introduced species.
2. Introduced pathogens can be considered logically as "biological pollutants" and, as such, should be subject to all regulations governing pollutant discharge control. This perception of introduced agents has received some support, but has not yet, to my knowledge, been incorporated directly into regulatory regimes concerned with ocean pollution.

3. Disease-causing organisms may seem benign and innocuous in adapted host populations, but they may become serious pathogens of related species when introduced into other geographic areas (for example, the eel nematode). Assessment of this potential risk can be done only in appropriately designed isolation laboratories where imported and native stocks can be reared and studied together for some reasonable period of time (preferably at least one complete life cycle).

4. The possible role of infected resistant carriers in transmitting a disease to a susceptible but geographically separate subpopulation of the same species must be considered in decisions about transfers within the total geographic range of the species.

5. Two forms of disease risks exist when animals are relocated outside their normal range:

   a. The infection of introduced stocks by an enzootic pathogen in the recipient country, to which native stocks are resistant — with the possible overwhelming of that resistance by increased infection pressure or increased virulence (see Figure 8.3)

   b. The introduction of a pathogen that affects susceptible native species but to which the introduced stock is resistant (see Figure 8.4).

6. These dual risks may be further complicated by interbreeding of a resistant introduced stock with remnants of a disease-ravaged native stock. This may lower the resistance of the new population to the pathogen. Another danger, particularly from introduced viral agents, is the rapid potentiation of virulence in intensive rearing facilities — as has been seen with infectious haematopoietic necrosis (IHN) virus in trout farms

The validity of any of these proposed concepts and generalizations about introduced pathogens must, of course, be scrutinized continuously as new information becomes available. The listing here merely represents my assessment of our present understanding.

CONCLUSIONS

It is important that any proposed introduction of a nonnative marine species should be preceded by adequate scientific attention to three considerations emphasized in this chapter:

Ecological considerations — including competition, predation, and community characteristics of species (diversity, carrying capacity)

Genetic considerations — including the potential for hybridization, change in gene frequency (genetic diversity), and change or modification in disease and parasite resistance

Pathological considerations — including the potential for unintentional introduction of diseases and parasites

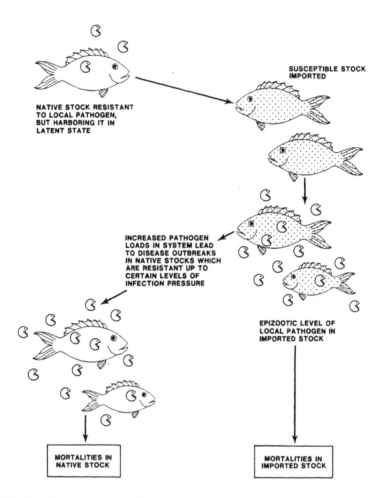

FIGURE 8.3 Effects of an enzootic pathogen on an introduced stock.

This chapter, with its case history approach, is built around these three principal categories: changes in ecosystems resulting from invasion by alien species, illustrated here by proliferation of the introduced macroalga *Caulerpa taxifolia* in the Mediterranean Sea and by the explosive population growth of the introduced ctenophore *Mnemiopsis leidyi* in the Black Sea; genetic changes in species characteristics, illustrated by results of extensive research on Atlantic salmon (*Salmo salar*) and the Pacific coho salmon (*Oncorhynchus kisutch*); and population effects of introduced pathogens, documented here for oysters, scallops, shrimp, eels, and pilchards.

The wealth of examples of negative effects of introduced marine species continues to accrue. Deliberate introductions for aquaculture purposes are often accompanied by problems with alien pathogens, and accidental introductions with ships' ballast water exchange or by hull-fouling organisms can result in unfavorable ecosystem modifications. Global control measures have been proposed, and some minimally effective actions have been taken by individual nations or regional regulatory

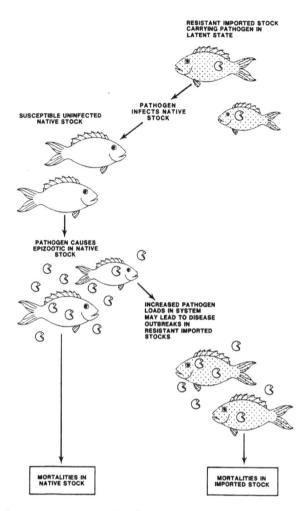

FIGURE 8.4 Effects of an introduced pathogen on native stock.

commissions, but the beat goes on. The following vignette is a sample of extreme thinking about the problem.

A Proposed Solution to Problems Created by Introductions of Nonindigenous Marine Species

It was in the very late 1970s — at a now long-forgotten conference on "Food and Drugs from the Sea" held in Mayaguez, Puerto Rico — that a radical new proposal about introductions of non-indigenous marine species was first presented to the scientific community. This was during the productive years of the much-respected NSF-funded International Decade of Ocean Exploration (IDOE), but was totally unrelated to that program. The proposal grew out of concern for the rapidly accelerating rate of unregulated global movements of marine organisms for aquaculture purposes, and especially about the consequences of the spread of infectious diseases in fish and shellfish by such movements.

It was also a time when the attention of international research groups such as the prestigious International Council for the Exploration of the Sea (ICES) was beginning to focus on the emerging problem of introduced species, by designating a working group to propose a course of action. One of the early products of that group was a suggested "Code of Practice Governing Introductions of Non-indigenous Marine Species" to be followed by member countries — a code that included rigid protocols governing importation of live marine species, for aquaculture or any other purpose. The code was approved by a few nations, but was widely ignored, especially by the then Soviet-bloc countries.

So it was from a sense of frustration over lack of progress in developing an effective intergovernmental regulatory framework for controlling introductions that I offered the new proposal at that obscure scientific meeting in Puerto Rico. It called for a new global initiative, tentatively labeled an "International Decade of Indiscriminate Ocean Transfers" (IDIOT). The core of the proposed program was to be <u>aggressive large-scale global transportation of marine organisms, in substantial numbers, from one estuary, bay, or ocean to another, worldwide, for an entire decade — for purposes of aquaculture, or for any other purpose, or for no purpose at all</u>. Then, after an estimated half-century or more of ecosystem dislocations and epizootics in all oceans, a plateau of relative stability might be achieved, without future international concerns for impacts of introduced species, and without need for any regulatory regimes.

Not surprisingly, the proposal met with disbelief, or (rarely) lukewarm but quizzical reception, at that meeting and subsequently. Some said that it was too advanced for its time, or that it was too large in scale, but since then, <u>the commerce of the world has followed, insofar as transport of marine organisms is concerned, a very similar path, with few non-economic guiding principles to be followed</u>. We humans are satisfying some of the objectives envisioned by that IDIOT proposal, but on a one-disaster-after-another slow-motion time scale!

> ***From Field Notes of a Pollution Watcher***
> *(C.J. Sindermann, 1999)*

I have already offered some concepts and generalizations about introductions of alien species, particularly those that concern introduced diseases. I will conclude this chapter with two "directives":

1. Promulgate and enforce, vigorously, new scientifically based international regulations about introductions. The ICES Code of Practice (ICES 1988, 1990; see Figure 8.5) is a good theoretical model, but it lacks necessary teeth, beyond voluntary compliance. A key provision of the code calls for quarantine and examination of a species proposed for introduction through one complete life cycle, followed by limited test

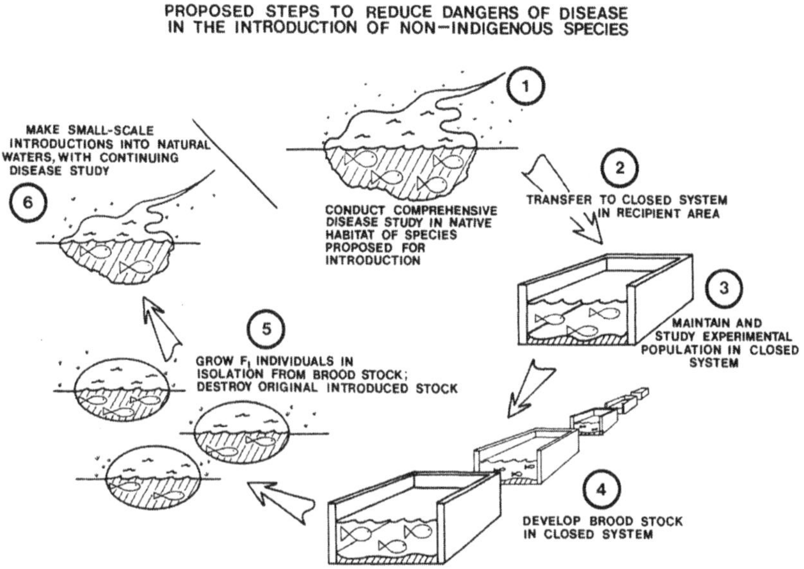

PROPOSED STEPS TO REDUCE DANGERS OF DISEASE
IN THE INTRODUCTION OF NON−INDIGENOUS SPECIES

FIGURE 8.5 The Code of Practice proposed by the International Council for the Exploration of the Sea concerning introductions of nonindigenous species.

introduction of F_1 offspring (but never the parent stock) into open waters, if no negative findings (ecological, genetic, or pathological) are reported. This Code of Practice has been widely, but not universally, accepted and followed by the world aquaculture community. It should be accepted and followed universally.

2. In developing national and international policies on introductions of nonnative aquatic species, consideration should also be given to adopting the "precautionary principle" proposed by Germany and accepted at the Second International Conference on the Protection of the North Sea in 1987. That principle "requires action to reduce pollution even in the absence of soundly established scientific proof for cause and effect relationships." The principle could be applied especially to control accidental introductions (including pathogens), which are clearly forms of biological pollution.

REFERENCES

Balouet, G., M. Poder, and A. Cahour. 1983. Haemocytic parasitosis: Morphology and pathology of lesions in the French flat oyster, *Ostrea edulis* L. *Aquaculture 34:* 1–14.

Burreson, E.M., N.A. Stokes, and C.S. Friedman. 2000. Increased virulence in an introduced pathogen: *Haplospridium nelsoni* (MSX) in the eastern oyster *Crassostrea virginica*. *J. Aquat. Anim. Health 12:* 1–8.

Caddy, J.F. 1993. Toward a comparative evaluation of human impacts on fishery ecosystems of enclosed and semi-enclosed seas. *Rev. Fish. Sci. 1:* 57–96.

Caddy, J.F. and R. Griffiths. 1990. A perspective on recent fishery-related events in the Black Sea. *General Fisheries Council for the Mediterranean (GFCM) Studies and Reviews 63:* 43–71.

Carlton, J.T. and J.B. Geller. 1993. Ecological roulette: The global transport and invasion of nonindigenous marine organisms. *Science 261:* 78–82.

Carlton, J.T., J.K. Thompson, L.E. Schemel, and F.H. Nichols. 1990. Remarkable invasion of San Francisco Bay (California, USA) by the Asian clam *Potamocorbula amurensis*. I. Introduction and dispersal. *Mar. Ecol. Prog. Ser. 66:* 81–94.

Center, T.D., J.H. Frank, and F.A. Dray, Jr. 1997. Biological control, pp. 245–263. In: D. Simberloff, D.C. Schmitz, and T.C. Brown (eds.), *Strangers in Paradise: The Impact and Management of Nonindigenous Species in Florida.* Island Press, Washington, DC.

Chew, K.K. 1990. Global bivalve shellfish introductions. *World Aquacult. 21:* 9–22.

Comps, M. 1970. Observations sur les causes d'une mortalité anomale des huîtres plate dans la Bassin de Marennes. *Rev. Trav. Inst. Pêches Marit. 34:* 317–326.

Comps, M. and J.L. Duthoit. 1976. Infection virale associés a la "maladie des branchies" de l'huitre portugaise *Crassostrea angulata* Lmk. *C.R. Hebd. Séances Acad. Sci. Ser. D. Sci. Nat. 283:* 1595–1596.

Comps, M., R. Bonami, C. Vago, and A. Campillo. 1976. Une virose de l'huître portugaise (*Crassostrea angulata* Lmk). *C.R. Hebd. Séances Acad. Sci. Ser. D. Sci. Nat. 282:* 1991–1993.

Comps, M., G. Tige, and H. Grizel. 1980. Pathologie animale — étude ultrastructurale d'une protiste parasite de l'huitre *Ostrea edulis* L. *C.R. Heb. Séances Sci. Ser. D 290:* 383–385.

Dalton, R. 2000. Researchers criticize response to killer algae. *Nature 406:* 447.

Devlin, R.H., T.W. Tesaki, C.A. Biagi, E.M. Donaldson, P. Swanson, and W.-K. Chan. 1994. Extraordinary growth in salmon. *Nature 371:* 209–210.

Devlin, R.H., T.Y. Yesaki, E.M. Donaldson, S.J. Du, and C.-L. Hew. 1995. Production of germline transgenic Pacific salmonids with dramatically increased growth performance. *Can. J. Fish. Aquat. Sci. 52:* 1376–1384.

Elston, R.A., C.A. Farley, and M.L. Kent. 1986. Occurrence and significance of bonamiasis in European flat oysters *Ostrea edulis* in North America. *Dis. Aquat. Org. 2:* 49–54.

Farley, C.A., P.H. Wolf, and R.A. Elston. 1988. A long-term study of "microcell" disease in oysters with a description of a new genus, *Mikrocytos* (g.n.), and two new species, *Mikrocytos mackini* (sp. n.) and *Mikrocytos roughleyi* (sp. n.). *Fish. Bull. 86:* 581–593.

Farrell, A.P., W. Bennett, and R.H. Devlin. 1997. Growth-enhanced transgenic salmon can be inferior swimmers. *Can. J. Zool. 75: 335–337.*

Ferguson, M.M. 1990. The genetic impact of introduced fishes on native species. *Can. J. Zool. 68:* 1053–1057.

Figueras, A.J. 1991. *Bonamia* status and its effects in cultured flat oysters in the Ria de Vigo, Galicia (N.W. Spain). *Aquaculture 93:* 225–233.

Fletcher, G.L., M.A. Shears, M.J. King, P.L. Davies, and C.L. Hew. 1988. Evidence for antifreeze protein gene transfer in Atlantic salmon (*Salmo salar*). *Can. J. Fish. Aquat. Sci. 45:* 352–357.

Gausen, D. and V. Moen. 1991. Large-scale escapes of farmed Atlantic salmon (*Salmo salar*) into Norwegian rivers threaten natural populations. *Can. J. Fish. Aquat. Sci. 48:* 426–428.

GESAMP (Group of Experts on Scientific Aspects of Marine Pollution). 1997. Opportunistic settlers and the problem of the ctenophore *Mnemiopsis leidyi* invasion in the Black Sea. Rep. Stud. GESAMP No. 58, International Maritime Organization, London, UK. 84 pp.

Grizel, H. and M. Héral. 1991. Introduction into France of the Japanese oyster (*Crassostrea gigas*). *J. Cons. Int. Explor. Mer 47:* 399–403.

Harbison, G.R. and S.P. Volovik. 1993. Annex I. Methods for the control of populations of the ctenophore, *Mnemiopsis leidyi,* in the Black and Azov Seas, pp. 42–44. In: FAO Fisheries Report No. 495, Second Technical Consultation on Stock Assessment in the Black Sea. General Fisheries Council for the Mediterranean. UN/FAO, Rome.

Harbison, G.R. and S.P. Volovik. 1994. The ctenophore, *Mnemiopsis leidyi,* in the Black Sea: A holoplanktonic organism transported in the ballast water of ships, pp. 25–36. In: *Nonindigenous Estuarine and Marine Organisms (NEMO),* Proceedings of the Conference and Workshop, Seattle, WA (1993). U.S. Dept. Commerce. National Oceanic and Atmospheric Admin., Washington, DC.

Hill, J.A., A. Kiessling, and R.H. Devlin. 2000. Coho salmon (*Oncorhynchus kisutch*) transgenic for a growth hormone gene construct exhibit increased rates of muscle hyperplasia and detectable levels of differential gene expression. *Can. J. Fish. Aquat. Sci.* 57: 939–950.

Hindar, K., N. Ryman, and F. Utter. 1991. Genetic effects of cultured fish on natural fish populations. *Can. J. Fish. Aquat. Sci. 48:* 945–957.

Hudson, E.B. and B.J. Hill. 1991. Impact and spread of bonamiasis in the U.K. *Aquaculture 93:* 279–285.

ICES (International Council for the Exploration of the Sea). 1988. Codes of practice and manual of procedures for consideration of the introductions and transfers of marine and freshwater organisms. Cooperative Research Report 159. EIFAC Occas. Pap. No. 23, 44 pp.

ICES (International Council for the Exploration of the Sea). 1990. Code of practice to reduce the risks of adverse effects arising from introductions and transfers of marine species. ICES, Copenhagen, Denmark.

Ivanov, L. and R.G.H. Beverton. 1985. The fisheries resource of the Mediterranean, Part II. Black Sea. *General Fisheries Council for the Mediterranean (GFCM) Studies and Reviews 60:* 1–135.

Jousson, O., J. Pawlowski, L. Zaninetti, A. Meinesz, and C.-F. Boudouresque. 1998. Molecular evidence for the aquarium origin of the green alga *Caulerpa taxifolia* introduced to the Mediterranean Sea. *Mar. Ecol. Prog. Ser. 172:* 275–280.

Jousson, O., J. Pawlowski, L. Zaninetti, F.W. Zechman, F. Dini, G. Di Giuseppe, R. Woodfield, A. Millar, and A. Meinesz. 2000. Invasive alga reaches California. *Nature 408:* 157–158.

Kulser, J. 2000. California algae may be feared European species. *Science 289:* 222–223.

Kalagayan, H., D. Godin, R. Kanna, O. Hagino, J. Sweeney, J. Wyban, and J. Brock. 1991. IHHN virus as an etiological factor in runt-deformity syndrome (RDS) of juvenile *Penaeus vannamei* cultured in Hawaii. *J. World Aquacult. Soc. 22:* 235–243.

Karlsson, J.D. 1990. Parasites of the bay scallop, *Argopecten irradians* Lamarck, pp. 180–190. In: S.E. Shumway (ed.), *International Compendium of Scallop Biology and Culture.* World Aquaculture Soc., Baton Rouge, LA.

Kennedy, C.R. and D.J. Fitch. 1990. Colonization, larval survival and epidemiology of the nematode *Anguillicola crassus,* parasitic in the eel, *Anguilla anguilla,* in Britain. *J. Fish Biol. 36:* 117–131.

Kideys, A.E., A Roohi, S. Bagheri, G. Finenko, and L. Kamburska. 2005. Impacts of invasive ctenophores on the fisheries of the Black Sea and Caspian Sea. *Oceanography 18:* 77–85.

Koie, M. 1991. Swimbladder nematodes (*Anguillicola* spp.) and gill monogeneans (*Pseudodactylogyrus* spp.) parasitic on the European eel (*Anguilla anguilla*). *J. Cons. Int. Explor. Mer 47:* 391–398.

Koops, H. and F. Hartmann. 1989. *Anguillicola* infestations in Germany and in German eel imports. *J. Appl. Ichthyol. 1:* 41–45.

Langar, H., A. Djellouli, K. Ben Mustapha, and A. El Abed. 2000. Première signalisation de *Caulerpa taxifolia* (Vahl) J. Agardh en Tunisie. *Bull. Inst. Natl. Sci. Technol. Mer 27:* 1–8.

Lee, C.G., R.H. Devlin, and A.P. Farrell. 2003. Swimming performance, oxygen consumption and excess post-exercise oxygen consumption in adult transgenic and ocean-ranched coho salmon. *J. Fish Biol. 62:* 753–766.

Lightner, D.V. 1990. Viroses section: Introductory remarks, pp. 3–6. In: F.O. Perkins and T.C. Cheng (eds.), *Pathology in Marine Science*. Academic Press, New York.

Lightner, D.V. and R.M. Redman. 1998. Emerging crustacean diseases, pp. 68–71. In: A.S. Kane and S.L. Poynton (eds.), *Proceedings of the Third International Symposium on Aquatic Animal Health*. APC Press, Baltimore, MD.

Lightner, D.V., R.M. Redman, and T.A. Bell. 1983. Infectious hypodermal and hematopoietic necrosis (IHHN), a newly recognized virus disease of penaeid shrimp. *J. Invertebr. Pathol. 42:* 62–70.

Lightner, D.V., R.M. Redman, T.A. Bell, and J.A. Brock. 1983. Detection of IHHN virus in *Penaeus stylirostris* and *P. vannamei* imported into Hawaii. *J. World Maricult. Soc. 14:* 212–225.

Lightner, D.V., T.A. Bell, and R.M. Redman. 1989. A review of the known hosts, geographical range and current diagnostic procedures for the virus disease of cultured penaeid shrimp, pp. 113–126. In: *Advances in Tropical Aquaculture, Workshop at Tahiti*. Actes de Colloque 9.

Lightner, D.V., T.A. Bell, R.M. Redman, and L.A. Perez. 1992a. A collection of case histories documenting the introduction and spread of the virus disease IHHN in penaeid shrimp culture facilities in northwestern Mexico. *Int. Counc. Explor. Sea, Mar. Sci. Symp. 194:* 97–105.

Lightner, D.V., R.M. Redman, R.M. Williams, T.A. Bell, and R.B. Thurman. 1992b. Geographic dispersion of the viruses IHHNV, MSV and HPV as a consequence of transfers and introductions of penaeid shrimp to new regions for aquaculture purpose, pp. 155–173. In: A. Rosenfield and R. Mann (eds.), *Dispersion of Living Organisms into Aquatic Ecosystems*. Sea Grant Publ. UM-SG-TS-92-04, Univ. Maryland, College Park, MD.

McArdle, J.F., F. McKiernan, H. Foley, and D.H. Jones. 1991. The current status of *Bonamia* disease in Ireland. *Aquaculture 93:* 273–278.

McGinnity, P., C. Stone, J.B. Taggart, D. Cooke, D. Cotter, R. Hynes, C. McCamley, T. Cross, and A. Ferguson. 1997. Genetic impact of escaped farmed Atlantic salmon (*Salmo salar* L.) on native populations: Use of DNA profiling to assess freshwater performance of wild, farmed, and hybrid progeny in a natural river environment. *ICES J. Mar. Sci. 54:* 998–1008.

McGladdery, S.E., R.J. Cawthorn, and B.C. Bradford. 1991. *Perkinsus karlssoni* n. sp. (Apicomplexa) in bay scallops *Argopecten irradians*. *Dis. Aquat. Org. 10:* 127–137.

Mee, L.D. 1992. The Black Sea in crisis: A need for concerted international action. *Ambio 21*(4): 278–286.

Meinesz, A. 1992. Mode de dissémination de l'algue *Caulerpa taxifolia* introduite en Méditerranée. *Rapp. P.V. Reun. Comm. Int. Explor. Sci. Mer Mediterr. 33B:* 44.

Meinesz, A. 1999. *Killer Algae*. Univ. Chicago Press, Chicago. 360 pp.

Meinesz, A. and C.F. Boudouresque. 1996. On the origin of *Caulerpa taxifolia* in the Mediterranean Sea. *C.R. Acad. Sci. Ser. III Sci. Vie 319:* 603–613.

Meinesz, A. and B. Hesse. 1991. Introduction et invasion de l'algue tropicale *Caulerpa taxifolia* en Méditerranée nord-occidentale. *Oceanol. Acta 14:* 415–426.

Meinesz, A., D. Pietkiewicz, T. Komatsu, G. Caye, J. Blachier, R. Lemee, and A. Renoux-Meunier. 1994. Notes taxinomiques préliminaires sur *Caulerpa taxifolia* et *Caulerpa mexicana*, pp. 105–114. In: C.-F. Boudouresque, A. Meinesz, and V. Gravez (eds.), *1st International Workshop on* Caulerpa taxifolia. GIS Posidonie, Marseille, France.

Meinesz, A., J.-M. Cottalorda, D. Chiaverini, N. Cassar, and J. Vaugelas. 1998. *Suivi de l'invasion de l'algue tropicale Caulerpa taxifolia en Méditerranée: Situation au 31 décembre 1997.* Laboratoire Environnement Marin Littoral, Univ. Nice, Sophia Antipolis, Nice, France.

Meusnier, G.I., J.L. Olsen, W.T. Stam, C. Destombe, and M. Valero. 2001. Phylogenetic analyses of *Caulerpa taxifolia* (Chlorophyta) and of its associated bacterial microflora provide clues to the origin of the Mediterranean introduction. *Mol. Ecol. 10:* 931–946.

Morales-Covarrubias, M.S., L.M. Nunan, D.V. Lightner, J.C. Mota-Urbina, M.C. Garza-Aquirre, and M.C. Chavez-Sanchez. 1999. Prevalence of infectious hypodermal and hematopoietic necrosis virus (IHNNV) in wild adult blue shrimp *Penaeus stylirostris* from the northern Gulf of California, Mexico. *J. Aquat. Anim. Health 11:* 296–301.

Neumann, W. 1985. Schwimmblasenparasit *Anguillicola* bei Aalen. *Fisch. Teichwirt. 11:* 322.

Nunan, L.M., B.T. Poulos, and D.V. Lightner. 1998. The detection of white spot syndrome virus (WSSV) and yellowhead virus (YHV) in imported commodity shrimp. *Aquaculture 160:* 19–30.

Oviatt, C.A. and P.M. Kremer. 1977. Predation on the ctenophore, *Mnemiopsis leidyi*, by butterfish, *Peprilus triacanthus*, in Narragansett Bay, Rhode Island. *Chesapeake Sci. 18:* 236–240.

Pantoja, C.R. 1993. Prevalencia del virus IHHNV en poblaciones silvestres de camarón azul (*Penaeus stylirostris*) en las costas de Sonora, México. Master's thesis, Instituto Tecnológico y de estudios superiores de Monterrey, Guaymas, Sonora, México.

Pantoja, C.R., D.V. Lightner, and K.H. Holtschmit. 1999. Prevalence and geographic distribution of infectious hypodermal and hematopoietic necrosis virus (IHHNV) in wild blue shrimp (*Penaeus stylirostris*) from the Gulf of California, Mexico. *J. Aquat. Anim. Health 11:* 23–34.

Peters, G. and F. Hartmann. 1986. *Anguillicola*, a parasitic nematode of the swim bladder spreading among eel populations in Europe. *Dis. Aquat. Org. 1:* 229–230.

Ray, S.M. and A.C. Chandler. 1955. *Dermocystidium marinum*, a parasite of oysters. *Exp. Parasitol. 4:* 172–200.

Shiganova, T.A., J.J. Dumont, A. Mikaelyan, D.M. Glazov, Yu. V. Bulgakova, E.I. Musayeva, P.Y. Sorokin, L.A. Pautova, Z.A. Mirzoyan, and E.I. Studenikina. 2004. Interaction between the invading ctenophores *Mnemiopsis leidyi* (A. Agassiz) and *Beroe ovata* Mayer 1912, and their influence on the pelagic ecosystem of the northeastern Black Sea, pp. 33–70. In: *Aquatic invasions in the Black, Caspian and Mediterranean Seas*, H. Dumont, T. Shiganova and U. Niermann, eds. Kluwer Academic Publishers, Dordecht, The Netherlands.

Skaala, O., G. Dahle, D.E. Jorstad, and G. Naevdal. 1990. Interactions between natural and farmed fish populations: Information from genetic markers. *J. Fish Biol. 36:* 449–460.

Stewart, J.E. 1991. Introductions as factors in diseases of fish and aquatic invertebrates. *Can. J. Fish. Aquat. Sci. 48*(Suppl. 1): 110–117.

Tsikhon-Lukanina, E.A., O.G. Reznichenko, and T.A. Lukasheva. 1992. Diet of the cteno-phore *Mnemiopsis* in inshore waters of the Black Sea. *Oceanology 32*(4): 496–500 (*Okeanologia 32*[4]: 733–738).

Tsikhon-Lukanina, E.A., O.G. Reznichenko, and T.A. Lukasheva. 1993. Predation rates on fish larvae by the ctenophore *Mnemiopsis* in the Black Sea inshore waters. *Oceanology 33*(6): 895–899. (In Russian)

Van Banning, P. 1991. Observations on bonamiasis in the stock of the European flat oyster, *Ostrea edulis*, in the Netherlands, with special reference to the recent developments in Lake Grevelingen. *Aquaculture 93:* 205–211.

Verspoor, E. 1997. Genetic diversity among Atlantic salmon (*Salmo salar* L.) populations. *ICES J. Mar. Sci. 54:* 965–973.

Vroom, P.S. and C.M. Smith. 2001. The challenges of siphonous green algae. *Am. Sci. 89:* 525–531.

Ward, T.M., F. Hoedt, L. McLeay, W.F. Dimmlich, M. Kinloch, G. Jackson, R. McGarvey, P.-J. Rogers, and K. Jones. 2001. Effects of the 1995 and 1998 mass mortality events on the spawning biomass of sardine, *Sardinops sagax*, in South Australian waters. *ICES J. Mar. Sci. 58:* 865–875.

Whittington, R.J., J.B. Jones, P.M. Hine, and A.D. Hyatt. 1997. Epizootic mortality in the pilchard *Sardinops sagax neopilchardus* in Australia and New Zealand in 1995. I. Pathology and epizootiology. *Dis. Aquat. Org. 28:* 1–16.

Wiedenmann, J., A. Baumstark, T.L. Pillen, A. Meinesz, and W. Vogel. 2001. DNA fingerprints of *Caulerpa taxifolia* provide evidence for the introduction of an aquarium strain into the Mediterranean Sea and its close relationship to an Australian population. *Mar. Biol. 138:* 229–234.

Zaika, V.E. 1993. The drop of anchovy stock in the Black Sea: Result of biological pollution? *GFCM/FAO Fish. Rep. 495*(Annex III): 54–59.

Zaitsev, Y.P. 1992a. Impacts of eutrophication on the Black Sea fauna, pp. 59–62. In: Fisheries and environmental studies in the Black Sea system. General Fisheries Council for the Mediterranean (GFCM) Studies and Reviews 64.

Zaitsev, Y.P. 1992b. Recent changes in the trophic structure of the Black Sea. *Fish. Oceanogr. 1*(2): 180–189.

Zaitsev, Y.P. 1993a. Biological aspects of Western Black Sea coastal waters. FAO Fish. Rep. No. 495.

Zaitsev, Y.P. 1993b. Impacts of eutrophication on the Black Sea fauna, pp. 63–86. In: Fisheries and Environmental Studies in the Black Sea System. General Fisheries Council for the Mediterranean (GFCM) Studies and Reviews 65.

Zaitsev, Y.P. and L.N. Polischuk. 1984. An increase in number of the jellyfish *Aurelia aurita* in the Black Sea. *Ecology of the Sea 17:* 35–46.

Section II

Effects of Coastal Pollution on Marine Animals

Thus far in this narrative I have touched lightly — in the introduction — on broad national and global pollution-related events and problems. After that, I discussed — briefly — my eight choices for pollution-induced undersea horrors. It is time now to focus on the first of two major objectives of the book: an examination of the effects of coastal/estuarine pollution on living marine animals.

The chapter sequence begins logically with consideration of some of the more significant sublethal effects of pollution on marine animals and then discusses some of their principal responses to pollution — responses that increase the likelihood of survival in contaminated habitats (Chapter 9). With this firm foundation of events at an individual level, we can move briskly to quantitative matters in Chapter 10 to effects of pollution at the population level — emphasizing possible impacts on fish and shellfish abundance. After that, I have included in Chapter 11 the special case of effects of pollution on survival and well-being of marine mammals. These three chapters seem to form a cohesive unit, contributing to an understanding of pollution effects on marine populations.

9 Sublethal Effects of Coastal Pollution on Marine Animals

Abnormal Pacific Oysters on the Coast of France — A Biological Detective Story

Pacific oysters <u>Crassostrea gigas</u> were introduced to the coastal waters of France by mass importations beginning in the late 1960s, after the native oysters had declined in abundance dramatically, due mostly to effects of epizootic diseases. Countless millions of seed oysters were airlifted from Japan during the period 1968 to 1974, with hope of reestablishing the industry in such traditional French oyster growing areas as Arcachon, Oleron, Marennes, and La Trinité. Initial results of the mass transplantation were very encouraging. The introduced species survived, grew, and even reproduced in some protected coastal waters. By the mid-1970s though, indications of a severe problem with these immigrant oysters were appearing in some of the bays. They were exhibiting poor growth and grossly malformed shells. Shell abnormalities that made the oysters unsalable reached an intolerable level of 90% in the Bay of Arcachon in 1980 to 1982 — just as an example.

Crisis response research conducted during the late 1970s and early 1980s in France and Britain eventually demonstrated that the cause of the deformities was an environmental pollutant — tributyltin — an organic compound used as an ingredient in antifouling paint for small boats, that was leaching out into growing areas. The problem, which was for several years a real threat to the successful reestablishment of the oyster industry in France was solved with the immediate imposition of a ban on the use of organotin compounds in antifouling paint for boats. Prevalences of the shell abnormalities fell to negligible levels soon after the ban took effect, and the oyster industry regained momentum in the growing areas that had been affected.

One of the fascinating aspects of this scientific detective story is the extreme sensitivity of the Pacific oyster to almost inconceivably minute concentrations of the specific environmental contaminant. The research demonstrated clearly that the presence of this single toxic chemical, in vanishingly small concentrations, could have a striking effect on the physiology of shell deposition in the oyster (probably by disrupting normal calcium metabolism), and ultimately on the marketability of the product. A more ominous finding was that those

extremely low concentrations of tributyltin could also kill oyster and crab larvae
— pointing to potential impacts on population abundance. (Later studies showed
that juvenile crabs were also very sensitive to the contaminant; experimental
exposures to tributyltin retarded limb regeneration, delayed molting, and pro-
duced deformities in regenerated appendages.)

From Field Notes of a Pollution Watcher
(C.J. Sindermann, 1986)

The tributyltin/oyster episode is only one of many examples of disabilities and
deformities that can be attributed to chemical contamination of inshore habitats. Too
often, though, such a clear cause-and-effect relationship of abnormalities with specific contaminants has not been demonstrated — being obscured by the simultaneous
presence of other suspect pollutants or other variable environmental factors.

This chapter examines some of the many ways in which animals can be affected
by pollutant chemicals, short of mortality. For ease of description, I have subdivided
it into four major sections:

1. Effects on reproduction and early development
2. Effects on juvenile and adult fish
3. How marine animals respond to chemical pollution
4. Stress from pollution

EFFECTS OF COASTAL POLLUTION ON
REPRODUCTION AND EARLY DEVELOPMENT OF FISH

Effects of pollution on reproduction and early development of fish have been investigated from physiological, biochemical, genetic, structural, and population perspectives. A general conclusion is that the most severe consequences of coastal/estuarine
contamination are to be found here, but that much remains to be learned — especially
about population level effects. I think that some insights can be gained about the
relative importance of pollution effects on reproduction and development by dis-
secting the topic into two somewhat distinct but still closely joined components:

1. Effects of pollutants on biochemical and structural (cellular) events in the
   adult fish prior to spawning
2. Effects of pollution on postspawning events — embryonic, larval, postlarval, and juvenile development

With the artificial compartments of this dissection clearly in mind, it is possible
to trace a descending spiral of contaminant-related departures from normal reproduction and early development as we move through successive phases in the maturation of the parental generation and then through the entire life cycle of the offspring
(as described in Table 9.1).

TABLE 9.1
Effects of Environmental Contaminants on Life History Stages of Fish

Life Stage	Effects of Pollutants on Life History Stages of Fish
Maturation	Delay or inhibition of gonad development in parent male or female
Spawning	Decreased fecundity (reduced numbers of eggs or sperm produced per adult)
Egg development (embryo)	Defective eggs or sperm, resulting in abnormal development and mortalities of embryonated eggs
Hatching	Reduced egg hatching success
Larval development	Abnormalities and mortalities of larvae
Postlarval development	Physiological/morphological abnormalities in postlarvae
Juvenile development	Further expression of physiological/morphological abnormalities, often producing disability and death
Adult	Genetic defects may be transmitted to offspring; contaminants derived from parent female or from polluted habitats may be transmitted to offspring; population abundance may be reduced by pollutants, especially if stocks are heavily exploited.

Before examining the life cycle stages — spawners, embryos and larvae, and juveniles — where pollutants can have severe effects, it seems relevant to review a few of the terms used by fisheries scientists to describe the reproductive process. Reproductive success — the production of viable offspring — can be affected by pollution before and after spawning. Success thus represents an *integration* of survival at early life history stages.*

EFFECTS OF POLLUTION ON BIOCHEMICAL AND STRUCTURAL (CELLULAR) EVENTS IN ADULT FISH PRIOR TO SPAWNING

Pollutants can have major disruptive effects on reproduction in fish — effects that may occur at multiple sites in the reproductive apparatus of maturing fish and in many developmental stages of their offspring (see Figure 9.1). An excellent review by Kime (1995) listed important pollution-induced structural and functional changes in reproductive capacities of spawners that included the following:

- Lesions or malformations of gonads, pituitary, brain, and liver
- Inhibition of production and release of hormones of the hypothalamus, pituitary, and gonads

* Measures of reproductive success, as described by Spies and Rice (1988), include the following descriptors: *Fecundity (N)* = total number of eggs spawned; *Viable eggs (V)* = % eggs that float (salmonids and certain other species); *% fertilization success* = # fertilized eggs (F) ÷ # eggs that float (V) × 100; *% embryological success* = # eggs that hatch (H) ÷ # fertilized eggs (F) × 100; *% normal larvae* = # normal larvae (L) ÷ # eggs that hatch (H) × 100; *Hatching success* integrates survival from spawning to hatching; and *Viable hatch* integrates survival from spawning to development of normal swimming larvae (negative descriptors that may be used include "reproductive depression" and "reproductive failure.")

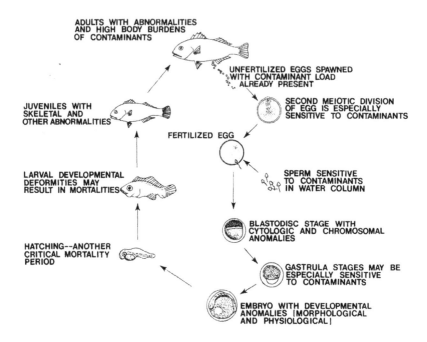

FIGURE 9.1 Points in the life cycle when fish are especially sensitive to pollutants.

- Altered metabolism of hormones by the liver
- Reduction in size of gonads
- Inhibition of egg production, vitellogenin (yolk precursor) synthesis, and growth of eggs
- Reduction in sperm production and motility
- Death or malformation of embryos and hatched larvae

As a result of his literature review, Kime concluded that "almost all pollutants may adversely affect the reproductive potential of [fish] species ... at concentrations below that at which significant mortality occurs" (p. 66). Based on the available data, he then made the rather dangerous generalization that "exposure to 0.0001 mgl⁻¹ (1 ppb) of pollutant is generally sufficient to produce harmful effects for long-term exposure, although some organochlorines show harmful effects even at one-thousandth of this level" (p. 66).

Kime then summarized his examination of the literature on effects of pollution on reproduction in this way:

> The overall impact of long-term environmental pollution can, therefore, decrease a population by decreasing fecundity, decreasing the numbers of reproductive cycles in the lifetime of each fish, and decreasing the survival of the offspring at early stages of their life cycle. (p. 68)

However, he then softened his statement, insofar as it involved population effects, with this caveat:

> Although the evidence points to decreased fecundity of fish populations resulting from pollution, hard evidence of this as a cause of decreasing fish stocks is lacking. In the marine environment overfishing and pollution probably both contribute to such a decrease but the relative contribution of each is not clear. (p. 67)

(He might have added that extremes of natural environmental factors can contribute to year class failures as well.)

Kime's review of effects of pollution on reproduction in fish was concerned with the ways in which the actual process of reproduction could be modified by pollutants, but there is an important closely integrated set of phenomena involved in reproductive success of fish populations — that of early development of offspring — emphasizing events that occur after hatching, when sensitive vulnerable larvae encounter the harsh external environment. The following section treats some of the problems encountered at that stage.

EFFECTS OF POLLUTION ON EMBRYONIC AND LARVAL DEVELOPMENT

The scientific literature from the past several decades has demonstrated the important role of organochlorine contamination in the reproductive process in adult fish, and this role continues during the early development of offspring. Associations of chlorinated hydrocarbon contamination of habitats with harmful effects on the earliest life history stages of marine fish — eggs, embryos, and larvae in particular — have been described abundantly. Among the persistent organic contaminants that are of major interest are the polychlorinated biphenyls (PCBs), dibenzo-p-dioxins (PCDDs), and dibenzofurans (PCDFs; Walker & Peterson 1992). These are members of a family of lipophilic, halogenated, aromatic hydrocarbons that persist in the environment and bioaccumulate in fish. Early life stages of many fish species are very sensitive to these synthetic hydrocarbons, which are transferred to maturing eggs from the contaminated tissues of parent females. Mortalities and abnormalities of embryos and larvae have been observed in several species of marine fish and have been correlated with high concentrations of organic pollutants.

The reproductive success of starry flounders (*Platichthys stellatus*) from polluted San Francisco Bay was compared with that of a reference population from an unpolluted site. The total PCB content of eggs correlated inversely with embryological success and hatching success, supporting the stated hypothesis that chronic contamination of reproductive tissues by relatively low PCB concentrations (<200 µg/kg) has a pervasive deleterious effect on the reproductive success of starry flounders in San Francisco Bay (Spies & Rice 1988).

Good evidence also came from European studies in which Baltic flounders (*Platichthys flesus*) with elevated levels of PCBs in their ovarian tissues were found to have a significant reduction in viable hatch of larvae (von Westernhagen et al. 1981). A threshold level of 120 ng/g (0.12 ppm) PCB (wet weight) in eggs and ovarian tissue was considered to be a contamination point above which reduced survival of developing eggs and larvae of that species could be expected. Levels of other chlorinated hydrocarbons or heavy metals could not be correlated with reductions in viable hatch. In a subsequent study of North Sea whiting (*Merlangius merlangus*), the same research team concluded that 0.2 ppm PCB in ovarian tissue constituted a threshold above which impaired reproductive success could be expected (Cameron et al. 1986, von Westernhagen et al. 1989).

Effects of PCBs and DDE on reproductive success of Baltic herring (*Clupea harengus*) were also investigated (Hansen, von Westernhagen, & Rosenthal 1985). Findings included these:

- Viable hatch was significantly reduced by ovarian PCB concentrations of more than 120 ng/g and by DDE concentrations of more than 18 ng/g (wet weight).
- A positive correlation existed between ovarian residues of PCBs and DDE.
- A linear relationship existed between ovarian residue levels of PCBs and DDE and viable hatch.
- The effects of PCBs and DDE on reproductive success were probably additive.

Levels of contaminants that reduced reproductive success in this study were low; the authors cautioned that other contaminants, not analyzed, may also have been involved.

Despite a 3-decade-old ban (in the United States) on its production and disposal, DDT and its metabolites are still implicated in reproductive impairment of marine fish. Attempts to spawn white croaker (*Genyonemus lineatus*) from contaminated San Pedro Bay in California were unsuccessful if the ovarian DDT concentration exceeded 4 ppm (36% of the sample exceeded this level; Cross & Hose 1988, Hose et al. 1989). Of those females with higher DDT tissue contamination, fecundity and fertilization success were lower, suggesting reduced reproductive success, although the authors pointed out that other contaminants, such as polycyclic aromatic hydrocarbons and metals, were present and are known to cause reproductive impairment (Hose et al. 1981; Brown, Gossett, & Jenkins 1982).

Results from earlier studies with estuarine-dependent species have provided additional evidence that high tissue concentrations of chlorinated hydrocarbons in spawning adults can result in mortalities of developing eggs and larvae. Reproductive failure of a sea trout (*Cynoscion nebulosus*) population in Texas was attributed to this phenomenon (Butler, Childress, & Wilson 1972). The sea trout population inhabited an estuary that was contaminated heavily with DDT, where DDT concentrations in ovaries reached a peak of 8 ppm prior to spawning compared to <0.5 ppm in sea trout from other, less contaminated estuaries. Spawning seemed normal, but eggs failed to develop.

Note that these are mostly field investigations, sometimes augmented by chemical analyses of water and tissue contaminant levels, and in a few cases by examination of maternal (ovarian) tissue burdens, accompanied infrequently by shipboard spawning and survival experiments, and even more rarely by concurrent laboratory experiments using environmental concentrations of selected chemicals. The correlations observed are highly suggestive of a relationship between environmental contamination and adverse effects on early life stages of fish. Many ecoepidemiological criteria have been satisfied, and experimental findings have been supportive — this we should consider as a close approximation of a causal relationship.

This consideration of the effects of pollution on reproduction and early development in fish should be an important aspect of any thinking about sublethal effects. So many examples of those effects cluster around the early life stages — either as eggs in the prespawning female or as vulnerable embryos and larvae in the external environment. Look at just a few of the problems created by contaminants: damage to DNA of sex products from male and female parents, abnormal cell divisions in embryos, and production of larvae that are structurally or functionally defective. The exit line, written in most of the life scripts for these larvae, calls for the sublethal — the abnormal — to become rapidly lethal in a harsh predator-filled environment, long before maturity is reached. Death, early and sudden, is a fundamental law of the coastal ocean, and such a fate is often enhanced by the added stressor of chemical pollution.

EFFECTS OF COASTAL POLLUTION ON JUVENILE AND ADULT FISH

We can easily grasp the concept of *acute effects* of chemical pollutants on individual fish — particularly that of death when physiological limits are exceeded and adaptive responses are overwhelmed. *Subacute* or chronic effects require more explanation, since they can be so varied and are often interrelated. Such sublethal effects can be expressed at any level of biological organization, and at any life history stage, but are most apparent as

1. Genetic and developmental abnormalities
2. Damage to cell metabolism, leading to progressive physiological disability of the animal
3. Disruptions of endocrine functions
4. Suppression of immune responses and concomitant reduction in disease resistance
5. Pathological changes in cells and tissues

Each of these categories deserves some investigation here.

Genetic Abnormalities

Pollutants can modify the genetic development of the animal, especially in the egg, embryo, and early larval stages — as we have just considered. Some modifications may take the form of chromosomal damage during early embryonic cell divisions, disruption of the normal mechanism of cell division, or effects on DNA-RNA transcription in the developing egg. Effects may be reflected in abnormalities that prevent hatching or failure of larvae to survive if hatching does occur. Additional genetically induced disorders — physiological or structural — may be expressed throughout the individual's life span.

Modifications in Cell Metabolism

Living organisms (most of them) can be characterized as integrated cellular systems, so cellular events and their modification by contaminants are fundamental to all that happens at higher levels of organization. Metabolic processes are controlled by intracellular enzymes, so the actions of contaminant chemicals at this organizational level are critical. Modification of enzyme activity within the cells results in disturbed metabolism, which is reflected at higher levels of organization. An excellent example of this would be sublethal effects of metal exposure on cellular enzymes. Such effects include energy-requiring chronic demand for compensatory induction of enzymes, or blocking of sensitivities by which enzyme reaction rates are regulated. This lessens the metabolic flexibility necessary for an animal's adaptation and survival during environmental challenge. Another example is seen in induction of so-called mixed function oxygenases (to be described later in this chapter) by chlorinated hydrocarbons; such induced enzymes have been implicated in disturbances of reproductive physiology, probably by altering liver steroid metabolism.

DISRUPTIONS OF ENDOCRINE FUNCTIONS

The synthesis and secretion of hormones are cellular processes under control either of other chemicals in the body fluids or of the nervous system. Contaminants can modify hormone production and activity through the following avenues:

- Blocking the synthesis of hormones
- Mimicking the natural hormones
- Providing receptors that inhibit cell synthesis of hormones (Arnold et al. 1993)

Undoubtedly, the most fascinating recent focus of attention in research on effects of aquatic pollutants has been on contaminant-induced hormonal disruption and its consequences. The role of pollutants as "endocrine disrupters" was explored in several series of studies of freshwater fish (brook trout, rainbow trout, and carp) and a few marine fish (cod, Atlantic croaker, and sole) beginning in the mid-1970s (Sangalang & Freeman 1974, Freeman & Idler 1975). The pace of investigations accelerated during the 1980s and the 1990s, so that today a substantial body of observational and experimental literature exists. It has been reviewed by Kime (1995) and the relationships to similar phenomena in other vertebrates have been emphasized by Colborn and Clement (1992); Colborn (1993); Colborn, von Saal, and Soto (1993); Colborn and Smolen (1996); and Rolland, Gilbertson, and Peterson (1997).

The ability of certain contaminants, especially some of the chlorinated hydrocarbons, to disrupt endocrine functions in fish and other vertebrates was the focus of a series of workshops (1991–1995) organized and supported principally by the World Wildlife Fund–U.S. Impetus for the workshops was described as the increasing number of reports of alterations in the development and function of reproductive, endocrine, nervous, and immune systems of fish and other vertebrates (including humans). The workshops have led to a series of position statements and three books. One book, edited by Colborn and Clement (1992), contains papers resulting from the 1991 conference and is titled *Chemically-Induced Alterations in Sexual and Functional Development: The Wildlife-Human Connection;* a second, authored by Colborn, Dumanoski, and Myers (1996), is titled *Our Stolen Future.* Technical papers in the first volume (1992) identify hormonal and associated developmental dysfunctions in fish and other vertebrates that seem to be induced by specific contaminants — especially organochlorines. Proceedings of the 1995 workshop, titled *Chemically-Induced Alterations in Functional Development and Reproduction of Fishes*, were published in an edited volume by Rolland, Gilbertson, and Peterson (1997), and other papers documenting research on effects of contaminants on marine fish have appeared in increasing numbers in an array of scientific journals.

Some of the findings (from the conferences and from other technical literature) include the following:

1. Dysfunctions in early stages of reproductive cycles of fish and other vertebrates that are thought to be associated with endocrine disruption include:
   - Reduced egg production
   - Delayed oocyte maturation
   - Decreased ovarian growth
   - Reduced vitellogenesis
   - Morphological abnormalities, especially of the brain and reproductive system (Reijnders & Brasseur 1992)
2. Gonadal activity of fish has been shown to be inhibited by pollutants in a number of studies, including the following:
   - Ovarian development and plasma estradiol were reduced in female English sole (*Parophrys vetulus*) from polluted estuarine waters. PCBs and polycyclic aromatic hydrocarbons (PAHs) were suspected (Johnson et al. 1988).
   - Testosterone synthesis in male Atlantic cod (*Gadus morhua*) was inhibited by PCBs (Freeman, Sangalang, & Flemming 1982).
   - Exposure of female Atlantic croaker to lead, benzo[a]pyrene and PCBs resulted in decreased plasma steroid levels, ovarian steroid secretion, and ovarian growth. Plasma testosterone levels in male croakers were also reduced (Thomas 1988).
   - In a later study (Thomas 1990), decreased pituitary gonadotropin secretion was found in croaker pituitaries maintained *in vitro* after *in vivo* PCB exposure.
3. Sewage effluents containing alkyl phenols — degradation products of detergents — were found to have estrogenic (feminizing) effects on male rainbow trout, inhibiting growth of testes and inducing production of vitellogenin (Jobling & Sumpter 1993, Jobling et al. 1995).
4. In early embryonic development, the brain is especially vulnerable to endocrine dysfunctions resulting from trace levels of certain contaminants, especially some synthetic chlorinated organic molecules. The thyroid gland and its secretions are intimately involved and can be affected by vanishingly small amounts of contaminants at specific developmental stages.
5. More than 50 synthetic chemicals (especially dioxins, furans, and chlorobiphenyls) have been found to disrupt endocrine function — as have cadmium and lead (Colborn, Dumanoski, & Myers 1996).

6. Chlorobiphenyls, such as the PCBs, may be metabolized by enzymes of the liver microsomal P450 system to create forms more toxic than the parent compound, causing a decrease in vertebrate thyroid hormone levels (Brouwer, Reijnders, & Koeman 1989; Brouwer, Murk, & Koeman 1990). Another pollutant-mediated effect on the P450 system is the inhibition of endogenous steroid synthesis.

From this brief examination of some of the literature, it seems obvious that greater consideration should be given in future fish-population studies to the potential role of contaminants as endocrine disrupters and, as such, as possible causes of reduced reproductive success and decreases in abundance of commercial species. A critical question in fish population dynamics remains: "Will the additional wastage of reproductive potential due to pollution have an effect on year class strength?" (This question will be addressed in the next chapter, on quantitative effects — don't miss it!)

SUPPRESSION OF IMMUNE RESPONSES

Another integrated cellular activity that finds full expression at the level of the organism is the synthesis of chemicals that protect the animal from invasion by toxic foreign substances (chemical, microbial, and others). Normal functioning of this internal protective system — the immune system — can be suppressed by the presence of contaminants in the tissues. The consequence is increase in vulnerability to toxins or microbial invasion.

PATHOLOGICAL CHANGES IN CELLS, TISSUES, AND ORGANS

The presence of sublethal concentrations of environmental contaminants within cells, tissues, and organs can result in the development of pathological changes, such as skeletal abnormalities, tumors, and skin lesions in fish (see Figure 9.2, Figure 9.3, and Figure 9.4). The onset of such abnormalities can occur at any life stage, beginning with deformed embryos and larvae, and can continue throughout growth and maturation.

SUMMARY

So here, then, is a window — offering an opportunity for a brief scrutiny of the kinds of sublethal effects of pollutants on marine fish that can be caused by contaminants added by humans to coastal/estuarine waters. Principal effects that have been identified and discussed here are:

- Genetic abnormalities
- Modifications of cell metabolism leading to progressive disability
- Disruptions of endocrine functions
- Suppression of immune responses and concomitant reduction in disease resistance
- Pathological changes

FIGURE 9.2 Ulcers and fin erosion in bluefish *Pomatomus saltatrix* (*above*) and sea trout *Cynoscion regalis* (*below*). (Photographs courtesy of Myron Silverman, National Marine Fisheries Service.)

Other effects, such as behavioral modifications, could be included, but the above list is long enough, and it encompasses the dominant sublethal effects.

HOW MARINE ANIMALS RESPOND TO CHEMICAL POLLUTION

Responses of aquatic animals to an environment changed by pollutants may take a number of forms. A generalization that is becoming increasingly apparent — but is often overlooked — is that *aquatic organisms are equipped with a wide variety of physiological/biochemical mechanisms that tend to preserve the status quo and permit survival in the presence of pollutants.* Individuals can tolerate or at least survive levels of contaminants that are within physiological limits, and tolerances may increase with continued exposure to sublethal doses of the contaminant (Bryan & Hummerstone 1971, Bryan 1976; in some instances, however, tolerances may decrease). Furthermore, individual animals may respond to organic contaminants by chemically or physically sequestering them, or by the induction of enzymes that detoxify the foreign chemicals. Detoxification of organic chemical pollutants through a number of metabolic pathways can be effective, although in some instances the transformed (metabolized) compound (for example, benzo[a]pyrene) can be more toxic or carcinogenic for the animal than the compound itself. Trace metal toxicity can be reduced by protein binding. Some specific methods of reducing the effects of environmental pollutants include heavy metal "traps," cytochrome P450 enzymes, modification of immune responses, and the selection of resistant strains.

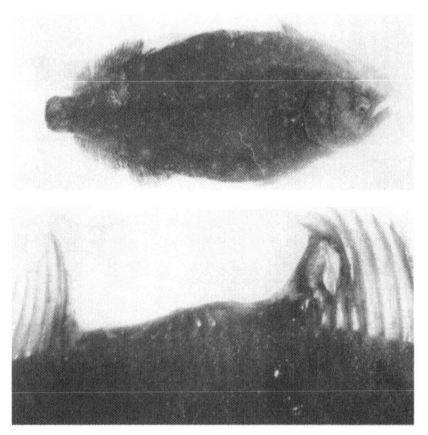

FIGURE 9.3 Extensive fin erosion in a flounder from the New York Bight (*above*), with closeup of an eroded area (*below*).

Heavy Metal "Traps"

Among the internal mechanisms mitigating the effects of heavy metals, an interesting early study (Nöel-Lambot, Bouquegneau, & Disteche 1980) demonstrated the presence in marine fish of mucus complexes with high copper-, zinc-, cadmium-binding capacity. The phenomenon is partly physical–chemical and partly an extension of biological manipulation of metal salts. In the normal environment, fish swallow seawater, and, aided by the pH of the intestinal contents, calcium and magnesium are precipitated out in mucus strands. In waters polluted by cadmium and other heavy metals, high concentrations of those metals are also precipitated out, and the granules of metal salts are incorporated into mucus complexes that are subsequently eliminated. With this process, levels of heavy metals in the intestinal lumen do not become excessive and thus are not absorbed by the fish. The mechanism, demonstrated in eels (*Anguilla anguilla*) and other species, consists of mucus secretion in the anterior intestine, creating extracellular "mucus traps" in which the heavy metal precipitates are incorporated and eliminated with the feces.

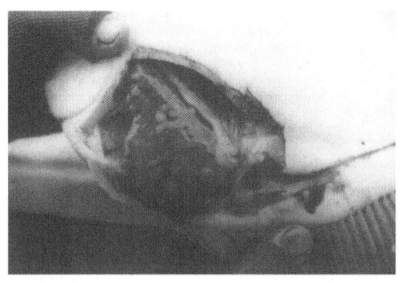

FIGURE 9.4 Gross preneoplastic or neoplastic lesions in the liver of a flounder from Boston Harbor. (Photograph courtesy of Dr. R.A. Murchelano, National Marine Fisheries Service.)

Another extracellular trap for heavy metals consists of increased mucus production by the gills in the presence of metal intoxication. Experimental studies have demonstrated fixation of mercury by gill mucus and its subsequent elimination, thus preventing high metal concentrations from contacting the gill epithelium.

An "intracellular trap" for heavy metals consists of binding of the metals (such as cadmium and mercury) to low molecular-weight proteins with a high content of the amino acid cysteine (metallothioneins). The presence of heavy metals induces biosynthesis of metallothioneins in tissues; the bound metal is toxicologically inert, providing tolerance to high contamination levels in chronic exposures (the metallothioneins appear when the high molecular-weight soluble proteins reach saturation with cadmium or certain other metals).

MIXED FUNCTION OXYGENASES (CYTOCHROME P450 SYSTEM)

A number of organic pollutants are known to induce so-called mixed function oxygenases, now collectively known as the cytochrome P450 system, in fish. These are enzymes that participate in metabolism and degradation of several categories of foreign compounds in the animal (Payne & Penrose 1975, Stegeman & Sabo 1976, Stegeman 1978). Oxidized metabolites of toxic foreign organic compounds can be eliminated by diffusion across membranes, or they can be conjugated with serum components and then excreted. The toxic compounds are eventually metabolized to less toxic ones and excreted, although, as noted earlier in this chapter, there are examples where metabolites are more toxic than the parent compound (Stegeman, Skopek, & Thilly 1979). Some compounds, such as the aromatic hydrocarbons, act as strong inducers of cytochrome P450 enzymes, whereas others, such as certain of the polychlorinated biphenyls, are poor inducers. The cytochrome P450 system is particularly useful as a mechanism for detoxifying short-term sublethal doses of organic contaminants, but it may not be as effective in ensuring long-term survival in chronically or heavily polluted habitats (Burns 1976).

Induction of other kinds of physiological changes that increase resistance in fish has also been demonstrated with several organochlorine and petroleum compounds. Mechanisms of resistance to two pesticides in one study included a membrane barrier that reduced uptake of the contaminants and a brain barrier in the form of insensitivity at the target site (Yarbrough & Wells 1971).

MODIFICATION OF IMMUNE RESPONSES

Although suppression of immune responses is often cited as one of the important effects on fish of exposure to pollutants, there are some aspects of internal resistance than tend to favor survival of fish in polluted habitats. In one experiment, exposure of cunner (*Tautogolabrus adspersus*), a small inshore species, to cadmium did not reduce humoral antibody production but did reduce the bactericidal capabilities of phagocytes (Robohm & Nitkowski 1974).

In a related field study, a survey of antibodies to a wide spectrum of bacteria in summer flounder (*Paralichthys dentatus*) from polluted and unpolluted habitats found significantly higher antibody levels and a greater diversity of antibodies in samples from polluted waters (the New York Bight; Robohm, Brown, & Murchelano 1979). Weakfish (*Cynoscion regalis*) also exhibited increased titers against many bacteria. The greatest proportion of increased titers was against *Vibrio* spp., although prominent titers against other fish pathogens were seen.

In a subsequent experimental study with summer flounder, a greater proportion of high antibody responders was found in fish taken from polluted waters than in those from unpolluted sites (Robohm & Sparrow 1981). Fish pathogens *Vibrio anguillarum* and *Aeromonas salmonicida* elicited particularly strong responses. The investigators suggested that fish surviving in polluted areas may be genetically selected as high antibody responders. Other recent studies have demonstrated that many organic contaminants (for example, dioxins and PCBs) are immunosuppressive and that some metals may selectively suppress critical parts of the immune system.

SELECTION OF RESISTANT STRAINS THROUGH DIFFERENTIAL MORTALITY OF SUSCEPTIBLE INDIVIDUALS

On a population level, there may be long-term adaptations to high environmental levels of naturally occurring contaminants such as heavy metals and petroleum hydrocarbons. Part of the process is the selection of resistant strains, which is feasible since many species have a high reproductive potential and high genetic plasticity. Some evidence exists for selective action of pollutants (in addition to selection for increased immunological competence). In one study of killifish (*Fundulus hetero-clitus*), some females from unpolluted coastal areas were found to produce eggs that were much more resistant to methylmercury than eggs from other females (as measured by percentages of developmental anomalies that followed exposure; Weis & Weis 1981). When a population from a heavily polluted coastal area was examined, a much higher percentage of the females produced "resistant" eggs. However, more recent studies by the same investigators indicated that even though embryos from polluted areas were more resistant to methylmercury toxicity, adults seemed less tolerant, as determined by mortality and rate of fin regeneration (Weis & Weis 1987).

A number of separate lines of investigation converge to illustrate other strategies employed by fish and invertebrates for survival in degraded habitats. Examination of scope for growth — the relative energy budget (energy available for growth and gamete production) — of animals in polluted zones as compared to unpolluted zones demonstrated that at a critical level of contamination animals were living at an energy deficit. Part of the survival strategy for species such as mussels was to reduce reproductive output during periods of high contamination, thereby conserving energy for growth (Bayne et al. 1978). Population replacement during such periods would depend on recruitment from populations outside the deficit zone — a good strategy so long as contamination is not uniformly or extensively distributed. Resistance strategies impose a burden on the animal in that they require energy, placing the animal at a physiological and perhaps survival disadvantage relative to other members of the species living in nonpolluted zones.

This kind of perspective on pollution effects suggests that *damage occurs as a consequence of exposure to pollutants, but important mechanisms for survival in modified habitats may be mobilized to mitigate the damage* — although at the expense of energy (see Figure 9.5). What emerges is, almost predictably, a limited system of checks and balances. Physiological mechanisms that compensate for environmental chemical imbalances function up to a point; beyond such limits, the phenomena of toxic effects, collapse, and death appear.

Sequestering, detoxifying, and excreting toxic chemicals are forms of protection that may be overloaded or overwhelmed. Examples of protective mechanisms include induced MFOs to reduce toxicity of synthetic organics, protein binding and mucus traps to reduce heavy metal toxicity, and the possible selection of high antibody responders as compensation for the immunosuppressive effects of some pollutants. One significant "downside" to these mechanisms may be, however, that survivors may build up high levels of contaminants in their tissues, thus posing chemical threats to predators, including human consumers.

Since the matter of adaptations for survival in degraded habitats is a critical one, it might be worth restating the central thesis: *Marine animals are equipped with a remarkable armamentarium of biochemical responses with which to confront chemical contaminants that are within physiological limits:*

1. They may have metabolic preadaptations, developed and retained in the species in response to earlier encounters with chemicals having some similarity to the contaminant.
2. They may adapt to the new chemical environmental factor through physiological or behavioral modifications.
3. They may function temporarily at an energy deficit if cell enzyme systems are affected.
4. They may reduce reproductive activities and growth in the presence of chronic pollution.

(A)

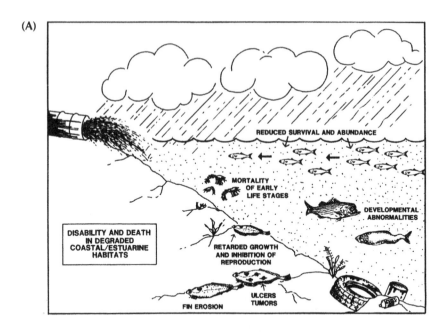

(B)

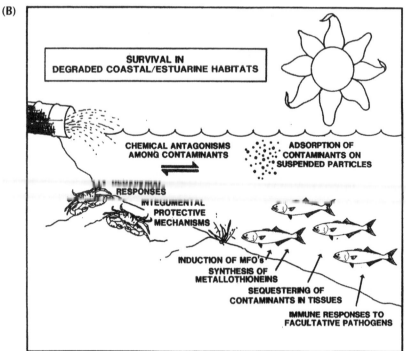

FIGURE 9.5 (A) Some examples of the effects of contaminants on marine organisms. (B) Some mechanisms that enhance the survival of marine organisms in degraded habitats.

5. They may transfer sequestered pollutants to offspring.
6. They may escape from zones of heaviest contamination.
7. They may modify migration pathways to avoid zones of heaviest contamination.
8. They may sequester or metabolize contaminants to reduce their effects.

It seems important to point out, somewhere in this discussion of adaptation and survival, that marine animals may have had no previous evolutionary experience with new synthetic organochlorine chemical formulations, but they are equipped with biochemical responses that enable enzymatic transformations and/or sequestering of these alien compounds (Forbes & Forbes 1994). In the words of Forbes and Forbes: "Through the course of evolution, organisms have repeatedly had to adjust to the addition of new 'natural' chemicals — so-called 'naturally existing pollutants'" (p. 14). As an example, the PCBs did not appear in the marine environment until the 1930s, but "earlier contacts by the species with natural organic compounds laid the evolutionary groundwork for sequestering and transforming the synthetic chemicals" (p. 15). Such responses are, of course, energy-requiring, imposing a degree of stress on the animal when exposed. We shall examine the important topic of stress in polluted coastal environments in the next section.

STRESS FROM POLLUTION

In examining, as we have just done, the brief descriptions of responses of marine animals to environmental changes resulting from pollution, it should be obvious that we have not yet probed deeply enough for underlying biological and biochemical mechanisms. What triggers these responses; how do they produce effects on the exposed animal; and how are they controlled? To approach these more fundamental questions, we need to explore the basic concept of stress, and ask how pollutants function as environmental stressors.

Environmental factors that affect fish and shellfish can be identified, and their principal elements are listed in Figure 9.6. The diagram includes most of the principal sources of stress for aquatic organisms and also illustrates the extent of the problem before us. Some factors are biological, whereas others are physical or chemical; single factors may dominate at any particular time. Some are directly lethal; others can lead to debilitation, physiological malfunction, or morphological abnormalities that render individuals more vulnerable to the effects of other factors.

It is obvious that pollutants and other man-made changes constitute only part of the total array of factors — physical, chemical, and biological — that impinge on marine populations. Pollutants and other nonoptimum environmental factors act as stressors, which, if extreme enough or prolonged enough, may affect survival. Stress is a significant but elusive concept in biology, so it is worth examining here. *Stress can be defined generally as the sum of morphological, physiological, biochemical, and behavioral changes in individuals that result from actions of stressors,* or (in its original sense), *the consequences of all the mechanisms whereby the organism attempts to maintain equilibrium in the face of environmental change* (Selye 1953, 1955).

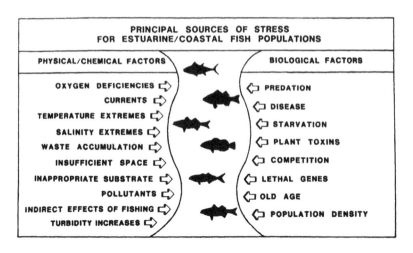

FIGURE 9.6 Sources of stress for marine animals.

The classical physiological/biochemical responses of vertebrates to stressors (accompanied by structural or behavioral changes) include the following (see Figure 9.7):

Primary phase: Increased output of regulatory chemicals (corticosteroids and catecholamines) — the initial or "alarm" phase

Secondary phase: A multitude of metabolic and osmoregulatory disturbances, among the most important of which are immunosuppression and decreased lymphocyte production (in fish, at least) — the stage of resistance or adaptation

Tertiary phase: Decreased resistance to disease and increasing physiological malfunctions or abnormalities, leading to effects on reproduction and growth, and (in extreme malfunctions) death — the stage of exhaustion and collapse of resistance mechanisms

These responses are distributed temporally; some occur immediately, whereas others may take months to develop, as illustrated in Figure 9.8.

Perspectives on responses to stressors will vary with the background of the observer. To the pathologist, many of the responses to stressors can be described as or result in "disease," if disease is defined as any departure from normal structure or function of the animal. The biochemist would see all of the above as expressions of altered cellular metabolism that produce changes at molecular and subcellular levels, mediated by hormonal and enzyme activity. The behaviorist and the physiologist would of course have quite different but still correct perceptions of stress responses — all of which lead to the observation that stress from environmental changes can be an excellent, all-encompassing concept in biology, and one that is vital to understanding of how pollution exerts effects on marine animals.

There is another small complication to this already complex story of stress responses. The *nonspecific* responses described by Selye (1953, 1955) as the "general

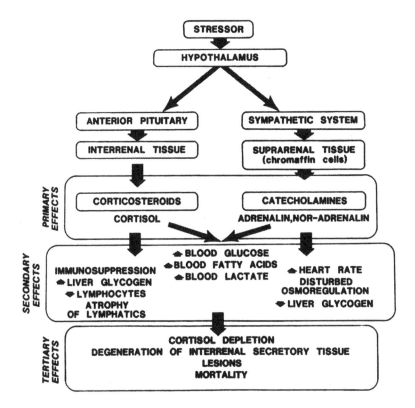

FIGURE 9.7 Pathways of the stress syndrome. (Modified from Mazeaud, M.M., F. Mazeaud, and E.M. Donaldson. 1977. *Trans. Am. Fish. Soc. 106:* 201–212. With permission.)

adaptation syndrome" have to be distinguished from *specific* stress responses — localized reactions, physiological and/or morphological, to injury or infection of a particular tissue or organ. These are superimposed on and may modify the nonspecific reactions to the same stressor. Responses to stressors will depend on the intensity and duration of environmental change. Each species has a series of physiological life zones with respect to variations in any stressor. The animal usually functions in a zone of normal adjustment and has a limit of compensation for changes in any environmental factor, as shown in Figure 9.9. Beyond this limit, the animal functions with increasing energy expenditure, and disabilities appear — some reversible if the environmental change is not too severe or too prolonged, and others irreversible and fatal if the change is drastic or prolonged. This diagram is worthy of attention; it encompasses the entire range of pollution effects on individuals.

A perceptive book by Forbes and Forbes (1994), referred to earlier, offers the following summarizing statement:

Biological responses to stressors, whether natural or anthropogenic, occur via individual organism acclimation, followed at increasing levels of stress by selective elimination of less tolerant genotypes within populations, and by selective reduction or elimination of less tolerant species. (p. 14)

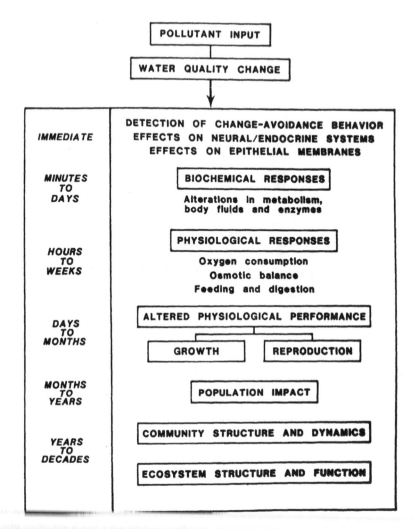

FIGURE 9.8 Temporal sequences of stress effects. (Modified from Sastry, A.N. and D.C. Miller. 1981. Application of biochemical and physiological responses to water quality monitoring, pp. 265–294. In: J. Vernberg, A. Calabrese, E.P. Thurberg, and W.B. Vernberg (eds.), *Biological Monitoring of Marine Pollutants.* Academic Press, New York. With permission.)

Concerning metal pollution, Forbes and Forbes state, "Most [species] assemblages occupying metal-polluted habitats showed increased resistance to metals, resulting from *some combination of physiological acclimation, changes in species composition, and genetic adaptation*" (p. 84; italics mine). (It should be noted that tolerance and resistance to heavy metal toxicity may be used interchangeably by some authors; see, for example, Weis and Weis, 1987.)

Although most studies of the physiological and morphological consequences of stress have emphasized the vertebrates, and particularly humans, there are indications that counterpart phenomena may exist in the lower animals as well (Bayne et al.

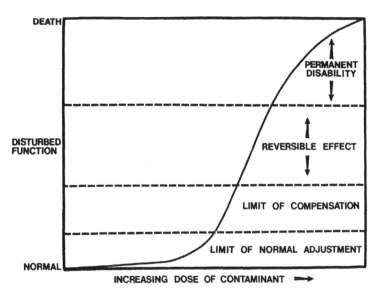

FIGURE 9.9 General life zones in the presence of a varying environmental factor. (Modified from Wilson, K.W. 1980. *Rapp. P.-V. Reun. Cons. Int. Explor. Mer 179:* 333–338. With permission.)

1985). In bivalve molluscs, signs of stress may include mantle recession, pale digestive gland, regression of digestive tubule epithelium, hemocyte infiltration of tissues, edema, lag in gametogenesis, shell abnormalities, and increased ceroid (brown bodies) — all or most of which constitute a stress syndrome, as illustrated in Figure 9.10. In the larger crustaceans, signs of stress include black gills (see Figure 9.11), abdominal muscle opacity, molt retardation, exoskeletal overgrowth with filamentous bacteria and protozoan epibionts, frequent occurrence of shell disease (see Figure 9.12), disoriented or inappropriate behavior, presence in the tissues of gram-positive bacteria, and clotting of the hemolymph (as a response to gram-negative bacterial endotoxin) — again constituting a stress syndrome, as summarized in Figure 9.13.

As a final point in this discussion of stress in marine animals, we should not overlook an emerging body of information about the protective role of stress proteins as part of the adaptive response of vertebrate and invertebrate organisms to potentially harmful environmental changes — chemical, physical, and biological. Exposure to a stressor results in the rapid synthesis of these stress proteins, which are in the range of 60 to 70 kDa and 80 to 90 kDa, and the suppression of synthesis of other proteins. Although the cellular mechanisms of induction and activity of these proteins are not fully understood, the result is an increase in tolerance of the animal, even to other types of stressors. The stress proteins thus represent a significant expansion of the concept of a stress response, beyond the physiological/biochemical changes that constitute the generally understood adaptation syndrome in fish and the counterpart adaptive responses in invertebrates (stress proteins can be induced in invertebrates as well as in the vertebrates).

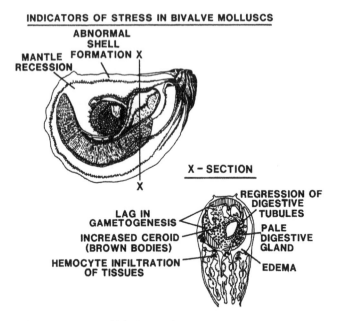

FIGURE 9.10 Some responses of bivalve molluscs to stressors.

To end this discussion of the effects of chemical stressors (pollutants) on marine animals, we can return to the effects of tributyltin and the vignette about abnormal French oysters that began this chapter. That pollutant chemical, acting as an environmental stressor, affected cellular enzyme systems involved in calcium metabolism; enzyme alteration resulted in abnormal shell deposition, producing an unmarketable oyster.

CONCLUSIONS

We have explored in this chapter some of the important physiological/biochemical response mechanisms that enable survival in degraded habitats, if the effects of the contaminant are not too severe or too prolonged. Polluting chemicals are being added in ever-increasing variety to coastal/estuarine waters, as effluents from inventive and seemingly insatiable human technologies. Fish and shellfish in affected habitats either live, survive marginally, or die, depending on their ability to adapt to the changed chemical environment. The methods of adaptation are marvelously varied; some principal biochemical devices include inducing cellular enzymes to metabolize the foreign chemical, blocking the entrance of the chemical at the cell membrane level, creating mucus traps in the digestive tract and gills for excess metals, sequestering fat-soluble toxic chemicals in fat cells, and linking metals with proteins to reduce their toxicity. Other strategies to counter toxic chemicals include energy conservation mechanisms, especially reduced growth and reproductive capacities, and exclusionary mechanisms such as increased mucus production in fish and prolonged shell closure in bivalve molluscs.

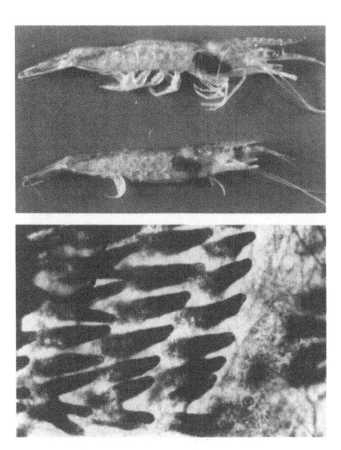

FIGURE 9.11 Black gills in shrimp (*above*) with closeup (*below*). (Photographs courtesy of Dr. D.V. Lightner, University of Arizona.)

Although the distinctions between "effects" of pollution and "adaptive responses" to pollution may sometimes not funny, I am much merit in trying to keep the terms separate in any preliminary discussion. *Effects* are to me the consequences — physiological, biochemical, structural, or behavioral — of exposure of marine animals to toxic chemicals. Effects may be lethal or sublethal, rapid or slow, and expressed at any life cycle stage. *Responses* of marine animals to toxic levels of pollutants may also be physiological, biochemical, structural, or behavioral — but they are easier to assign to specific pigeonholes. So, for example, biochemical responses include elaboration of mucus heavy metal traps and other methods of sequestering toxic contaminants, or induction of enzyme systems (mixed function oxygenases, for example) to metabolize such contaminants. Behavioral responses could include escaping from toxic habitats, altering migration routes, or (in the case of shellfish) exclusion of the toxic environment by shell closure. Responses that enhance survival in polluted habitats include such strategies as reduced growth and reproductive output, or sequestering toxic contaminants in body tissues.

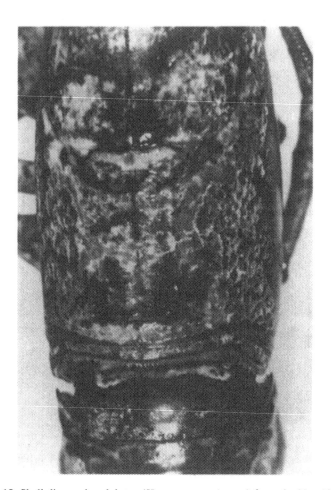

FIGURE 9.12 Shell disease in a lobster (*Homarus americanus*) from the New York Bight.

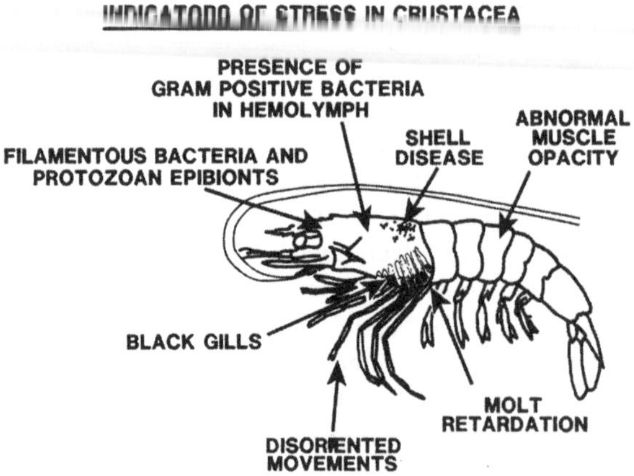

INDICATORS OF STRESS IN CRUSTACEA

PRESENCE OF
GRAM POSITIVE BACTERIA
IN HEMOLYMPH

ABNORMAL
MUSCLE
OPACITY

SHELL
DISEASE

FILAMENTOUS BACTERIA AND
PROTOZOAN EPIBIONTS

BLACK GILLS

MOLT
RETARDATION

DISORIENTED
MOVEMENTS

FIGURE 9.13 Some responses of crustaceans to stressors.

Interwoven throughout any discussion of responses to pollution must be a consideration of stress. Chemical contaminants act as stressors for marine animals, eliciting an array of biochemical, physiological, and behavioral responses that constitute (in the vertebrates) a "stress syndrome" with an initial (alarm) stage, followed by adaptation to or resistance to the stressor, and then a third stage of exhaustion or collapse of resistance in the presence of prolonged or overwhelming levels of the stressor. Invertebrates give evidence of different responses to stressors, but specific syndromes have not been described completely. Essential to an understanding of the consequences of stress from pollutants is a tour of the complex biochemical pathways that are induced or altered by the presence of contaminant chemicals — which has been attempted in abbreviated form in this chapter.

REFERENCES

Arnold, S.F., D.M. Klotz, B.M. Collins, P.M. Vonier, L.J. Guillette, Jr., and J.A. McLachlan. 1993. Synergistic activation of estrogen receptor with combinations of environmental chemicals. *Science 272:* 1489–1492.

Bayne, B.L., D.L. Holland, M.N. Moore, D.M. Lowe, and J. Widdows. 1978. Further studies on the effects of stress in the adult on the eggs of *Mytilus edulis. J. Mar. Biol. Assoc. U.K. 58:* 825–841.

Bayne, B.L., D.A. Brown, K. Burns, D.R. Dixon, A. Ivanovici, D.R. Livingstone, D.M. Lowe, M.N. Moore, A.R.D. Stebbing, and J. Widdows. 1985. *Effects of Stress and Pollution on Marine Animals.* Greenwood Press, Westport, CT. 384 pp.

Brouwer, A., P.J.H. Reijnders, and J.H. Koeman. 1989. Polychlorinated biphenyl (PCB)-contaminated fish induces vitamin A and thyroid hormone deficiency in the common seal, *Phoca vitulina. Aquat. Toxicol. 15:* 99–106.

Brouwer, A., A.J. Murk, and J.H. Koeman. 1990. Biochemical and physiological approaches in ecotoxicology. *Funct. Ecol. 4:* 275–281.

Brown, D.A., R.W. Gossett, and K.D. Jenkins. 1982. Contaminants in white croakers *Genyonemus lineatus* (Ayers, 1855), from the Southern California Bight. II. Chlorinated hydrocarbon detoxification/toxification, pp. 197–213. In: W.B. Vernberg, A. Calabrese, F.P. Thurberg, and F.J. Vernberg (eds.), *Physiological Mechanisms of Marine Pollution Toxicity.* Academic Press, New York.

Bryan, G.W. 1976. Some aspects of heavy metal tolerance in aquatic organisms, pp. 7–34. In: A. Lockwood (ed.), *Effects of Pollutants on Aquatic Organisms.* Cambridge Univ. Press, New York.

Bryan, G.W. and L.G. Hummerstone. 1971. Adaptation of the polychaete *Nereis diversicolor* to estuarine sediments containing high concentrations of heavy metals. *J. Mar. Biol. Assoc. U.K. 51:* 845–863.

Burns, K.A. 1976. Microsomal mixed function oxidases in an estuarine fish, *Fundulus heteroclitus,* and their induction as a result of environmental contamination. *Comp. Biochem. Physiol. 53B:* 443–446.

Butler, P.A., R. Childress, and A.J. Wilson. 1972. The association of DDT residues with losses in marine productivity, pp. 262–266. In: M. Ruivo (ed.), *Marine Pollution and Sea Life.* Fishing News Ltd., London.

Cameron, P., H. von Westernhagen, V. Dethlefsen, and D. Janssen. 1986. Chlorinated hydrocarbons in North Sea whiting (*Merlangius merlangus*) and effects on reproduction. *Int. Counc. Explor. Sea* E:25, pp. 1–10.

Colborn, T. 1993. The wildlife/human connection: Modernizing risk decisions. *Environ. Health Perspect. Suppl. 102*(12): 55–59.

Colborn, T. and C. Clement, Editors. 1992. *Chemically Induced Alterations in Sexual and Functional Development: The Wildlife-Human Connection.* Advances in Modern Environmental Toxicology, Vol. 21, Princeton Science Publishing, Princeton, NJ.

Colborn, T. and M.J. Smolen. 1996. Epidemiological analysis of persistent organochlorine contaminants in cetaceans. *Rev. Environ. Contam. Toxicol. 146:* 91–171.

Colborn, T., F.S. von Saal, and A.M. Soto. 1993. Developmental effects of endocrine-disrupting chemicals in wildlife and humans. *Environ. Health Perspect. 101:* 378–384.

Colborn, T., D. Dumanoski, and J.P. Myers. 1996. *Our Stolen Future.* Dutton Publ., New York.

Cross, J.N. and J.E. Hose. 1988. Evidence for impaired reproduction in white croaker (*Genyonemus lineatus*) from contaminated areas off southern California. *Mar. Environ. Res. 24:* 185–188.

Forbes, V.E. and T.L. Forbes. 1994. *Ecotoxicology in Theory and Practice.* Chapman and Hall, London.

Freeman, H.C. and D.R. Idler. 1975. The effect of polychlorinated biphenyl on steroidogenesis and reproduction in the brook trout (*Salvelinus fontinalis*). *Can. J. Biochem. 53*(6): 666–670.

Freeman, H.C., G.B. Sangalang, and B. Flemming. 1982. The sublethal effects of polychlorinated biphenyl (Aroclor 1254) diet on the Atlantic cod *Gadus morhua. Sci. Tot. Environ. 24:* 1–11.

Hansen, P.D., H. von Westernhagen, and H. Rosenthal. 1985. Chlorinated hydrocarbons and hatching success in Baltic herring spring spawners. *Mar. Environ. Res. 15:* 59–76.

Hose, J.E., J.B. Hannah, M.L. Landolt, B.S. Miller, W.T. Iwaoka, and S.P. Felton. 1981. Uptake of benzo(a)pyrene by gonadal tissue of flatfish (family Pleuronectidae) and its effects on subsequent egg development. *J. Toxicol. Environ. Health 7:* 991–1000.

Hose, J.E., J.N. Cross, S.G. Smith, and D. Diehl. 1989. Reproductive impairment in a fish inhabiting a contaminated coastal environment off southern California. *Environ. Pollut. 57:* 139–148.

Jobling, S. and J.P. Sumpter. 1993. Detergent components in sewage effluent are weakly oestrogenic to fish: An *in vitro* study using rainbow trout (*Oncorhynchus mykiss*) hepatocytes. *Aquat. Toxicol. 27:* 361–372.

Jobling, S., T. Reynolds, R. White, M.G. Parker, and J.P. Sumpter. 1995. A variety of environmentally persistent chemicals, including some phthalate plasticizers, are weakly estrogenic. *Environ. Health Perspect. 103:* 582.

Johnson, L.L., E. Casillas, T.K. Collier, B.B. McCain, and U. Varanasi. 1988. Contaminant effects on ovarian development in English sole (*Parophrys vetulus*) from Puget Sound, Washington. *Can. J. Fish Aquat. Sci. 45:* 2133–2146.

Kime, D.E. 1995. The effects of pollution on reproduction in fish. *Rev. Fish Biol. Fish. 5:* 52–95.

Mazeaud, M.M., F. Mazeaud, and E.M. Donaldson. 1977. Primary and secondary effects of stress in fish: Some new data with a general review. *Trans. Am. Fish. Soc. 106:* 201–212.

Nöel-Lambot, F., J.M. Bouquegneau, and A. Disteche. 1980. Some mechanisms promoting or limiting bioaccumulation in marine organisms. Int. Counc. Explor. Sea, Doc. C.M.1980/E:39, 25 pp.

Payne, J.F. and W.R. Penrose. 1975. Induction of aryl hydrocarbon (benzo[a]pyrene) hydroxylase in fish by petroleum. *Bull. Environ. Contam. Toxicol. 14:* 112–116.

Reijnders, P.J.H. and S.M.J.M. Brasseur. 1992. Xenobiotic induced hormonal and associated developmental disorders in marine organisms and related effects in humans: An overview, pp. 159–174. In: T. Colborn and C. Clement (eds.), *Chemically-Induced Alterations in Sexual and Functional Development: The Wildlife/Human Connections. Adv. Modern Environ. Toxicol. 21.*

Robohm, R.A. and M.F. Nitkowski. 1974. Physiological response of the cunner, *Tautogolabrus adspersus*, to cadmium. IV. Effects on the immune system. U.S. Dept. Commerce, NOAA Tech. Rep. NMFS SSRF-681, pp. 15–20.

Robohm, R.A. and D.S. Sparrow. 1981. Evidence for genetic selection of high antibody responders in summer flounder (*Paralichthys dentatus*) from polluted areas, pp. 273–278. In: *Proceedings of the International Symposium on Fish Biologics, Serodiagnostics, and Vaccines.* S. Karger, Basel.

Robohm, R.A., C. Brown, and R.A. Murchelano. 1979. Comparison of antibodies in marine fish from clean and polluted waters in the New York Bight: Relative levels against 36 bacteria. *Appl. Environ. Microbiol. 38:* 248–257.

Rolland, R.M., M. Gilbertson, and R.E. Peterson, Editors. 1997. *Chemically Induced Alterations in Functional Development and Reproduction of Fishes.* Society of Environmental Toxicology and Chemistry (SETAC), Pensacola, FL. 194 pp.

Sangalang, G.B. and H.C. Freeman. 1974. Effects of sublethal cadmium on maturation and testosterone and 11-ketotestosterone production *in vivo* in brook trout. *Biol. Reprod. 11*(4): 429–435.

Sastry, A.N. and D.C. Miller. 1981. Application of biochemical and physiological responses to water quality monitoring, pp. 265–294. In: J. Vernberg, A. Calabrese, E.P. Thurberg, and W.B. Vernberg (eds.), *Biological Monitoring of Marine Pollutants.* Academic Press, New York. 559 pp.

Selye, H. 1953. *The Story of the Adaptation Syndrome.* Acta, Montreal, Quebec. 255 pp.

Selye, H. 1955. Stress and disease. *Science 122:* 625–631.

Spies, R.B. and D.W. Rice, Jr. 1988. The effects of organic contaminants on reproduction of starry flounder, *Platichthys stellatus* (Pallas) in San Francisco Bay. Part II. Reproductive success of fish captured in San Francisco Bay and spawned in the laboratory. *Mar. Biol. 98:* 191–200.

Stegeman, J.J. 1978. Influence of environmental contamination on cytochrome P450 mixed function oxygenases in fish: Implications for recovery in the Wild Harbor marsh. *J. Fish. Res. Board Can. 35:* 668–674

Stegeman, J.J. and D.J. Sabo. 1976. Aspects of the effects of hydrocarbons on intermediary metabolism and xenobiotic metabolism in marine fish, pp. 423–436. In: *Proceedings of the Symposium on Sources, Effects and Sinks of Hydrocarbons in the Aquatic Environment.* American Institute of Biological Sciences, Washington, DC.

Stegeman, J.J., T.R. Skopek, and W.G. Thilly. 1979. Bioactivation of polynuclear aromatic hydrocarbons to cytotoxic and mutagenic products by marine fish. In: *Carcinogenic Polynuclear Aromatic Hydrocarbons in the Marine Environment.* Ann Arbor Science Publ., Ann Arbor, MI.

Thomas, P. 1988. Reproductive endocrine function in female Atlantic croaker exposed to pollutants. *Mar. Environ. Res. 24:* 179–183.

Thomas, P. 1990. Effects of Aroclor 1254 and cadmium on reproductive endocrine function and ovarian growth in Atlantic croaker. *Mar. Environ. Res. 28:* 499–503.

von Westernhagen, H.D., H. Rosenthal, V. Dethlefsen, W. Ernst, U. Harms, and P.D. Hansen. 1981. Bioaccumulating substances and reproductive success in Baltic flounder, *Platichthys flesus. Aquat. Toxicol. 1:* 85–99.

von Westernhagen, H., P. Cameron, V. Dethlefsen, and D. Janssen. 1989. Chlorinated hydrocarbons in North Sea whiting (*Merlangius merlangus* L.) and effects on reproduction. *Helgol. Meeresunters. 43:* 45–60.

Walker, M.K. and R.E. Peterson. 1992. Toxicity of polychlorinated dibenzo-p-dioxins, dibenzofurans, and biphenyls during early development in fish, pp. 195–202. In: T. Colborn and C. Clement (eds.), *Chemically-Induced Alterations in Sexual and Functional Development:The Wildlife/Human Connection. Adv. Modern Environ. Toxicol. 21.*

Weis, J.S. and P. Weis. 1981. Methylmercury tolerance in killifish. *COPAS 1*(3): 35–36.

Weis, J.S. and P. Weis. 1987. Pollutants as developmental toxicants in aquatic organisms. *Environ. Health Perspect. 71:* 77–85.

Wilson, K.W. 1980. Monitoring dose–response relationships. *Rapp. P.-V. Reun. Cons. Int. Explor. Mer 179:* 333–338.

Yarbrough, J.D. and M.R. Wells. 1971. Vertebrate insecticide resistance: The *in vitro* endrin effect on succinic dehydrogenase activity on endrin-resistant and susceptible mosquitofish. *Bull. Environ. Contam. Toxicol. 6:* 171–177.

10 Effects of Coastal Pollution on Yields from Fish and Shellfish Resources

Mackerel Migrations in Coastal Waters of the Western North Atlantic

Rachael Carson, in her first book <u>Under the Sea Wind</u> published in 1941, wrote in almost poetic but still technically impeccable language about the annual spring migrations of Atlantic mackerel <u>Scomber scombrus</u> into the coastal waters of the Middle Atlantic States, and described in poignant terms the fates of billions of offspring produced during that mass movement — the many ways that death can and does come to far more than 99% of the helpless eggs and larvae in a hostile ocean environment.

But Ms. Carson published her first book about the sea in a simpler time, well before toxic industrial chemicals were found to be so lethal to marine organisms, and more than two decades before she published her major work <u>Silent Spring</u> (1962) that aroused public sensitivities to the menace of pesticides in the terrestrial environment. So in the spring of 2006, as in every other spring, the mackerel will again move in great masses over the continental shelves of the Middle Atlantic States. As they migrate, they will as usual produce uncountable numbers of offspring — except that now the beleaguered defenseless young will encounter an additional source of disability and death in the form of increasing levels of chemicals synthesized by man and discarded from his wasteful industrial processes. Developing embryos within the floating eggs and hatched larvae in the surface waters will encounter strange and often lethal man-made chemical creations — PCBs, DDT, polycyclic aromatic hydrocarbons, and many others — with which marine animals have had no previous evolutionary experience, and to which the younger life stages are particularly vulnerable.

The precise percentage of all the mackerel offspring that will be disabled or killed this coming spring by the relatively new environmental stressor — pollutants at lethal levels — is not easily determined, but it is substantial, and the estimates are improving in quality. Some indications of the magnitude of mortalities have been obtained from percentages of dead eggs and abnormal embryos in samples that have been collected from coastal waters by the National

163

Marine Fisheries Service intermittently since the mid-1970s. In one series of studies, a direct correlation was found between dead or abnormal young and environmental levels of contaminants throughout the New York Bight from Montauk, Long Island to Cape May, New Jersey. In some samples from severely contaminated inshore stations, as many as 40% of the eggs were found to be dead or to contain abnormal embryos. Fortunately, the spawning area for mackerel in the Middle Atlantic Bight extends well out into continental shelf waters, to areas where levels of contaminants are much lower than those in the immediate coastal zones, and where percentages of abnormalities are correspondingly smaller.

Scientists have learned much about mackerel during the past half-century. An extensive long-term data base, beginning in the 1930s, on size of mackerel stocks, spawning areas and intensities, and annual survival rates can now be combined with more recent information on geographic distribution of pollutants in surface waters and prevalences of abnormalities and mortalities in eggs and larvae in those same areas. All these pieces of information provide an unusually strong opportunity for computer simulations of population impacts, and for realistic estimates of actual impacts of pollution.

Those estimates will not improve the odds for survival of any of the tiny vulnerable offspring from the 2005 spawning migration of mackerel in Middle Atlantic waters — but they <u>will</u> contribute to the gradual accretion of evidence that may someday convince a very destructive terrestrial species, <u>Homo sapiens</u>, to clean up its act, insofar as abuse of the oceans is concerned.

From Field Notes of a Pollution Watcher
(C.J. Sindermann, 1995)

The introduction to this book promised "undersea horrors." Here is a massive one, repeated year after year. Millions upon millions of fish larvae less than a quarter inch long are being bred in polluted waters from eggs already contaminated by toxic chemicals. They struggle for survival in a harsh coastal environment, despite structural and functional abnormalities that are almost certain precursors of early death. The most important mitigating factor is that the killing zone of highest contamination is restricted, when compared with the total spawning area of the species.

To make this litany of disaster a little more realistic, we could trace the early encounters of a single mackerel larva, born in the polluted waters 20 mi seaward of New York City in the spring of 2005. The example has to be hypothetical to some degree, since all other environmental factors that lead to disability and death — predation, starvation, disease, storms, temperature changes — which are likely causes of mortality must be excluded here to enable a focus on effects of pollution.

In the long evolutionary history of the species, habitat degradation from human sources has been a very recent phenomenon, but one that could have severe localized effects (as, for example, in the immediate vicinity of a minor oil spill, or in coastal waters contaminated by outflow of polluted rivers, such as the Hudson). Fortunately, these heavily impacted waters constitute only a small fraction of the total spawning range of adult mackerel as they make their annual northward migration through the Middle Atlantic Bight. Depending on chance alone, the eggs can be deposited in contaminated waters of the New York Bight, or they may be spawned in less polluted waters of the outer continental shelf (although no Middle Atlantic coastal area can be described any longer as "pristine" or "unpolluted").

The larva that we have chosen was unlucky, being spawned in the inner New York Bight, an area contaminated by PCBs and other industrial chemicals from the Hudson River, by toxic chemicals and other repulsive debris from municipal sewage treatment plants, by non–point source outwash of fertilizers from farmland and from the lawns and gardens of more than 11 million people in the New York metropolitan area, and from deposition of airborne contaminants from the industries of the north-eastern states.

Already carrying a burden of PCBs and other toxic organic chemicals in its tissues — a legacy passed on to offspring by the parent female through incorporation of fat-soluble pollutants in the yolk of her eggs — the helpless larva hatches in heavily contaminated waters of the Bight. But the helplessness is relative; the rudiments of adaptive biochemical systems that favor survival are already in place and quickly become functional. Again purely by chance, our hypothetical larva may or may not have escaped the pollution-induced genetic damage that cripples many of its cohorts during early developmental stages, but as the young fish develops, environmentally induced abnormalities are still threats. Metamorphosis of the larva into the early juvenile stage does provide some degree of escape from total vulner-ability, with increased mobility and schooling behavior to reduce predation. Death, early and instantaneous, is still the rule at any stage, but never more so than in those first hours, days, and weeks of existence.

As this anecdote about Atlantic mackerel suggests, to isolate and then to quantify specific pollutant effects on population abundance — as distinct from a maelstrom of other environmental influences on survival — is pushing the current state of the art in marine population dynamics too far. As a consequence, the approach that quantitative biologists usually resort to is to develop simulation models based on available data, to simulate and then predict potential effects of various levels of pollution (as additional causes of mortality) on the life stages of fish and shellfish under varying conditions of exploitation by man (Boreman 1997).

In an attempt to understand pollution effects on abundance of marine species, I have tried a "case history" approach, in which several coastal/estuarine species of fish at high risk from pollution have been examined in detail — made possible by the extent of available published information about those species and the popula-tion-modeling efforts that have been used.

CASE HISTORIES OF POLLUTION IMPACT STUDIES

Probably the most critical problem in assessing pollution effects on fish stocks is that of separating natural and fishing mortality from pollution-induced mortality. The nature of the problem was explored thoroughly by Jones (1982), whose illus-tration of the complexity of population responses is presented in Figure 10.1.

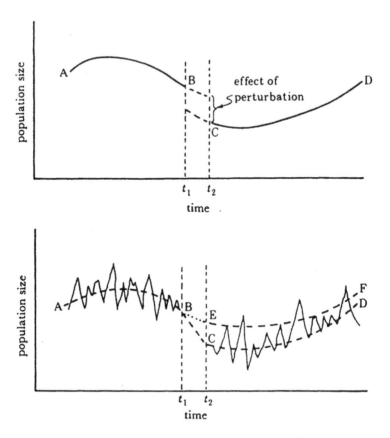

FIGURE 10.1 (*Top*) Simplified response in which a population subject to long-term cyclical variation at level AB is perturbed from time$_1$ to time$_2$, displacing the two parts of the cycle; and (*bottom*) response of a population subject to long-term cyclical variation and short-term random variation (AB) that is perturbed from time$_1$ to time$_2$. The result of perturbation causes a change in population size (BC) to be different than had the perturbation not occurred. (From Jones, R. 1982. *Philos Trans R Soc Lond* B297: 353–368. With permission from The Royal Society.)

The extent of continuing frustration about inability to distinguish causation can be detected in extreme statements such as "biological monitoring programs that cannot separate pollution-induced change from natural change should be terminated and those regulatory requirements which require such programs changed" (Segar & Stamman 1986, p. 876). A reasonable rebuttal to such a subjective conclusion might be (in my opinion) that we should intensify research leading to better delimitation of pollution-induced from natural changes in marine populations, and in the interim should adopt a "precautionary principle" developed recently by Germany (Dethlefsen & Tiews 1985, Dethlefsen 1986) and accepted at the Second International Conference on the Protection of the North Sea in 1987. The principle *requires action to reduce pollution even in the absence of soundly established scientific proof for cause-and-effect relationships.* A proposed elaboration of the principle states that

Only those reduction measures will be applicable which are technically and economically feasible. But the decision whether xenobiotic substances are introduced into the marine environment or not should not be based on considerations of the assimilative capacity of the recipient water but on technically available reduction options.

Additionally, the precautionary concept always has to be accompanied by intensive research on the effects of pollutants in the marine environment. The principle is a policy-making strategy, acknowledging that scientific evidence is often inconclusive, but providing for action to protect the environment even without that elusive scientific certainty.

Cause-and-effect relationships — demonstrations of population impacts of pollutants — have been subjects of searches by many investigators, especially during the past 3 decades. The approach selected for this chapter is to consider a few species of economic importance that may be at risk from coastal/estuarine pollution and to use these as case histories illustrating the current status of technical information about pollution effects on abundance. The species chosen are Atlantic menhaden, *Brevoortia tyrannus;* striped bass, *Morone saxatilis;* and winter flounder, *Pleuronectes americanus.*

ATLANTIC MENHADEN

Estimates of mortality rates in fish stocks from fishing and from natural causes can be made through the use of a variety of models developed for assessment and management purposes. Such models may be adapted to assessing the effects of pollution by integrating pollution mortality terms into them. The modified models can then provide estimates of varying pollution mortality on fish stocks at various levels of exploitation. Such a model has been used to demonstrate population responses of menhaden (*Brevoortia tyrannus*), a relatively short-lived species, to simulated pollution events (Kanciruk, Breck, & Vaughan 1982; Vaughan, Kanciruk, & Breck 1982; Vaughan et al. 1982). The investigators who developed the model pointed out that all sources of mortality — natural, fishing, and pollution — at all life stages of a species throughout its range must be quantified for effective prediction of pollution effects. Using the extensive (> 30 yr) menhaden database of the Beaufort (NC) Laboratory of the National Marine Fisheries Service, and imposing both one-time pollution-related catastrophic mortalities and chronic mortalities on one year-class, the response of the entire population for the next 30 yr was simulated.

For a simulated catastrophic event (an oil well blowout in coastal waters), and assuming a one-time 50% reduction in survival of 0-age-group menhaden, the researchers estimated that the total biomass would be reduced by about 12% over 30 yr. Based on the fact that stocks are heavily exploited already and thus may have little compensatory reserve, the simulation indicated a permanent reduction in stock size of 9% as a consequence of the event. For a comparable simulation of chronic effects, assumptions were made of a slow but continuous decline in estuarine water quality and a substantial increase in ocean dumping. The simulation predicted a 40% decrease in total population biomass in 30 yr.

In subsequent papers by the Beaufort scientists, estimates of acute and chronic pollution effects derived from simulation modeling were made for Atlantic menhaden as well as for seven other inshore species (Schaaf et al. 1987, 1993). The studies were designed not only to examine population effects, but also to compare the relative vulnerability of several marine fish stocks, at different life stages, to pollution. Findings included these:

- The modeled stocks responded to a simulated catastrophic event (a one-time 50% reduction in first year survival) by taking, on average, 10 years to equilibrate at 88% of preimpact abundance.
- Species that are impacted the most by coastal/estuarine pollution seem to stabilize most rapidly following acute stress.
- Stocks most susceptible to acute stress are even more susceptible to chronic stress for at least up to 20 yr.
- Estimates of first-year survival are critical to any attempts at simulation modeling.
- Estimates of a species' susceptibility to pollution stress (in terms of stabilization time after impact) can be made from life history data, including age-specific survival and fecundity rates.

As the biologists pointed out, simulations of this kind can provide estimates of the magnitude and time duration of pollution impacts that should be useful to resource managers.

In a related modeling effort exploring the possible effect of a single man-induced catastrophic event (such as an oil spill) on stock abundance in future years, the staff of the Beaufort Laboratory examined the effect of mass mortalities of young menhaden that occurred in 1984. The study was designed to test the ability to detect reductions in populations following acute pollution events, because of variability in young-of-the-year survival (Vaughan, Merriner, & Schaaf 1986). They concluded that a catastrophic loss to the Atlantic menhaden 1984 year-class (e.g., >50% loss in abundance of the 1984 menhaden year-class from the entire Atlantic coast) would have to occur to be detectable at reasonable levels of statistical power (e.g., >70% chance of detection), but more subtle reductions (e.g., <25% loss to the Atlantic menhaden 1984 year-class from the entire Atlantic coast) would undoubtedly go undetected (e.g., chance of detection <12%). *Such difficulties in detecting reductions are typical of most fish stocks having comparable or larger inherent variability in recruitment or landings* (p. 128; italics are mine).

These studies with menhaden in the 1980s were preceded by similar attempts to evaluate the ability to detect reduction in year-class strength of white perch (*Morone americana*) in the Hudson River (Van Winkle 1977, Van Winkle et al. 1981, Vaughan & Van Winkle 1982). General conclusions from the white perch analysis were that at least 20 yr of data collection would be required to detect an actual 50% reduction in mean year-class strength and that annual fluctuations can mask major reductions in mean year-class strength.

Simulation models developed for predicting impacts of pollution and other man-induced habitat changes can be designed to do two things:

1. Simulate the annual effect on recruitment of young-of-the-year (YOY) fish into the adult population (YOY model)
2. Simulate the long-term effect of reduced recruitment on population size (life cycle model)

The YOY models predict either the number of YOY surviving to age 1 or the percent reduction in the YOY. These outputs from YOY models are then used in conjunction with life cycle models to predict long-term yields or adult population reductions (Swartzman, Deriso, & Cowan 1977).

Development and application of simulation models such as these to predict the effects of pollution on fish stocks is of course only one item in the battery of approaches that can be and are being applied to sorting out the causes of mortality in resource species. Other sources of information are:

- Data from fisheries landings and from fisheries-independent surveys, useful in estimating stock size, recruitment, and mortality
- Data on physical/chemical changes in the oceans
- Data from chemical analyses of contaminant levels in the habitat and in fish tissues
- Data on physiological responses and biochemical transformations of contaminants by marine organisms
- Data from experimental exposures to single or multiple contaminants
- Data from experimental field exposures of resource species in polluted zones

The level of effort required for a program that includes all these elements listed would appear to be overwhelming, but problems of this complexity can be addressed with modern computer power and a major long term commitment of of research resources.

STRIPED BASS

The striped bass (*Morone saxatilis*) has had a long history of extensive fluctuations in catches (and presumably abundance); its center of abundance is Chesapeake Bay, with lesser centers in the Hudson and Delaware estuaries. During their life cycles, striped bass occupy a variety of aquatic habitats, from the lower freshwater zones of spawning rivers to the open sea (see Figure 10.2). For most of their lives, but especially during the first year, they are exposed to an array of chemical contaminants, as well as to the highly variable estuarine and riverine environments. It is generally accepted that survival during the first 60 d of life determines the size of a year-class (Figure 10.3). It is also well accepted that certain of the early life stages are particularly vulnerable to environmental toxicants.

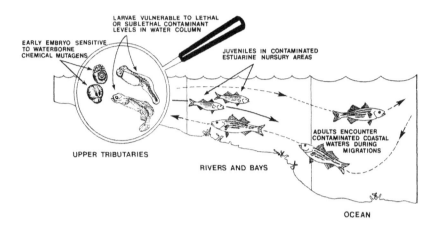

FIGURE 10.2 Life cycle of the striped bass (*Morone saxatilis*), with potential pollutant impact points.

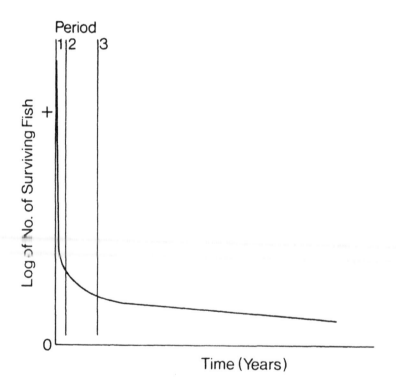

FIGURE 10.3 Hypothetical survival curve of one year-class of fish. The survival curve is divided into three periods: (1) eggs and larval fish; (2) prerecruit fish; (3) fished stocks. (Redrawn from Munro, A.L.S., A.H. McVicar, and R. Jones. 1983. *Rapp. P.-V. Reun. Cons. Int. Explor. Mer 182:* 21–32. With permission from ICES.)

Two aspects of human interference with the natural order of things (other than overfishing) have been responsible for extensive quantitative studies of striped bass on the Atlantic coast of the United States: (1) possible population impacts of power plant siting and operation on the Hudson River, and (2) the possible role of toxic contaminants in reducing abundance of Chesapeake Bay stocks. The Hudson River studies peaked in the mid-1970s; the Chesapeake Bay studies culminated in extensive documentation in the late 1970s and early 1980s. Investigators were particularly interested in quantitative information linking human activities with the effects on striped bass survival.

The most recent decline in striped bass stocks in the Chesapeake Bay, beginning in 1973 and continuing to 1985, was attributed to various causes, especially over-fishing and reduced survival of larvae because of chemical pollution. A major research effort in the late 1970s and early 1980s confirmed that fishing effort had increased since the mid-1960s and might have affected recruitment (Merriner 1976; Goodyear 1978, 1980, 1984a, 1984b). Evidence was also found of elevated levels of PCBs and other contaminants in tissues of young-of-the-year fish and in their habitats, and experimental exposure of yolk-sac larvae in the laboratory to environmental levels (but artificial mixtures) of selected contaminants resulted in increased mortality (Mehrle et al. 1982, Hall et al. 1987). Yolk-sac larvae exposed in *in situ* chambers to Nanticoke River water (a tributary of Chesapeake Bay) also died differentially, and high aluminum and low pH were thought to be implicated. However, a later experimental study of yolk-sac larvae, also using *in situ* chambers placed in their natural habitat in the upper Chesapeake Bay (Chesapeake and Delaware Canal), did not indicate acute harmful effects, although sublethal gill abnormalities were seen in yearlings (Hall et al. 1985).

In addition to the possible effects of toxic chemicals on larvae, effects on reproduction were also investigated as a possible cause of the population decline. Concentrations of PCBs up to 26 ppm were reported in adult striped bass from the Hudson River, and elevated concentrations of PCBs, DDT, and dieldrin were found in striped bass eggs (Mehrle et al. 1982). The organochlorine residues were associated with the failure of cleavage in fertilized eggs, but correlations were not considered significant. In an earlier study of striped bass in California, eggs were found to contain 5 to 10 ppm DDT, but reproductive depression was not demonstrated (Hunt & Linn 1969).

In other studies, the survival of larvae from eggs spawned by contaminated female striped bass was found to be inversely related to concentrations of chlorinated hydrocarbons (hexachlorobenzene, DDT, PCBs, and chlordane) in the eggs. It was also found that parental sources of these contaminants had greater effects on survival than did dietary sources. However, in an earlier investigation, eggs with PCB content from 1.1 to 8.1 μg/g (ppm) wet weight did not differ from controls in survival and growth after yolk absorption (Westin, Olney, & Rogers 1983, 1985).

In still other research, tissue contaminants were found to produce abnormalities that might affect stock abundance. Young-of-the-year striped bass from the Hudson River had high levels of PCBs in their tissues and had vertebrae that were fragile and that ruptured under minimal force (Mehrle et al. 1982). The authors referred to other laboratory studies indicating that contaminants such as PCBs, cadmium, and lead could weaken vertebral structure and contribute to mortality.

Simulation models were used to examine the relative influence of two factors — increased fishing mortality and contaminant toxicity — in producing the observed decline in striped bass abundance (Goodyear 1985a, 1985b). Principal findings were:

- [A]t low levels of density-dependent mortality, an increase in fishing mortality, or an equivalent decrease in early life-stage survival caused by the toxic effect of a contaminant would cause similar declines in the stock.
- At high levels of density-dependent mortality, the effects on the yield are similar if the contaminant-induced mortality precedes the density-dependent mortality. However, if the contaminant-induced mortality occurs after the period of density-dependent mortality, the decline in yield will be more severe than that caused by an equivalent increase in fishing mortality.

However, despite the impressive amount of data and analyses available from these and other studies, the investigator concluded that "the actual level of any excess mortality that is imposed on the striped bass population from toxic substances is unknown, and it will probably remain so for some time" (Goodyear 1985x, p. 111). But he then went on in a more optimistic vein, stating that reduced fishing mortality could result in a 20- to 30-fold increase in population fecundity, which could "offset even rather severe losses due to contaminant toxicity, and could halt or reverse the decline in stock" (p. 112).

It is very interesting, and no doubt significant, that conclusions based on this modeling effort were supported by the behavior of the fishery and the population in the late 1980s. Beginning in 1985, the states of Maryland and Virginia imposed a moratorium on striped bass fishing that remained in effect until 1990. A dramatic increase in spawning stocks occurred (an estimated fivefold increase in spawning females in 1989 as compared to 1984), as well as a high average young-of-the-year index. Based on available evidence for the resurgence of striped bass stocks, the Atlantic States Marine Fisheries Commission (ASMFC) adopted a conservative management plan for limited harvest and a quota system beginning in 1990. The plan called for closure of federal waters (3 to 200 mi) to striped bass fishing, with catch and size limitations in state waters. In 1995 the ASMFC declared that the population was "fully restored," and at present (2005) that organization has recommended lifting the federal ban.

What can be concluded from the results of this research on Chesapeake Bay striped bass, followed by implementation of management measures and the resurgence of the population in the late 1980s? It now begins to appear that high fishing mortality was the principal culprit in the decline in stocks from 1973 to 1985. (It might be noted parenthetically that to some quantitative biologists it has appeared that way all along, especially because fishing mortality rates for immature striped bass were very high.) It also appears that high larval mortality in some spawning tributaries may have been contributory, if stressful environmental conditions (pH, hardness) occurred coincident with toxic levels of specific contaminants (as, for example, aluminum did in the studies conducted in 1985 in the Nanticoke River).

These conclusions support the concept that a heavily exploited population may have limited compensatory reserve and may be particularly sensitive to pollution. In the present case of striped bass, a drastic reduction in fishing mortality during the period 1985 to 1990 because of the moratorium allowed population expansion, even if other factors, such as low larval survival due to pollution, had contributed to the previous decline. An important caution was offered by Boreman (1995): "Reduction of fishing mortality can be an effective short-term measure while habitats are being restored, but unless pollution is reduced, the measure may only delay the decline in fish populations."

WINTER FLOUNDER

Over the past 4 decades, the winter flounder (*Pleuronectes americanus*) has been the subject of many studies that have emphasized pollution effects on early life history stages (Figure 10.4). The average female produces about 600,000 eggs, of which an estimated 10 to 16% hatch. Only an estimated 18 individuals per 100,000 hatched larvae survive to age 1 (Saila 1962). The species is estuarine-dependent, both for nursery areas and for overwintering sites for adults. Many estuaries are polluted, and much of the life history of the winter flounder is spent in close association with contaminated bottom sediments in those estuaries. Adults lie partially buried in bottom sediments; spawning occurs near the bottom; eggs sink to the bottom, where they aggregate in clusters; and larvae, after hatching, alternately swim upward and then sink to the bottom (Hughes et al. 1986). Fin erosion (a good pollution indicator) is a condition commonly seen in adults from polluted habitats, and tissue levels of pollutants in juveniles and adults can be significantly elevated.

The accumulation of chlorinated hydrocarbons in the tissues of adults and their transfer by females to eggs has been found in several studies to contribute substantially to larval mortality. Evidence was found in studies conducted in the 1960s that high mortalities of larval winter flounder in a Massachusetts estuary (a tributary of Buzzards Bay) could be related to pesticide pollution (Smith & Cole 1970). Adult females concentrated DDT, DDF, and heptachlor epoxide in their ovaries as spawning approached, and mortality of post–yolk-sac larva was estimated to approximate 100%. The authors pointed out the similarity of this pattern of reduced hatchability and larval mortality to that reported for several salmonid species and considered it to be the result of DDT contamination (Johnson & Pecor 1969). The explanation offered was that DDT was translocated to the maturing eggs, where, bound to yolk fats, it remained inactive biologically until such fats were metabolized by the developing fry; DDT was then released with lethal results.

Corollary to this study of mortality of winter flounder larvae, juveniles of age 2 and younger, year-round residents of the same polluted Massachusetts estuary, contained higher tissue levels of pesticide residues than did the migratory adults, but mortalities were not observed (Smith & Cole 1970).

In a related experimental study of DDT effects on developing eggs, adult female winter flounders were exposed to sublethal concentrations of DDT and dieldrin. Dieldrin exposure did not affect survival to hatching, but DDT exposure of females

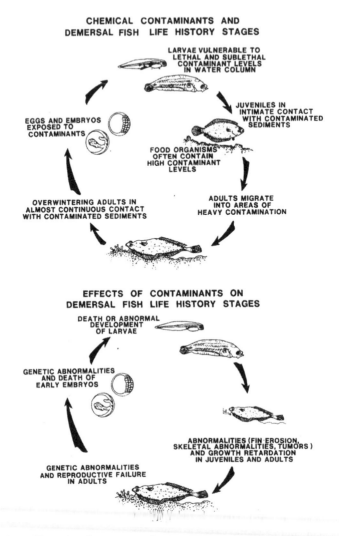

**CHEMICAL CONTAMINANTS AND
DEMERSAL FISH LIFE HISTORY STAGES**

LARVAE VULNERABLE TO
LETHAL AND SUBLETHAL
CONTAMINANT LEVELS
IN WATER COLUMN

JUVENILES IN
INTIMATE CONTACT
WITH CONTAMINATED
SEDIMENTS

EGGS AND EMBRYOS
EXPOSED TO
CONTAMINANTS

FOOD ORGANISMS
OFTEN CONTAIN
HIGH CONTAMINANT
LEVELS

OVERWINTERING ADULTS IN
ALMOST CONTINUOUS CONTACT
WITH CONTAMINATED SEDIMENTS

ADULTS MIGRATE
INTO AREAS OF
HEAVY CONTAMINATION

**EFFECTS OF CONTAMINANTS ON
DEMERSAL FISH LIFE HISTORY STAGES**

DEATH OR ABNORMAL
DEVELOPMENT
OF LARVAE

GENETIC ABNORMALITIES
AND DEATH OF
EARLY EMBRYOS

ABNORMALITIES (FIN EROSION,
SKELETAL ABNORMALITIES, TUMORS)
AND GROWTH RETARDATION
IN JUVENILES AND ADULTS

GENETIC ABNORMALITIES
AND REPRODUCTIVE FAILURE
IN ADULTS

FIGURE 10.4 Life cycle of the winter flounder, *Pleuronectes americanus*, with potential pollutant impact points (*above*) and the effects of pollutants (*below*).

resulted in abnormal gastrulation and mortality of eggs after fertilization and severe vertebral deformities in 39% of the larvae at hatching. Experimentally induced gonad levels of the insecticides in spawning females duplicated levels found in wild fish (Smith & Cole 1973).

More recent studies of the effects of polluted habitats on the reproductive success of winter flounder have produced conflicting results. Examination of samples collected along a composite pollution gradient in Long Island Sound found no significant differences in the percentage of viable hatch from eggs taken from females at each site. In cytogenetic studies, however, increased percentages of chromosomal anomalies and reduced mitotic rates characterized embryos from the more polluted sites (western Long Island Sound). Analysis of PCBs in eggs did not indicate a correlation between contaminant levels and the cytogenetic data (Longwell et al. 1983).

Another pollution gradient, this one a composite of stations in Narragansett Bay, Rhode Island, and Buzzards Bay, Massachusetts, was exploited in a study of the effects of inherited contaminants on eggs and larvae of resident winter flounder (Black, Phelps, & Lapan 1988). Progeny from flounders captured in Buzzards Bay, an area noted for long-term PCB contamination, contained significantly higher levels of PCBs (averaging 39.6 µg/g dry weight), and hatched larvae were smaller in length and weight than progeny from reference site adults. No information was given on the percentage of viable hatch, but as larvae grew to metamorphosis the length/weight differences disappeared; the compensatory growth of larvae from contaminated parents but grown in clean water was attributed by the authors to biotransformation and detoxification of contaminants via mixed-function oxidase systems of the embryos and larvae.

Petroleum contamination can affect the survival of winter flounder larvae. Experimental exposures of mature female winter flounders and their developing eggs and larvae to low concentrations of No. 2 fuel oil produced results that should be useful in population analyses. Exposure to 100 ppb throughout the gonad maturation of parents, and during fertilization and embryogenesis, resulted in a 3- to 9-d delay in hatching, a 19% reduction in viable hatch, and a 4% prevalence of spinal defects in hatched larvae. Larvae produced from gametes contaminated during parental gonadal maturation but then reared in clean water had a mortality coefficient of 0.130, considered by the authors to be much higher than the calculated mortality coefficient for untreated, laboratory-reared winter flounder larvae of 0.036 to 0.059 (Kühnhold et al. 1978). Growth of larvae hatched from gametes from oil-exposed spawners was also slower. (It might be noted here that the topic of acute and chronic oil pollution in the sea was examined in detail from a fishery perspective by McIntyre [1982]. He concluded a detailed review by stating that no long-term adverse effects on fish stocks can be attributed to oil, but local impacts can be extremely damaging in the short term).

The Northeast Fisheries Science Center of the National Marine Fisheries Service (Milford, Connecticut, Laboratory) has carried on an extensive interdisciplinary study of the effects of pollution on winter flounder populations of Long Island Sound. Among findings were these (Hughes et al. 1986):

- Fish from a severely polluted site had a smaller proportion of females actually spawning (whether naturally or artificially induced), a smaller proportion of live eggs at the time of extrusion, and a much smaller proportion of successful egg cultures than fish from a relatively clean reference site.
- The same polluted site had the highest level of developmental abnormality and mortality in the early life stages.
- In decreasing order of importance, pesticides, aromatic hydrocarbons, and polychlorinated biphenyls are the chemical body burdens that most limit the reproductive success of female winter flounder.

- In laboratory experiments, young-of-the-year winter flounder held under constant low levels (2.2 mg/l of dissolved oxygen for 11 to 12 weeks grew only about half as much as flounder held under high levels (6.7 mg/l).
- Yearling winter flounder died during a 20-h exposure at 20°C to a dissolved oxygen (DO) range of 1.1 to 1.5 ppm. They withstood, however, an 8-h exposure to a DO range of 1.2 to 1.4 ppm at 20°C.

Another study by Milford Laboratory staff members of winter flounder from selected stations in Long Island Sound and Boston Harbor disclosed a low percentage of viable hatch, small larvae, and delayed embryogenesis in offspring from fish taken at the most polluted sites (New Haven Harbor and Boston Harbor). Such findings indicate low larval survival and reproductive impairment at the most heavily degraded sites (Nelson et al. 1991). A related study found the greatest prevalences of chromosomal abnormalities and mitotic disruptions in developing embryos from spawning adults taken at the most polluted sites (Perry, Hughes, & Hebert 1991).

Results from studies with flatfish species other than the winter flounder have provided additional evidence that high tissue concentrations of chlorinated hydrocarbons in spawning adults can result in mortalities of developing eggs and larvae. The reproductive success of starry flounders (*Platichthys stellatus*) from polluted San Francisco Bay was compared with that of a reference population from an unpolluted site. The total PCB content of eggs correlated inversely with embryological success and hatching success, supporting the stated hypothesis that chronic contamination of reproductive tissues by relatively low PCB concentrations (<200 μg/kg) has a pervasive deleterious effect on the reproductive success of starry flounders in San Francisco Bay. Good evidence also came from European studies, in which Baltic flounders (*Platichthys flesus*) with elevated levels of PCBs in their ovarian tissues were found to have a significant reduction in viable hatch of larvae (von Westernhagen et al. 1981, 1989). A threshold level of 120 ng/g (0.12 ppm) PCB (wet weight) in eggs and ovarian tissue was considered to be a contamination point above which reduced survival of developing eggs and larvae of that species could be expected. Levels of other chlorinated hydrocarbons or heavy metals could not be correlated with reductions in viable hatch.

Thus far in this case history on winter flounder (augmented by data from other flounders), the effects of pollution on early life history stages have been emphasized, but effects on adults should not be ignored. Detailed studies of pollution effects on winter flounder began 35 yr ago, with one of the first experimental studies of tissue lesions resulting from exposure to copper (Baker 1969). Principal effects were hemolytic anemia, fatty degeneration of the liver, and renal necrosis — all potentially lethal. Sublethal physiological effects of other heavy metals, such as cadmium and mercury, were reported by other investigators, and the distribution, metabolism, and excretion of DDT and Mirex were examined, with the interesting observation that the winter flounder stores its pesticide burden primarily in body muscle (Calabrese et al. 1975).

A study published in 1986 of contaminants in winter flounder from a number of polluted sites on the northeast coast of the United States disclosed that levels of polycyclic aromatic and chlorinated hydrocarbons in stomach contents were higher

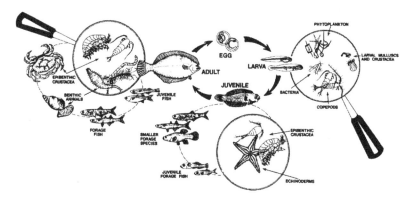

FIGURE 10.5 Effects of contaminants on food chains of winter flounder life history stages.

than those in bottom sediments, indicating that the compounds were being accumulated by prey organisms (see Figure 10.5; Zdanowicz, Gadbois, & Newman 1986). Relating these findings to biological effects, the investigators described several kinds of necrotic and degenerative liver and kidney lesions that were relatively more prevalent in fish from severely polluted sites. Inferences were not made about possible pollution-induced mortality, but the correlation of certain sublethal pathological conditions with contaminated habitats and elevated contaminant levels in tissues is suggestive.

Related experimental studies, exposing adult winter flounders to oil-contaminated sediments for 4 to 5 months, resulted in mortalities in summer months, possibly because the oil was acting as a nonspecific stressor at a time when temperatures approached an incipient lethal level (Fletcher, Kiceniuk, & Williams 1981).

The association of progressively severe liver pathology and several types of liver tumors with badly degraded coastal/estuarine waters is becoming apparent from recent reports. Among them, winter flounder from several degraded areas on the East Coast of the United States (New Haven Harbor, upper Narragansett Bay, Boston Harbor) had prevalences of 3.4 to 7.5% tumors (Murchelano & Wolke 1985) What seems to be emerging from examinations of tumors in winter flounders and other species in different polluted estuaries is a sequence of histopathological changes in livers, beginning with fatty deposits and preneoplastic changes in liver parenchyma cells. The progression of pathological changes also seems roughly correlated with the extent of estuarine degradation and the length of residence of fish in the estuary. In studies of West Coast flatfish species, positive correlations were obtained between neoplasm prevalence in bottom-dwelling fish and levels of "certain individual groups of sediment-associated chemicals" (aromatic hydrocarbons, chlorinated hydrocarbons, and heavy metals; Malins et al. 1984, 1988). Other studies of East Coast winter flounder have identified an oncogene derived from tumorous liver and possibly indicating a specific interaction of flounder DNA with polycyclic aromatic hydrocarbons (PAHs; McMahon et al. 1988).

A persistent question, still not resolved, is whether the liver tumors kill the fish or if they regress when the fish moves from the heavily polluted habitat. If the tumors do not regress, the survival of individual fish, as described by one investigator

(Murchelano 1988) depends on many factors, including the extent of liver damage, the degree of toxicity resulting from decreased hepatic function, the impairment of other essential organs, whether metastasis occurs, and the degree of behavioral modification (the ability to capture prey and avoid predators may be compromised). Population effects are possible if large numbers of fish with hepatic carcinoma die.

Limited and somewhat conflicting information is available about a possible reduction in immune responses and hence a reduced survival potential of winter flounder as a consequence of exposure to pollutants. In one study, a 4-month exposure to oiled sediments resulted in a statistically significant reduction in pigmented liver macrophage aggregates, believed to be important components of the cellular immune system of fish and possibly primitive analogues of the mammalian lymph nodes (Payne & Fancey 1989). However, in an earlier investigation (Wolke, George, & Blazer 1984; Wolke et al. 1985) the numbers of aggregates in fish from polluted areas were not greater than those from reference sites, although their size was. In still other studies with different fish species, aggregates were found to be more numerous and larger in samples from polluted areas than in those from unpolluted sites (Peters, Köhler, & Kranz 1987). It seems plausible, as some investigators have pointed out, that at low chronic levels of pollution the cellular defense system may function effectively, whereas at higher, more toxic levels of pollution, phagocytosis may be impaired, leading to a decrease in melanomacrophage aggregates.

One more study of effects of pollution on adult winter flounders — this one demonstrating synergism between chronic oil pollution and protozoan (trypanosome) parasitization — was reported (Khan 1987). Infected fish exposed for 6 weeks to oil-contaminated sediments had higher mortality rates than uninfected individuals; intensity of infection was higher in oil-exposed fish than in untreated controls; and retardation of gonad development was more pronounced in the oil-treated, parasitized fish than in the other experimental fish.

So here then, with winter flounder, we have a species that has been examined extensively, from pathological, immunological, physiological, and biochemical perspectives, in an effort to understand effects of pollution. Probably the most significant findings, from a population point of view, are the demonstrated negative impacts on larval survival of exposure to chlorinated hydrocarbons, acting either on pre-spawning females or on eggs and larvae. Next in order of significance might be the severe morphological changes — liver tumors and fin erosion in particular — that have been seen in samples from badly degraded habitats such as the New York Bight apex, western Long Island Sound, and Boston Harbor. Not to be ignored, however, is the dissenting observation by one team of investigators (Haedrich & Haedrich 1974) that chronically polluted environments seem to have little influence on winter flounder populations.

The one missing ingredient in this scrutiny of effects of pollution on the species is a serious attempt to quantify the observations made and to provide numerical estimates of the extent of population reduction that may result from exposure to pollutants. Some narrowly focused efforts have been made, such as Smith and Cole's early study of the effects of DDT on larval mortality, but we are left with little published information that can be extrapolated to the entire population of winter flounder on the northeast coast of the United States.

STATUS OF KNOWLEDGE ABOUT EFFECTS OF
COASTAL POLLUTION ON ABUNDANCE OF FISH

The case histories of quantitative pollution effects on populations of menhaden, striped bass, and winter flounder presented here are of course only examples. Some information is available for certain other species, such as salmon and shad. If this paper were to be written from a European perspective, the long-term research of German workers (Kinne, Rosenthal, Dethlefsen, von Westernhagen, and their colleagues) examining pollutant effects on larval survival and development would certainly be emphasized. That series of studies began in 1967 (Kinne & Rosenthal 1967) and continues to the present. Among the many significant general findings reported by those investigators are these:

- Exposure of maturing females to low concentrations of contaminants — especially those that are bioaccumulated — can affect gonad tissue, with effects expressed in the next generation.
- Life cycle stages most vulnerable to contaminants are maturing females, early embryos, early hatched larvae, and larvae in transition from yolk-sac to feeding.
- A wide range of morphological, behavioral, and physiological abnormalities in larvae result from exposure to contaminants, in rough proportion to the environmental level of the particular contaminant.
- Common morphological abnormalities include malformed lower jaw, eye deformities, anomalies in the vertebral column, and reduced size at hatching.
- Common physiological abnormalities include reduced heart rate, reduced swimming ability, disturbance in equilibrium, and reduced feeding.
- Early developmental stages showed the highest malformation rates.

Of course the basic question that must be asked is, "Can defective embryonic development and high embryo mortality due to pollution affect recruitment?" Some observations relative to this question have been proposed by German investigators (Dethlefsen et al. 1987, von Westernhagen et al. 1988) According to their reasoning, total mortality during the embryonic stage of development of marine fish has been estimated to be high — 95 to 99% for species such as Baltic cod and plaice, for example. At this mortality level, decreases in survival rates due to embryo abnormalities, at observed levels from 22 to 33%, would be too small to detect in unexploited populations, but in overexploited populations in which spawning stocks have been reduced severely, the added impact of abnormal embryonic development and high embryo mortality could result in reduced recruitment. The investigators also pointed out that, for the North Sea, the highest prevalences of embryonic malformations occurred in highly polluted areas (off the mouths of the Rhine and Elbe rivers and in the vicinity of the dumping zone for titanium dioxide wastes).

From the perspectives of the research scientist and the resource manager, three rather clear responsibilities exist when considering pollution effects on fisheries: one is to demonstrate that coastal/estuarine pollution is really affecting fish and shellfish stocks, the second is to define the local or regional stocks (especially shellfish) that are affected, and the third is to propose management measures to mitigate damage if it exists.

These responsibilities can be satisfied in four stages, which are easy to state but difficult to accomplish:

1. Isolate and quantify pollution's effects on resource species — as distinct from effects of natural environmental variations.
2. Conduct critical examinations of pollution effects at levels of the individual, the local population, and the species.
3. Encourage the identification and quantification of sensitive early-warning indicators of environmental degradation
4. Attempt to reduce pollutant inputs, where damage to living resources has been or can be demonstrated.

Item 4 is particularly difficult to achieve; it points out the pressing need for aquatic scientists to interact with those responsible for managing and regulating terrestrial sources of pollution.

Case histories of the quantitative effects of pollution on populations of menhaden, striped bass, and winter flounder, as presented in this chapter, illustrate a series of generalizations useful in stock management:

- In assessing pollution impacts, the *entire range* of the species should be considered, and precise information about levels of all contaminants throughout that range should be available. Furthermore, migratory characteristics must be considered, since some species may move rapidly into or out of areas of severe pollution, while others may become semipermanent residents of such degraded areas.
- Particular attention must be paid to controlling pollution levels in spawning/nursery areas, since most pollution-associated mortality will occur during the first year of life. It is generally accepted by fish population biologists that survival during the first 60 d of life determines the size of the year-class of many estuarine-dependent species; factors affecting early survival (such as pollution) should be of concern and subjects of management action.
- "[V]ery little impact [from pollution] should be tolerated at low stock size, because it would prevent recovery to a ... maximum sustained yield" (Cushing 1979, p. 109).
- Simulation models can supply useful information, but the degree of reliability of such models depends on the extent of the database employed.
- Marine/anadromous fish populations are characterized by aperiodic dominant year-classes, which may form the basis for a fishery for many years. (An example would be the 1970 striped bass year-class in Chesapeake Bay.) Existence of these dominant year-classes and examination of factors responsible for their production can lead to important insights about environmental influences on abundance.

Previous sections of this chapter included examinations of case histories that indicated points in the life cycles of fish where pollutant stressors can exert lethal and sublethal influences. From experimental studies, it seems obvious that, with all these potential impact points throughout life cycles, populations in contaminated waters should dwindle and disappear; yet from experience on the east coast of North America, this has not happened (at least not yet). Only in severely degraded local waters, which are a small part of the total range of most fish species, have there been localized disappearances; even in those areas, other species that might be expected to be affected are still present and are in some instances abundant. This is true particularly of a number of coastal/estuarine-dependent fish species, many of which spend much of their life cycles in waters that are to some extent contaminated. It may also be true for migratory species that can move into or through degraded zones, although any possible impacts of transient exposures to contaminants are more difficult to recognize and much more difficult to quantify.

Two major attempts have been made to assess pollution impacts on fish stocks off portions of the U.S. coasts — one by Prager and MacCall (1993), examining data from the southern California bight, and the other by Summers et al. (1986), using historical data for a number of fish species from five estuaries of the northeastern states. Using two modeling methods — a biomass-based model and a recruitment model — Prager and MacCall looked for significant effects of climatic conditions and contaminant loadings on the spawning success of three Pacific species: northern anchovy (*Engraulis mordax*), Pacific sardine (*Sardinops sagax*), and chub mackerel (*Scomber japonicus*). Using the models, no climate or contaminant influences on the spawning success of northern anchovies were detected, but the spawning success of chub mackerel seemed to be strongly influenced by climatic variability, and (with the recruitment model) the spawning success of Pacific sardines was strongly negatively correlated with contaminant loadings — being consistent with the hypothesis that the stock, which had been overfished and had collapsed in the 1950s, was stressed beyond its limits by poor larval survival caused by ambient contaminant concentrations. The study of fish populations of five Atlantic coast estuaries by Summers et al., using a biostatistical modeling approach based on catch statistics, disclosed consistent patterns of pollution effects among similar species across different estuaries.

Another analysis — this one of recent changes in abundance of North Sea fish stocks — failed to find any effect from pollution. Tiews (1983, 1989) examined the abundance trends of 25 species for 35 yr (1954 to 1988) from trawl catches on the German North Sea coast. He found that some species had declined, some had increased, and others had fluctuated irregularly, but that no consistent long-term decline of commercial species could be attributed to deteriorating habitat conditions. That noted investigator had earlier (1983) expressed suspicion that, at least for some species, declines might be due to pollution effects, but his later analysis concluded with the comment that

> this study shows that the majority of species studied seems to be able to tolerate the present status of environmental deterioration of the ecosystem. Furthermore, the study does not indicate any fundamental impairment of the fishery biological situation of the

area during the last seven years in comparison to the preceding period (from 1954 to 1981). Six species have even substantially increased in abundance. (p. 11)

Tiews cautioned, however, that 35 yr may be too short a period to reveal natural population fluctuations, so that a very conservative interpretation of the data is necessary.

A well-known quantitative scientist from the United States, C.P. Goodyear, has described the dimensions of the problem of interpreting population responses to direct (lethal) and indirect (sublethal) levels of contaminants very succinctly:

- For cases where contaminants are directly lethal, two types of information are required: "the timing and extent of the excess mortality must be determined. ... [In addition,] the nature, timing, and intensity of the density-dependent processes that regulate the size of the population must be understood" (1984b, p. 418). (The latter, as Goodyear pointed out, is "a key, largely unsolved problem in fishery research and management in general" [1984b, p. 418].)
- For cases where indirect (sublethal) levels of contaminants exist, "The interpretation of the population response to the indirect effects of the contaminants on the species of interest requires quantification of the response in terms of a change in survival probability or a change in reproductive rates, and knowledge of the density-dependent processes that control population size" (1984b, p. 419). (Goodyear then pointed out with great aplomb that none of the required information is particularly amenable to measurement [1984b, p. 419.])

Experimental demonstrations of the negative effects of chemical contaminants on the survival and well-being of marine fish and shellfish are abundant, but the variability that results from the influences of natural factors is so great and so incompletely understood that experimental findings cannot usually be applied directly to assessments of exploited populations — despite repeated attempts to do so. Data extrapolations from single-contaminant experimental studies can be useful, however.

One paper (Føyn & Serigstad 1989) described calculations of year-class reductions in cod, herring, and saithe populations of the Norwegian coast that might result from a major oil spill. Detailed field studies of seasonal herring larval distribution and concurrent laboratory studies of effects of oil on larvae, when presented in a worst-case scenario, enabled calculation of what turned out to be a low percentage of potential reduction in recruitment. Field experimental studies, in which fish are exposed *in situ* to contaminated waters, have also produced meaningful results when combined with adequate chemical analyses of environmental and tissue samples.

The appearance of reports of sublethal effects of contaminants on marine animals was described 25 yr ago as of "avalanche" proportions (Lewis 1980, p. 458). In a thought-provoking discussion of options in environmental management and the deployment of future research efforts concerned with the population effects of pollution, the author of that report made the point (in 1980) that

The more subtle threats of chronic pollution well away from the hot-spots, and which because of their subtlety have always been seen as potentially the most dangerous, do not appear to have developed and produced effects on the scale that was feared a decade ago. (p. 464)

The author then asked difficult questions:

If after all the recent, intensive effort there is still difficulty in finding chronic effects upon communities does this not suggest that such effects are negligible? That while the initial concern over chronic effects was justified, is it not now time to acknowledge that it is only acute pollution that matters?

Most of us in pollution research would detect a strong odor of heresy in this superficially logical line of thought, probably rejecting it out of hand, since it tends to overestimate human ability to sort out the effects of natural vs. man-induced population changes, and it tends toward too easy acceptance of the significance of acute pollution events in reducing overall abundance of marine populations. However, we would have to admit that our arguments would not have a strong base in adequately demonstrated population impacts. We might also admit that, just as with certain terrestrial species, a threshold may be reached where contaminants in the marine environment could achieve a level, or operate over sufficient time, to greatly affect population abundance.

Assessment of the quantitative effects of pollution and other human-induced environmental perturbations on fish populations depends on an understanding of the stock recruitment relationships of fish populations of concern and on knowledge of the density-dependent and density-independent factors influencing those relationships. Of particular concern is mortality imposed by toxic levels of pollutants on eggs, larvae, and juveniles of economically important fish species. However, models predicting the effects of human intrusions (other than fishing) must be used conservatively, especially when impacts on early life stages are included. The concern has been expressed with precision as follows:

[T]he potential for error in numerical predictions of the effect of proposed levels of increased mortality on pre-recruit stages is large, while the biologically acceptable range of error is small. ... [U]ntil a better understanding is achieved of the interacting roles of density-dependent and density-independent factors in regulating population size and stability, and until a much better data base is available for the majority of the stocks subject to mortality from industrial activity, precise numerical predictions of the impact of this incremental mortality on adult stocks should be interpreted with great caution. (Leggett 1977, p. 316).

Assuming the correctness of the generally held principle that recruitment variability is a consequence of events that occur in egg, larval, and postlarval stages, it is logical to focus pollution studies more directly on reproductive success of populations at risk. This has been done in a number of studies, but one disturbing conclusion that can be reached after examining some of the accumulated literature is that investigators rarely if ever carry their findings to the point of actually estimating quantitative effects on fish populations and species.

Let's consider a few examples. German researchers (von Westernhagen et al. 1987a, 1987b) examined the contaminant content of ovaries of Baltic flounder (*Platichthys flesus*) and established a threshold level of 0.12 ppm PCBs beyond which reproductive impairment would occur. These authors found that 8.5% of the sample exceeded that level (range 0.05 to 3.17), and they reported that viable hatch was lower than 15% if gonad PCBs exceeded 0.25 ppm. They did *not* take the final quantitative steps. The Baltic flounder catches are known; the extent of PCB contamination in the Baltic is known. Why not make a rough estimate from these data of the possible impact of PCB contamination on recruitment of Baltic flounders? (And then even make an economic evaluation as well.) As another example, American scientists (Cross & Hose 1988) found reproductive impairment and high levels of ovarian DDT and PCBs in white croakers (*Genyonemus lineatus*) from a contaminated site near Los Angeles. Again, white croaker catch statistics are available, and environmental levels of chlorinated hydrocarbons on the California coast have been examined extensively. Why not take another step, armed with the data, and provide a rough estimate of the possible impact on recruitment?

Only once, to my knowledge, has even the penultimate step — estimating the effects of pollution on recruitment — been taken. This was in a report on developmental defects in pelagic fish embryos from the western Baltic (von Westernhagen 1988). Plankton net catches of cod, plaice, and flounder eggs and larvae were examined for mortality and abnormalities. Prevalences of defective embryos and abnormal larvae in various locations in the western Baltic were determined, and the effects on survival of larvae of two year-classes (1983 and 1984) were estimated. Decreases (18 to 44%) were considered too small in terms of biological significance to cause a detectable impact on recruitment, although the authors pointed out that even a small impact on larval production could, in the case of overexploited Baltic stocks, lead eventually to reduced recruitment. It is puzzling, considering the availability of excellent fishery statistics and environmental information for the Baltic, some of it extending back for more than a century, why the ultimate step — of estimating total possible population effects of existing pollution levels — was not taken, since it is only at this stage that pollution information would become meaningful to resource managers.

Part of the problem may be that many professionals interested in problems created by coastal/estuarine pollution do not communicate effectively with population dynamics specialists, so full exploitation of available data does not happen. Only in rare instances, such as the mid-Atlantic striped bass program discussed earlier, when simulations based on good resource and environmental data sets were attempted, has some measure of integration been achieved. The assessment scientist is often loath to use the minimalist approach necessary with the environmental scientist's data.

To do an effective job of quantifying pollution impacts on fish stocks, a large interdisciplinary research and monitoring program is required. The principal ingredients are shown in Figure 10.6. A triumvirate of population assessments, environmental assessments, and experimental studies constitutes the basic information source for the required modeling effort. Such a program must be long-term as well as geographically broad. Until now, only the mid-Atlantic striped bass program has

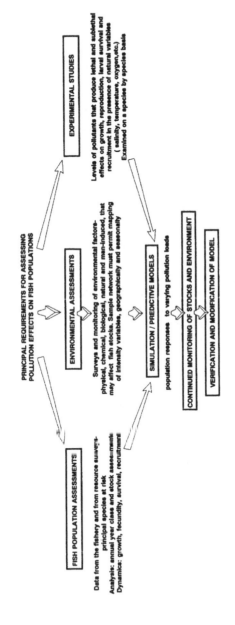

FIGURE 10.6 Principal requirements for assessing the effects of pollution on fish stocks.

approached the level of research commitment required, and even this major effort must face an annual struggle for adequate funding, in spite of its considerable importance and high levels of lobbying.

Mathematical models useful in assessing the status of fish stocks, as well as the effects of fishing on those stocks, were developed and refined during the latter half of the 20th century and are being used worldwide by quantitative biologists. Until the 1960s, though, their perspective on human-induced perturbations was mostly limited to the effects of fishing, with little attention to other man-influenced habitat variables — including pollution. Gradually, rising pressures resulting from new government regulations and public concern about impacts of environmental degradation have encouraged an examination by those same quantitative biologists (and their disciples) of coastal/estuarine pollution on fish populations. After such a late start, impact assessment models that focus on pollution-induced mortality have been and are being developed; fish population dynamics specialists and toxicologists are finally beginning to communicate and to produce findings useful to fisheries managers, environmental managers, and a concerned public.

Numerical models can be used to simulate the consequences of manipulating environmental variables, and then to predict outcomes of regulatory actions or inactions. Reliability of the simulations depends to a major degree on the extent of the pollution-related database that is available (and for many coastal areas it is less than adequate). Simulations may also lead to conclusions and correlations that can be tested in field and laboratory investigations.

Modeling approaches have had utility in comparing pollution-induced mortality in fish to mortality caused by fishing. The methods are (relatively) easy to apply and give results that can be understood by resource managers (Boreman 1997). Four principal methods have been proposed and tested are:

1. "Equivalent adult" (Horst 1977)
2. "Production foregone" (Rago 1984),
3. "Reproductive potential" (Goodyear 1988, 1993; Boreman et al. 1993)
4. "Individual based" (Rose et al. 1993)

Each method has advantages and disadvantages, and, as pointed out by Boreman (1997), all suffer from the difficulty of estimating survival rates for early life stages and from the possible existence of mechanisms that counterbalance effects of the increased mortality. Despite their shortcomings, these methods

allow managers to weigh tradeoffs between pollution-induced mortality and fishing-induced mortality in terms of their relative impacts on population measures such as abundance, harvestable biomass, spawning biomass, fishing opportunity foregone, and ecosystem production foregone. For species in which many ages are vulnerable to fisheries, such as striped bass, population-level impacts caused by a reduction in age 0 survival can be offset by a relatively smaller reduction in fishing mortality, as demonstrated by Goodyear (1988). For species in which fewer ages are vulnerable to fishing, such as winter flounder, the population-level impact from a reduction in age 0 survival may be more comparable to the impact from a proportionally equivalent increase in fishing mortality, as demonstrated by Boreman et al. (1993).

Boreman and colleagues' study of winter flounder (1993) provides an excellent example of the "reproductive potential" (eggs per recruit) method in evaluating options for restoring population abundance. A counterpart modeling study of winter flounder by Rose et al. (1996) provided approaches to understanding causes of long-term fluctuations in flounder populations. Then Landahl et al. (1997) examined with simulation modeling the potential impact of contaminant-related mortality and reproductive impairment on the population growth rate of English sole (*Pleuronectes vetulus*) from Puget Sound, Washington.

So the population/pollution effects literature is gradually expanding, and quantitative biologists are contributing significantly to that expansion.

CONCLUSIONS

Definitive conclusions about quantitative effects of pollution on abundance of coastal fish species require much more substantiating data of the kind summarized in the three case histories presented here (Atlantic menhaden, striped bass, and winter flounder). Subject to this caveat, and recognizing that large areas of disagreement may exist, I can nonetheless offer some statements in the proper form of hypotheses about such effects.

First, habitat destruction in estuaries and ocean disposal of contaminants have had localized adverse effects on resource organisms, but the overall effects at the species level have not been determined. Lack of success in assessing effects of pollution on population abundance may be due in part to the fact that the scale of impact usually is much smaller than the geographical range of the species.

Three decades ago, a British author (Cole 1972) pointed out a critical guideline for evaluating conclusions reached about marine pollution: that we must be careful to differentiate large-scale changes due to pollution from purely local, almost parochial, situations, usually in a few square kilometers of coastal/estuarine waters. This admonition was augmented by other authors (Cross, Peters, & Schaaf 1985), who stated that

Improvement in our capacity to predict the effects of pollution on fisheries requires an understanding of the environmental factors that control variability in fish populations and the effect of multiple stresses on these stocks over their entire geographical range. Increased predictive capability can be gained most cost-effectively through closer integration of the disciplines of population dynamics and toxicology.

A larger case should be made for reduced reproduction as a consequence of both chemical contamination and coastline modifications (loss of spawning and nursery areas through land reclamation, dredge and fill projects, and loss of seaweed and sea grass beds).

Second, significant negative impacts of pollution on commercial fish stocks have not been demonstrated, even for areas such as the North Sea, where statistical information about fish stocks has been collected for many decades. One British author (Lee 1978), for example, stated, "There is no evidence either way as to whether or not contamination of North Sea waters by metals, pesticide/residues, etc.

has affected the well being of the fish stocks." This pronouncement may be correct, but it seems to be an oversimplification; a more rational viewpoint was expressed by German researchers (Dethlefsen & Tiews 1985), who concluded that

> at present it is impossible to define the role of pollution on fish stocks of the North Sea as a whole. This is largely due to the fact that only drastic changes in marine ecosystems would be detectable and could be interpreted as manmade. Normally, chronic and sublethal changes are taking place very slowly and it will be impossible to separate natural fluctuations from [those that are] anthropogenically caused. Long-term research over decades might be necessary to clearly distinguish between man-made and natural fluctuations especially in offshore waters. (p. 118)

Third, marine resource populations are subject to large natural fluctuations whose causes are incompletely understood. Turning again to the North Sea, where resource and environmental data are probably better than almost anywhere else in the world, one noted German scientist's detailed report (Hempel 1978) on changes in fisheries and fish stocks concluded with this remarkably nebulous statement:

> It seems that direct and indirect effects of changes in the fisheries as well as climatic changes and their consequences for the biotic environment caused the recent changes in the fish stocks of the North Sea. It is not possible to quantify the effects of man-made and natural factors separately because of the complexity of interactions between the various fish stocks and the stages of their early life history. (p. 165)

Death, early and sudden, is the norm in marine populations, but unusual levels of mortality at a given time can have significance to future abundance. Mass mortalities result from many natural phenomena — disease, anoxia, storms, volcanism, toxic algal blooms, and so forth. Those directly attributable to pollution are usually rare and localized. The following snippet from my personal history illustrates how the natural factor of disease can affect population abundance.

Mass Mortalities of Herring in the Gulf of Saint Lawrence

Each spring, sometimes even before the ice has disappeared from the Gulf of Saint Lawrence, great silvery masses of sea herring Clupea harengus move westward into the gulf to spawn off the coasts of the Magdalen Islands, Quebec, and northern New Brunswick. Following ancient rhythms, the huge aggregations crowd into shallow waters to reproduce, and then disperse to the eastward later in the season.

But in this April of 1954, something was desperately wrong. The fish were dying in huge numbers in waters of the southern gulf. Masses of them littered the bottom, and washed up on the beaches. I was there at the request of the Canadian fisheries agency as a so-called "visiting expert" to offer an opinion on the cause of this major environmental event. My intellectual baggage at the time was remarkably slim (I was a fresh new Ph.D. with a few graduate courses in mycology and a few in parasitology) — too inexperienced to be asked the

"Why are they dying?" question by worried people who earned their living on the water.

I did know that three times before this — in 1898, 1914, and 1932 — these spawning migrations had become transformed into what seemed like predestined appointments with death. The incoming herring schools appear to encounter in the shallow waters large concentrations of a virulent fungal pathogen Ichthyophonus hoferi. Spores of the fungus cause systemic infections in herring (and a few other species, like alewives and mackerel) that are lethal within a few weeks. Herring die in truly overwhelming numbers; carcasses float on the surface, litter the bottom, or are washed up in windrows on extensive stretches of gulf beaches. In peak mortality years, a significant part of the entire adult herring population of that major body of water can be killed by the disease, and commercial catches are reduced drastically for several years following each outbreak. Then, a year or two after the peak of mortalities, just as mysteriously as it has appeared, the disease retreats. Infection rates decline rapidly and mortalities cease. Annual spawning migrations then go on for several decades unhindered by the threatening presence of fatal disease.

This extraordinarily complex interweaving of major ecological events — spawning aggregations clouded by epizootic disease, reproduction in the immediate presence of mass deaths — illustrates the limited distance that we have traveled toward understanding and controlling mortalities in marine fish populations. We still have a very marginal comprehension of these disease-induced phenomena; we still stand bewildered on the shores — almost as did our ancestors from the dark ages, when confronted by the great human pestilences of those far-off times.

I estimated from field observations made during that exciting period spent in Canada that over half of the adult herring and many of the alewives and mackerel of that entire gulf were killed during the three-year outbreak period (1954–1956), and the immediate decline of the herring fishery to less than half its previous level supported the estimates. This was an epic environmental event that had to be investigated. It was then that my long search to understand mass deaths in the sea began — a search that continues to the present time.

From Field Notes of a Pollution Watcher
(C.J. Sindermann, 2000)

Fourth, because of the inability to distinguish population changes due to pollution, such changes might become catastrophic before they are noted. Effects of pollutants on reproduction and on survival until recruitment into the fished stocks may be particularly critical. Survival may be reduced by parentally transmitted or dietary contaminants, such as PCBs. Sperm viability and egg fertilizability may be reduced. Larval behavior may be affected, reducing competencies in food capture and predator avoidance. Pollution may result in increased susceptibility to disease in early life stages. The reproductive rate — usually considered by environmental toxicologists to be a more sensitive indicator than survival — may be severely affected. Toxic chemicals may reduce reproductive rates by causing diversion of energy to metabolic deactivation of harmful chemicals or to tissue repair, by directly impairing protein synthesis and hence growth, by affecting digestion and assimilation, or by interfering with gonad maturation and gamete production. These physiological responses of individuals under chemical stress can contribute to reproductive failure of the population as a whole, although other factors may be involved, as summarized by a Canadian author:

> (1) The fish may be unable to reach its spawning grounds because of unfavourable ecological conditions, or its own weak physical state, and goes unspawned; (2) the eggs may never be released by the female owing to some unsuitable physiological condition; (3) the eggs and larvae may die because of their unhealthy state or poor conditions on the spawning grounds; (4) the eggs and larvae may be poisoned by a substance bioaccumulated in the gonads of the parent, e.g., DDT; or (5) the eggs and larvae may be poisoned by toxic substances in the environment. (Waldichuk 1979, p. 410).

Fifth, offshore fish stocks are not immune to damage from pollutants, since year-class abundance is determined early in the life history, which for many marine species is spent in coastal/estuarine waters.

Finally, many major fish stocks that are already overfished may be more vulnerable to additional stresses (such as pollution) than unexploited or lightly exploited populations. A population that is heavily fished may have limited "compensatory reserve" (in the form of increased growth rates and greater egg production) and thus may be particularly sensitive to pollution (Figure 10.7). A critical question to be asked about each stressed population concerns this compensatory response, especially the amount of reserve that is left in an exploited population.

The point about differential effects of exploitation on stressed populations was made forcefully by a noted British researcher (Cushing 1979). Defining the term

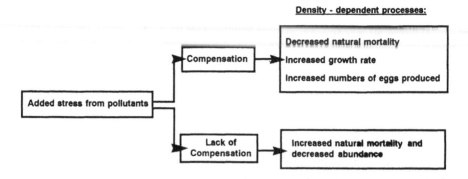

FIGURE 10.7 Population responses to pollution stress. Note that the gains from compensation are added (in an algebraic sense) to the losses from the added pollution stress. The net result could indeed be decreased M, but it could instead be constant M (if compensation precisely offsets the contaminant stress) or reduced M (if compensation is incomplete).

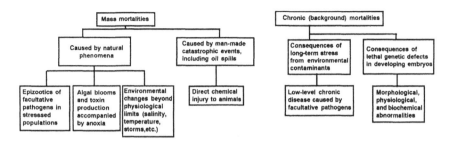

FIGURE 10.8 Causes of mortality in marine fish.

impact as "the loss of eggs, larvae, or juveniles to the recruitment of fish stocks," he stated that

> Variability of recruitment in marine and anadromous fish stocks is high — one to two orders of magnitude — and an impact due to pollution would be difficult to detect. However, the effect of impact depends on the degree of exploitation, so if stocks are low and heavily exploited, no impact can be tolerated, until the stock has been returned to the desired objective, the maximum sustainable yield or any lesser quantity that might be optional. (p. 607)

This is consistent with a related statement about environmental effects by another British scientist:

> [I]n general, recruitment [annual addition to a fishable stock] is not directly dependent on the size of the spawning stock, but *appears to be dependent on other factors, frequently environmental in origin*, that affect the fish during the early larval and juvenile stages. Fluctuations in stock size are therefore very largely due to events that occur during the early life stages and are generally less dependent on events that influence the size of the adult stock." (Jones 1982, p. 354; italics mine)

Coastal/estuarine pollution can of course affect any life stage of fish, but it is during their first year of life — and more specifically during their first few months of life — that fish can be particularly sensitive to toxic contaminants. *Factors affecting early mortality are therefore of great significance in determining the causes of long-term population trends, even though death may occur at any point in the life cycle.* When translated into population terms, mortality may be chronic or catastrophic, as shown in Figure 10.8. Pollution impacts may occur at numerous points in this process.

More than 2 decades ago, in a major symposium titled "Protection of Life in the Sea," held at the Biologische Anstalt Helgoland, a series of conclusions was proposed about pollution effects on fisheries — conclusions that still seem relevant today:

> Pollution effects on fisheries have received some scrutiny in recent decades, and information is accumulating, but is still insufficient to be very useful in resource management decisions — except as they involve local areas. Evidence exists for localized effects of pollutant stress on fisheries, but *as yet there is little specific evidence*

for widespread damage to major fisheries resource populations resulting from coastal/estuarine pollution. This may well be because we are unable to separate clearly the effects of pollutant stress from effects of the many other forms of environmental stresses to which marine populations are subject. Other factors, such as shifts in geographic distribution of fish populations, epizootic disease, changes in productive ecosystems, or overfishing, may cause pronounced changes in fisheries — changes which could obscure any effects of localized habitat degradation. *It seems, with the evidence presently available, that factors other than pollution are overriding in determining fish abundance, but we lack sufficient quantitative data to make positive statements about cause and effect relationships of abundance and pollution.*

It may be, of course, that coastal/estuarine pollution is exerting some overall influence on certain resource species, but that this influence may be masked by increased fishing effort, or by favorable changes in other environmental factors which create a positive effect on abundance, outweighing any negative effects of pollutants. Many experimental studies, particularly more recent ones concerned with long-term exposure of fish and shellfish to low levels of contaminants, suggest that some long-term effects on abundance should be felt, but our statistics, our monitoring, and our population assessments are not yet adequate to detect them.

Effective long-term monitoring of stocks and environment must be the basis of any attempt to isolate and identify pollutant effects. *A continuous integrated effort in stock assessment, environmental assessment, and experimental studies will be required to understand the role of all environmental stressors — natural and man-induced — in determining abundance of resource populations.* However, we do not yet have the principal pieces of the puzzle in place, so in the absence of full understanding of the phenomena involved, management decisions affecting coastal/estuarine pollution must be made on the basis of "best available scientific information," just as decisions about allowable resource exploitation are made. In both types of decision processes, a conservative action provides a lower risk of damage and loss than does a more extreme action.

Conservatism can be especially significant when decisions are made that might permit pollution to continue or increase, since long-term effects of existing levels on abundance of resource populations are largely undetermined. In addition to advocating conservatism, we must persist in attempts to quantify the effects of pollution, and to determine the precise pathways through which fishery resources are affected. (Sindermann 1980, pp. 684–686)

Freely admitting to a conservative mind-set (and a mild infatuation with my own prose), there is little about these conclusions that I would change today — more than 2 decades after their original publication. Research in the intervening years has added *substantially* but only *incrementally* and not *conceptually* to our understanding of pollution effects on resource species; some of that information has been summarized in the case histories just presented.

The major problem — being able to distinguish adequately the effects of pollution from all the other influences on marine fish population abundance — still confronts us, even though the boundaries of our knowledge have expanded. Recent advances include the creation and implementation of new pollution monitoring and assessment programs, developments in simulation modeling, annual lengthening of critical resource and environmental data sets, and findings from field and laboratory experimental studies.

Some pessimists feel that we will never have adequate data to distinguish clearly the quantitative effects of pollution on fish stocks; I think the accumulation of analyses and insights relevant to the problem forecasts a brighter future than that.

As a final personal note about the relative effects of humans on abundance of fish — and to cast my vote for massive overexploitation of stocks as a principal offender rather than coastal pollution — I offer the following vignette.

Destruction of the Georges Bank Sea Herring Stocks by Overfishing, 1950–1977

Sea herring Clupea harengus *are prized commercial fish in many North Atlantic countries — in fact the economic development of nations such as Norway, Sweden, Denmark, and Britain was shaped in part by major herring fisheries, beginning as early as the sixteenth century. Seemingly limitless stocks of this species existed in northern temperate waters, including those adjacent to North American shores. One such aggregation — of undetermined size — occurred on Georges Bank, located one hundred miles east of Cape Cod, but was only lightly exploited by U.S. fishermen, who were more interested in other species such as cod, haddock, and flounders.*

After World War II, ocean fisheries were recognized as a major global economic resource, and stocks were exploited by technologically advanced and ever-expanding foreign distant water fleets, able to fish in any productive location worldwide. I was an unwitting but still culpable participant in events that led to the almost total elimination of the Georges Bank herring population by such fishing methods in the period 1960 to 1977.

For a few years, beginning in the mid-1950s, I was a member of a government research group in Maine, concerned with the biology and population dynamics of herring. One aspect of the program was the conduct of seasonal fishery surveys of the western North Atlantic, including the Gulf of Maine and Georges Bank. Our cruise reports described herring larvae produced by vast seasonal spawning aggregations on that bank. How dreadfully naïve supposedly intelligent scientists can be. Those larval distribution charts that we published so proudly in the 1960s and early 1970s (and that indicated precise locations of spawning adults) were eagerly devoured and exploited by fisheries interests in communist eastern European countries — Russia, Poland, East Germany, Romania, and all the others. Their huge factory trawlers, some over 300 feet long, appeared off our coasts in awesome numbers in the late 1960s and the 1970s. Georges Bank at night looked like a floating city, with all the deck lights of those enormous fishing vessels. And the fish stocks disappeared into their holds.

Catches of adult herring increased geometrically, reaching an astounding peak of 200,000 tons *in 1977 (Sindermann 1979). The stocks collapsed in the following year, and herring became very scarce (a survey cruise in 1984*

recorded a dismal, almost unbelievable total of only three individual fish of that
species taken in all trawling stations on that Georges Bank cruise)!

(As a footnote: The Georges Bank herring stock recovered following the
imposition of a 200-mile U.S. Economic Zone and the forced departure of foreign
fleets in 1979; in the 1990s limited fishing on regenerating stocks resumed; see
Overholtz and Friedland 2002.)

From Field Notes of a Pollution Watcher
(C.J. Sindermann, 2003)

REFERENCES

Baker, J.T.P. 1969. Histological and electron microscopical observations on copper poisoning
    in the winter flounder (*Pseudopleuronectes americanus*). *J. Fish. Res. Board Can.*
    26: 2785–2793.

Black, D.E., D.K. Phelps, and R.L. Lapan. 1988. The effect of inherited contamination on
    egg and larval winter flounder, *Pseudopleuronectes americanus. Mar. Environ. Res.*
    25: 45–62.

Boreman, J. 1995. Pollution versus overfishing: Finding a cause for declining abundance of
    fish. (Abstract). In: *Pollution and Fisheries: A Conference on Population-level Effects*
    of Marine Contamination. Univ. Maryland Biotechnical Institute, Baltimore, MD.

Boreman, J. 1997. Methods for comparing the impacts of pollution and fishing on fish
    populations. *Trans. Am. Fish. Soc. 126:* 506–513.

Boreman, J., S.C. Correia, and D.B. Witherell. 1993. Effects of changes in age 0 survival on
    egg production of winter flounder in Cape Cod Bay, pp. 39–45. In: L.A. Fuiman,
    (ed.), *Water quality and the early life stages of fishes.* American Fisheries Society,
    Symposium 14, Bethesda, MD.

Calabrese, A., F.P. Thurberg, M.A. Dawson, and D.R. Wenzloff. 1975. Sublethal physiological
    stress induced by cadmium and mercury in the winter flounder, *Pseudopleuronectes*
    americanus, pp. 15–21. In: J.H. Koeman and J.J.T.W.A. Strik (eds.), *Sublethal Effects*
    of Toxic Chemicals on Aquatic Animals. Elsevier, Amsterdam.

Cole, H.A. 1972. North Sea pollution, pp. 3–9. In: M. Ruivo (ed.), *Marine Pollution and Sea*
    Life. Fishing News Ltd., London.

Cross, J.N. and J.E. Hose. 1988. Evidence for impaired reproduction in white croaker (*Geny-*
    onemus lineatus) from contaminated areas off southern California. *Mar. Environ. Res.*
    24: 185–188.

Cross, F.A., D.S. Peters, and W.E. Schaaf. 1985. Implications of waste disposal in coastal
    waters on fish populations, pp. 383–399. In: R.D. Cardwell, R. Purdy, and R.C. Bahner
    (eds.), *Aquatic Toxicology and Hazard Assessment.* ASTM-STP 854. American Soci-
    ety for Testing and Materials, Philadelphia, PA.

Cushing, D.H. 1979. The monitoring of biological effects: The separation of natural changes
    from those induced by pollution, pp. 597–609. In: H.A. Cole (ed.), *The Assessment*
    of Sublethal Effects of Pollutants in the Sea. Philos. Trans. R. Soc. Lond. B286.

Dethlefsen, V. 1986. Marine pollution mismanagement: Towards the precautionary concept.
    Mar. Pollut. Bull. 17: 54–57.

Dethlefsen, V. and K. Tiews. 1985. Review of the effects of pollution on marine fish life and
    fisheries in the North Sea. *Z. Angew. Ichthyol. 1:* 97–118.

Dethlefsen, V., P. Cameron, H. von Westernhagen, and D. Janssen. 1987. Morphologische und chromosomale Untersuchungen an Fischembryonen der südlichen Nordsee in Zusammenhang mit der Organochlorkontamination der Elterntiere. *Veröff. Inst. Küsten Binnenfisch. 97:* 1–57.

Fletcher, G.L., J.W. Kiceniuk, and U.P. Williams. 1981. Effects of oiled sediments on mortality, feeding and growth of winter flounder *Pseudopleuronectes americanus. Mar. Ecol. Prog. Ser. 4:* 91–96.

Føyn, L. and B. Serigstad. 1989. How can a potential oil pollution affect the recruitment to fish stocks? Int. Counc. Explor. Sea, Cod. C.M. 1989/Minisymp. No. 5, 23 pp.

Goodyear, C.P. 1978. Management problems of migratory stocks of striped bass, pp. 75–84. In: H. Clepper (ed.), *Marine Recreational Fisheries 3.* Sport Fishing Institute, Washington, DC.

Goodyear, C.P. 1980. Oscillatory behavior of a striped bass population model controlled by a Ricker function. *Trans. Am. Fish. Soc. 109:* 511–516.

Goodyear, C.P. 1984a. Analysis of potential yield per recruit for striped bass produced in the Chesapeake Bay. *N. Am. J. Fish. Manage. 4:* 488–496.

Goodyear, C.P. 1984b. Measuring effects of contaminant stress on fish populations, pp. 414–424. In: *Aquatic Toxicology: Sixth Symposium.* Spec. Tech. Publ. No. 802, American Society for Testing and Materials, Philadelphia, PA.

Goodyear, C.P. 1985a. Relationship between reported commercial landings and abundance of young striped bass in Chesapeake Bay, Maryland. *Trans. Am. Fish. Soc. 114:* 92–96.

Goodyear, C.P. 1985b. Toxic materials, fishing, and environmental variation: Simulated effects on striped bass population trends. *Trans. Am. Fish. Soc. 114:* 107–113.

Goodyear, C.P. 1988. Implications of power plant mortality for management of the Hudson River striped bass fishery. *Am. Fish. Soc. Monogr. 4:* 245–254.

Goodyear, C.P. 1993. Spawning stock biomass per recruit in fisheries management: Foundation and current use, pp. 67–81. In: S.J. Smith, J.J. Hunt, and D. Rivard (eds.), *Risk Evaluation and Biological Reference Points for Fisheries Management.* Can. Spec. Publ. Fish. Aquat. Sci. 120.

Haedrich, R.L. and S.O. Haedrich. 1974. A seasonal survey of the fishes of the Mystic River, a polluted estuary in downtown Boston, Massachusetts. *Estuar. Coast. Mar. Sci. 2:* 59–73.

Hall, L.W., Jr., A.E. Pinkney, L.O. Horseman and S.E. Finger. 1985. Mortality of striped bass larvae in relation to contaminants and water quality conditions in a Chesapeake Bay tributary. *Trans. Am. Fish. Soc. 114.* 861–868.

Hall, L.W., Jr., A.E. Pinkney, R.L. Herman, and S.E. Finger. 1987. Survival of striped bass larvae and yearlings in relation to contaminants and water quality in the upper Chesapeake Bay. *Arch. Environ. Contam. Toxicol. 16:* 391–400.

Hempel, G. 1978. North Sea fisheries and fish stocks — A review of recent changes. *Rapp. P.-V. Reun. Cons. Int. Explor. Mer 173:* 145–167.

Horst, T.J. 1977. Use of the Leslie matrix for assessing environmental impact with an example for a fish population. *Trans. Am. Fish. Soc. 106:* 253–257.

Hughes, J.B., D.A. Nelson, D.M. Perry, J.E. Miller, G.R. Sennefelder, and J.J. Periera. 1986. Reproductive success of the winter flounder (*Pseudopleuronectes americanus*) in Long Island Sound. Int. Counc. Explor. Sea, Doc. C.M. 1986/E:10, 11 pp.

Hunt, E. and J. Linn. 1969. Fish kills by pesticides, pp. 44–59. In: J.W. Gillette (ed.), *Proceedings of the Symposium on the Biological Impact of Pesticides in the Environment.* Oregon State Univ., Corvallis.

Johnson, H.E. and C. Pecor. 1969. Coho salmon mortality and DDT in Lake Michigan. *Trans. 34th N. Am. Wildl. Nat. Resour. Conf.*, pp. 159–166.

Jones, R. 1982. Population fluctuations and recruitment in marine populations. *Philos. Trans. R. Soc. Lond. B297:* 353–368.

Kanciruk, P., J.E. Breck, and D.S. Vaughan. 1982. Population-level effects of multiple stresses on fish and shellfish. Publ. No. ORNL/TM–8317, Oak Ridge Natl. Lab., Oak Ridge, TN.

Khan, R.A. 1987. Effects of chronic exposure to petroleum hydrocarbons on two species of marine fish infected with a hemoprotozoan, *Trypanosoma murmanensis. Can. J. Zool.* 65: 2703–2709.

Kinne, O. and H. Rosenthal. 1967. Effects of sulfuric water pollutants on fertilization, embryonic development and larvae of the herring, *Clupea harengus. Mar. Biol. (Berl.) 1:* 65–83.

Kühnhold, W.W., D. Everich, J.J. Stegeman, J. Lake, and R.E. Wolke. 1978. Effects of low levels of hydrocarbons on embryonic, larval and adult winter flounder (*Pseudopleu-ronectes americanus*), pp. 677–711. In: *Proceedings of the Conference on Assessment of Ecological Impacts of Oil Spills.* American Institute of Biological Science, Washington, DC.

Landahl, J.T., L.L. Johnson, J.E. Stein, T.K. Collier, and U. Varanasi. 1997. Approaches for determining effects of pollution on fish populations of Puget Sound. *Trans. Am. Fish. Soc. 126:* 519–535.

Lee, A. 1978. Effects of man on the fish resources of the North Sea. *Rapp. P.-V. Reun. Cons. Int. Explor. Mer 173:* 231–240.

Leggett, W.C. 1977. Density dependence, density independence, and recruitment in the American shad (*Alosa sapidissima*) population of the Connecticut River, pp. 3–17. In: W. Van Winkle (ed.), *Proceedings of the Conference on Assessing the Effects of Power-Plant–Induced Mortality on Fish Populations.* Pergamon Press, New York.

Lewis, J.R. 1980. Options and problems in environmental management and evaluation. *Helgol. Meeresunters. 33:* 452–466.

Longwell, A.C., D. Perry, J.B. Hughes, and A. Herbert. 1983. Frequencies of micronuclei in mature and immature erythrocytes of fish as an estimate of chromosome mutation rates — Results of field surveys on windowpane flounder, winter flounder and Atlantic mackerel. Int. Counc. Explor. Sea, Doc. C.M. 1983/E:55.

Malins, D.C., B.B. McCain, D.W. Brown, U. L. Chan, M.S. Myers, J.T. Landahl, P.G. Prohaska, A.J. Friedman, L.D. Rhodes, D.G. Burrows, W.D. Gronlund, and H.O. Hodgins. 1984. Chemical pollutants in sediments and diseases of bottom-dwelling fish in Puget Sound, Washington. *Environ. Sci. Technol. 13:* 705–713.

Malins, D.C., B.B. McCain, J.T. Landahl, M.S. Myers, M.M. Krahn, D.W. Brown, S.-L. Chan, and W.T. Roubal. 1988. Neoplastic and other diseases in fish in relation to toxic chemicals: An overview. *Aquat. Toxicol. 11:* 43–67.

McIntyre, A.D. 1982. Oil pollution and fisheries, pp. 401–411. In: R.B. Clark (ed.), *The Long Term Effects of Oil Pollution on Marine Populations, Communities, and Ecosystems.* Philos. Trans. R. Soc. Lond. B297.

McMahon, G., L.J. Huber, J.J. Stegeman, and G.N. Wogan. 1988. Identification of a C-Ki-ras oncogene in a neoplasm isolated from winter flounder. *Mar. Environ. Res. 24:* 345–350.

Mehrle, P.M., T.A. Haines, S. Hamilton, J.L. Ludke, F.L. Mayer, and M.A. Ribick. 1982. Relationship between body contaminants and bone development in east-coast striped bass. *Trans. Am. Fish. Soc. 111:* 231–241.

Merriner, J.V. 1976. Differences in management of marine recreational fisheries, pp. In: H. Clepper (ed.), *Marine Recreational Fisheries.* Sport Fishing Institute, Washington, DC.

Munro, A.L.S., A.H. McVicar, and R. Jones. 1983. The epidemiology of infectious disease in commercially important wild marine fish. *Rapp. P.-V. Reun. Cons. Int. Explor. Mer 182:* 21–32.

Murchelano, R.A. 1988. Fish as sentinels of environmental health. U.S. Dept. Commerce, NOAA Tech. Memo. NMFS-F/NEC-61, 16 pp.

Murchelano, R.A. and R.E. Wolke. 1985. Epizootic carcinoma in the winter flounder, *Pseudopleuronectes americanus. Science 228:* 587–589.

Nelson, D.A., J.E. Miller, D. Rusanowsky, R.A. Greig, G.R. Sennefelder, R. Mercaldo-Allen, C. Kuropat, E. Gould, F.P. Thurberg, and A. Calabrese. 1991. Comparative reproductive success of winter flounder in Long Island Sound: A 3-year study (biology, biochemistry, and chemistry). *Estuaries 14:* 318–331.

Overholtz, W.J. and K.D. Friedland. 2002. Recovery of the Gulf of Maine — Georges Bank Atlantic herring (*Clupea harengus*) complex: Perspectives based on bottom trawl survey data. *Fish. Bull. 100:* 593–608.

Payne, J.F. and L.F. Fancey. 1989. Effect of polycyclic aromatic hydrocarbons on immune responses in fish: Change in melanomacrophage centers in flounder (*Pseudopleuronectes americanus*) exposed to hydrocarbon-contaminated sediments. *Mar. Environ. Res. 28:* 431–435.

Perry, D.M., J.B. Hughes, and A.T. Hebert. 1991. Sublethal abnormalities in embryos of winter flounder, *Pseudopleuronectes americanus*, from Long Island Sound. *Estuaries 14:* 306–317.

Peters, N., A. Köhler, and H. Kranz. 1987. Liver pathology in fishes from the lower Elbe as a consequence of pollution. *Dis. Aquat. Org. 2:* 87–97.

Prager, M.H. and A.D. MacCall. 1993. Detection of contaminant and climate effects on spawning success of three pelagic fish stocks off southern California: Northern anchovy *Engraulis mordax*, Pacific sardine *Sardinops sagax*, and chub mackerel *Scomber japonicus. Fish. Bull. 91:* 310–327.

Rago, P.J. 1984. Production forgone: An alternative method for assessing the consequences of fish entrainment and impingement losses at power plants and other water intakes. *Ecol. Model. 24:* 79–111.

Rose, K.A., J.H. Cowan, Jr., E.D. Houde, and C.C. Coutant. 1993. Individual based modelling of environmental quality effects on early life stages of fishes. A case study using striped bass. *Am. Fish. Soc. Symp. 14:* 125–145.

Rose, K.A., J.A. Tyler, R.C. Chambers, G. Klein-McPhee, and D.J. Danila. 1996. Simulating winter flounder population dynamics using coupled individual-based young-of-the-year and age-structured adult models. *Can. J. Fish. Aquat. Sci. 53:* 1071–1091.

Saila, S.B. 1962. The contribution of estuaries to the offshore winter flounder fishery in Rhode Island. *Proc. Gulf Caribb. Fish. Inst. 1961:* 95–105.

Schaaf, W.E., D.S. Peters, D.S. Vaughan, L. Coston-Clements, and C.W. Krouse. 1987. Fish population responses to chronic and acute pollution: The influence of life history strategies. *Estuaries 10:* 267–275.

Schaaf, W.E., D.S. Peters, L. Coston-Clements, D.S. Vaughan, and C.W. Krouse. 1993. A simulation model of how life history strategies mediate pollution effects on fish populations. *Estuaries 16:* 697–702.

Segar, D.A. and E. Stamman. 1986. Monitoring in support of estuarine pollution management needs, pp. 874–877. In: *Proceedings of Oceans '86, Vol. 3.* Marine Technological Society, Washington, DC.

Sindermann, C.J. 1979. Status of Northwest Atlantic herring stocks of concern to the United States. Tech. Ser. Rep. No. 23, National Marine Fisheries Service, Sandy Hook Lab., Highlands, NJ.

Sindermann, C.J. 1980. Pollution effects on fisheries — Potential management activities. *Helgol. Meeresunters. 33:* 674–686.

Smith, R.M. and C.F. Cole. 1970. Chlorinated hydrocarbon insecticide residues in winter flounder, *Pseudopleuronectes americanus*, from the Weweantic River estuary, Massachusetts. *J. Fish. Res. Board Can. 27:* 2374–2380.

Smith, R.M. and C.F. Cole. 1973. Effects of egg concentrations of DDT and dieldrin on development in winter flounder (*Pseudopleuronectes americanus*). *J. Fish. Res. Board Can. 30:* 1894–1898.

Summers, J.K., T.T. Polgar, K.A. Rose, R.A. Cummins, R.N. Ross, and D.G. Heimbuch. 1986. Assessment of the relationships among hydrographic conditions, macropollution histories, and fish and shellfish stocks in major northeastern estuaries. U.S. Dept. Commerce, NOAA Tech. Memo. NOS-OMA 31, 223 pp.

Swartzman, G., R. Deriso, and C. Cowan. 1977. Comparison of simulation models used in assessing the effects of power-plant–induced mortality on fish populations, pp. 333–361. In: W. Van Winkle (ed.), *Proceedings of the Conference on Assessing the Effects of Power-Plant–Induced Mortality on Fish Populations.* Pergamon Press, New York.

Tiews, K. 1983. On the changes of fish and crustacean stocks in the German North Sea coast during the years 1954–1981 and the hypothetical role of pollution as a causative factor. Int. Counc. Explor. Sea, Doc. C.M.1983/E:16, 18 pp.

Tiews, K. 1989. 35 years' abundance trends (1954–1988) of 25 fish and crustacean stocks on the German North Sea coast. Int. Counc. Explor. Sea, Doc. C.M.1989/E:28, 11 pp.

Van Winkle, W., editor. 1977. *Proceedings of the Conference on Assessing the Effects of Power-Plant–Induced Mortality on Fish Populations.* Pergamon Press, New York. 380 pp.

Van Winkle, W., D.S. Vaughan, L.W. Barnthouse, and B.L. Kirk. 1981. An analysis of the ability to detect reduction in year-class strength of the Hudson River white perch (*Morone americana*) population. *Can. J. Fish. Aquat. Sci. 38:* 627–632.

Vaughan, D.S. and W. Van Winkle. 1982. Corrected analysis of the ability to detect reductions in year-class strength of the Hudson River white perch (*Morone americana*) populations. *Can. J. Fish. Aquat. Sci. 39:* 782–785.

Vaughan, D.S., F. Kanciruk, and J.E. Breck. 1982. Research needs to assess population-level effects of multiple stresses on fish and shellfish. Publ. No. ORNL/TM-8375, Oak Ridge Natl. Lab., Oak Ridge, TN.

Vaughan, D.S., R.M. Yoshiyama, J.E. Breck, and D.L. DeAngelis. 1982. Review and analysis of existing modeling approaches for assessing population-level effects of multiple stresses on fish and shellfish. Publ. No. ORNL/TM-8342, Oak Ridge Natl. Lab., Oak Ridge, TN.

Vaughan, D.S., J.V. Merriner, and W.E. Schaaf. 1986. Detectability of a reduction in a single year class of a fish population. *J. Elisha Mitchell Soc. 102:* 122–128.

Waldichuk, M. 1979. Review of the problems, pp. 399–424. In: H.A. Cole (ed.), *The Assessment of Sublethal Effects of Pollutants in the Sea.* Philos. Trans. R. Soc. Lond. B286.

Westernhagen, H. von. 1988. Sub-lethal effects of pollutants on fish eggs and larvae, pp. 253–346. In: W.S. Hoar and D.J. Randall (eds.), *Fish Physiology.* Academic Press, New York.

Westernhagen, H. von, H. Rosenthal, V. Dethlefsen, W. Ernst, U. Harms, and P.-D. Hansen. 1981. Bioaccumulating substances and reproductive success in Baltic flounder *Platichthys flesus. Aquat. Toxicol. 1:* 85–99.

Westernhagen, H. von, K.R. Sperling, D. Janssen, V. Dethlefsen, P. Cameron, R. Kocan, M. Landolt, G. Fürstenberg, and K. Kremling. 1987a. Anthropogenic contaminants and reproduction in marine fish. *Ber. Biol. Anst. Helgol. 3:* 1–70.

Westernhagen, H. von, V. Dethlefsen, P. Cameron, and D. Janssen. 1987b. Chlorinated hydrocarbon residues in gonads of marine fish and effects on reproduction. *Sarsia 72:* 419–422.

Westernhagen, H. von, V. Dethlefsen, P. Cameron, J. Berg, and G. Fürstenberg. 1988. Developmental defects in pelagic fish embryos from the western Baltic. *Helgol. Meeresunters. 42:* 13–36.

Westernhagen, H. von, P. Cameron, V. Dethlefsen, and D. Janssen. 1989. Chlorinated hydrocarbons in North Sea whiting (*Merlangius merlangus* (L.)) and effects on reproduction. *Helgol. Meeresunters. 43:* 45–60.

Westin, D.T., C.E. Olney, and B.A. Rogers. 1983. Effects of parental and dietary PCBs on survival, growth, and body burdens of larval striped bass. *Bull. Environ. Contam. Toxicol. 30:* 50–57.

Westin, D.T., C.E. Olney, and B.A. Rogers. 1985. Effects of parental and dietary organochlorines on survival and body burdens of striped bass larvae. *Trans. Am. Fish. Soc. 114:* 125–136.

Wolke, R.E., C.J. George, and V.S. Blazer. 1984. Pigmented macrophage accumulations (MMC; PMB): Possible monitors of fish health, pp. 93–97. In: W. Hargis (ed.), *USA–USSR Symposium on Pathogens and Parasites of the World Oceans.* U.S. Dept. Commerce, NOAA, Natl. Mar. Fish. Serv. Tech. Rep. 25.

Wolke, R.E., R.A. Murchelano, C.D. Dickstein, and C.J. George. 1985. Preliminary evaluation of the use of macrophage aggregates (MA) as fish health monitors. *Bull. Environ. Contam. Toxicol. 35:* 222–227.

Zdanowicz, V.S., D.F. Gadbois, and M.W. Newman. 1986. Levels of organic and inorganic contaminants in sediments and fish tissues and prevalences of pathological disorders in winter flounder from estuaries of the northeast United States, 1984, pp. 578–585. In: *IEEE Oceans '86 Conference Proceedings.* IEEE, Piscataway, NJ.

11 Mass Mortalities of Marine Mammals

Died at Sea of Unknown Causes: Dolphin Mortalities on the Atlantic Coast of United States in 1987–1988

The early morning mist was disappearing from the beach at Asbury Park, New Jersey, on a late summer day in 1988, and as it dissipated it revealed the bloated bodies of three sea mammals that had floated in with the tide during the night. They were young adult bottlenose dolphin (Tursiopsis truncatus) — a species noted for their graceful antics at sea and for crowd-pleasing trained behavior in seaside commercial aquaria. But these specimens were, unaccountably, dead. They were also statistics, because during the summers of 1987 and 1988 more than 700 other members of their species had washed up on Atlantic shores — an unprecedented event.

What killed these intelligent marine mammals? Had they encountered fatal levels of some industrial chemical, or some natural algal toxin? Had a new viral disease reached epizootic levels in the population? Or was some parasitic disease affecting equilibrium or respiration to the point where drowning could occur?

A task force of experts in marine diseases was assembled to examine the problem soon after the deaths began in 1987 — headed by a professor imported temporarily from a Canadian veterinary college. The group established field headquarters in 1987 and 1988 in an oceanfront motel and pursued various hypotheses about causes as specimens were dissected and examined. News media representatives asked every day for answers, and tourists gathered in subdued clusters around dead and decomposing animals on the beaches. After the various analyses pathological, microbiological, and chemical were completed, a tentative (and controversial) diagnosis was made and a final report prepared. According to the report, deaths during the outbreak period seemed associated with high body tissue levels of an algal toxin ("brevetoxin," known as a "red tide" toxin) ingested with food. Affected animals seemed less resistant to the microbial infections that were the immediate cause of death in most cases. Pathological effects may have been exacerbated also by exposure to industrial pollutants such as PCBs, which could create physiological stress and reduce disease resistance.

At any rate, the somewhat inconclusive results of the investigations did not seem to please many observers, and as an interesting but equally baffling sequel to these mortalities on the Atlantic coast, dolphins also died in 1990 in large numbers, estimated to be in excess of 200 — but this time in the Gulf of Mexico.

*Most marine mammals — especially the dolphins — have a strange hold
on the human conscience. We seem to actually care about their well-being. So
it should behoove us, in rare moments of introspection, to examine our possible
role in despoiling their habitats — to the degree that they can die in great
numbers, as they did in 1987, 1988, and 1990.*

From Field Notes of a Pollution Watcher
(C.J. Sindermann, 1992)

[Update 2003: A viral disease was recognized by a number of investigators as
being at epizootic levels in the dolphin population during the 1987–1988 period,
and was possibly responsible for the mortalities seen.]

The bottlenose dolphin deaths on the North Atlantic coast of the United States
in 1987 and 1988 — just described briefly — are one example of a more general
phenomenon. Mass deaths have occurred sporadically during the past century, but
with increasing frequency in the past 2 decades among populations of the two major
groups of marine mammals — *cetaceans* (dolphins, porpoises, whales) and *pinnipeds*
(seals, sea lions, fur seals, walrus; Reynolds & Rommel 1999).

Early in the 19th century a massive "red tide" bloom was thought to have killed
large numbers of Cape fur seals (*Arctocephalus pusillus;* Wyatt 1980). Then in the
summer of 1918 more than 1000 harbor seals (*Phoca vitulina*) died in Icelandic
waters, about 3000 crabeater seals (*Lobodon carcinophagus*) died in Antarctica in
1955 (Laws & Taylor 1957), and more than 1200 walruses (*Odobenus rosmarus*)
died in the Bering Strait in 1978.

Viral infections have been reported to be primary causes of more recent mass
deaths in striped dolphins and harbor seals, and algal toxins have also been implicated
in some mortalities of bottlenose dolphins and humpback whales on the Atlantic
coast of North America. Many reports allude to, but do not demonstrate, a possible
role for pollution, particularly in reducing immune responses or causing reproductive
failures in some species of marine mammals.

DOLPHIN AND WHALE MORTALITIES

Mass deaths of some cetaceans are matters of record. Mortalities of bottlenose
dolphins (*Tursiops truncatus*) and humpback whales (*Megaptera novaeangliae*)
occurred on the Atlantic coast of the United States in 1987 and 1988. Dolphin
mortalities began in the summer of 1987 in New Jersey waters and persisted along
the Atlantic seaboard for 11 months. During that period, more than 700 dolphins
died. Estimates were that over 50% of the inshore bottlenose dolphin population of
the Middle Atlantic coast may have died during the 1987–1988 outbreak (Lipscomb
et al. 1994a).

The humpback whale mortality occurred in late 1987 off the Massachusetts coast
and involved 14 animals killed in a little over 1 month. Detailed examinations and
chemical analyses of tissues led to the conclusion that mortalities were associated

with the presence in the animals of the algal neurotoxins brevetoxin and saxitoxin (Geraci 1989). The source of the toxins was hypothesized to be food fish, especially menhaden and mackerel, and effects included reduced physiological fitness and thus increased susceptibility to secondary bacterial infections that were the immediate cause of death (Anderson & White 1989).

Questions about the causes of the dolphin mortalities were at least partially resolved in the early 1990s, with findings of morbilliviruses (now labeled CeMV or cetacean morbillivirus) by several investigators (Lipscomb et al. 1994a, 1994b; Krafft et al. 1995; Duignan et al. 1995a, 1995b; Schulman et al. 1997). In a recent review of cetacean viruses, Van Bressem, Waerebeek, and Raga (1999) concluded that CeMV caused epizootics in bottlenose dolphin along the Atlantic USA coast in 1982, 1987–1988, and 1993–1994.

Mass mortalities of dolphins have also occurred recently in the Mediterranean. In 1990 and 1991 more than 1000 striped dolphins (*Stenella coeruleoalba*) died in the western and central Mediterranean; in 1992 striped dolphins died in large numbers in the eastern Mediterranean (Van Bressem et al. 1991, 1993). These events in the Mediterranean seemed to be the result of infections by the morbillivirus CeMV that had killed bottlenose dolphins a few years earlier in the North Atlantic (Van Bressem, Jepson, & Barrett 1998). Population impact of mortalities on Mediterranean striped dolphins was indicated by a 70% reduction in school size in the affected areas (Aguilar & Raga 1993, Forcada et al. 1994). Effects of the viral epizootics may have been exacerbated by reduced immune competence (lower disease resistance) induced by high levels of PCBs and other chlorinated hydrocarbon contaminants in food and in the environment, but no unusual mortalities have been detected in Mediterranean striped dolphins since 1992 (Van Bressem, Waerebeek, & Raga 1999). Those authors pointed out, though, that morbilliviruses in other animals do not induce a carrier state, and infection confers lifelong immunity, so recurrent epizootics, with significant population reductions, may be expected in the future (Harwood & Hall 1990).

SEAL MORTALITIES

Among the pinnipeds, seal mortalities have been especially dramatic both on the northern Atlantic coast of the United States and in northern European waters. During 1979 to 1980, more than 500 harbor seals (*Phoca vitulina*) died in New England waters; an influenza virus, probably of avian origin, was identified as the primary cause of death (Geraci et al. 1982). Then, in the spring and summer of 1988, massive numbers of dead harbor seals — some estimates exceed 20,000 — were washed ashore in northern Europe, especially in Norway, Sweden, Denmark, Britain, and Germany. A newly recognized virus, called phocine distemper virus (PDV), was identified as the primary cause of the mortalities (the virus belongs to the morbillivirus group, which also contains the organism responsible for distemper in dogs and measles in humans; Osterhaus et al. 1990).

The "seal plague" in northern Europe in 1988 was extensive enough to cause drastic reductions in population size, ranging from 40 to 60% of existing regional stocks (Tougaard 1989, Harkönen & Heide-Jorgensen 1990). Disease effects declined sharply in 1989. One fascinating aspect of this outbreak, as pointed out by Dietz, Heide-Jorgensen, and Harkönen (1989), was that North European waters experienced in the spring of 1988 massive blooms of a toxin-producing microalga, *Chrysochromulina polylepis* — just at the time when the harbor seal deaths were beginning. The bloom caused major ecosystem disturbances, killing fish, shellfish, and macroalgae (Rosenberg, Lindahl, & Blanck 1988), but its role (if any) in the seal mortalities remains unknown. The statement made by Dietz and colleagues (1989) seems to be most relevant:

[I]t cannot be entirely ruled out that the dramatic development of the seal epizootic was not in some way related to the ongoing changes in nutrients and microbial activity along the heavily urbanized and extensively cultivated coasts of Northern Europe. (p. 262)

It is difficult to accept that an extensive toxic bloom of this nature would not have had some effect on marine mammals in the area.

Another fascinating aspect of the outbreak in seals of northern Europe was that it was preceded (in the autumn of 1987) by a similar disease among Baikal seals (*Phoca sibirica*) in Lake Baikal in southern Siberia (Osterhaus et al. 1989). Several thousand seals died, and the virus that was isolated was also a morbillivirus but was more closely related to canine distemper virus than to the phocine virus that killed harbor seals in Europe in 1988 (Grachev et al. 1989).

ROLE OF POLLUTANTS IN MASS MORTALITIES

Marine mammals die from many causes, both natural and man made. They may be killed on the ice as pups in the Gulf of Saint Lawrence, they may drown in fishermen's nets, or they may be killed by predators such as polar bears or sharks. Occasionally, they may die in great numbers at roughly the same time — mass mortality — a reason for concern among scientists and humans in general.

From this brief account, we can discern that possible contributors to mass mortality of marine mammals (as identified in the literature) include:

- Pathogens
- Natural toxins
- Contaminants from human sources

Pathogens, particularly viruses, have clearly been causes of mortality in several species of marine mammals. Dolphins in the Mediterranean and Caspian seas and harbor seals on the New England and North European coasts were killed in large numbers during viral epizootics in the late 1980s and early 1990s. Other mass mortalities, of dolphins and whales off the New England coast, were attributed initially to toxins from algal blooms, transferred through food fish — although subsequent research implicated virus epizootics as the primary cause of mortalities. But the role of pollutants — especially chlorinated pesticides, PCBs, and heavy metals — in mass deaths of marine mammals is not clear, except for reduction in physiological condition, reproductive failure, and decreased resistance to disease as a consequence of exposure (for which there is some experimental information). O'Shea (1999), in a review of environmental contaminants, concluded that "there is mixed evidence for linkages with increased susceptibility to disease" (p. 520) and, "Although there is a lack of absolute scientific certainty in linking the presence of specific organochlorine contaminants to detrimental impacts on marine mammal populations, the body of indirect circumstantial evidence ... continues to grow" (p. 520).

Because this is a book about coastal pollution and its effects, it seems entirely fitting to focus on the involvement of pollutants in the mortalities advertised in the chapter title. One obvious linkage between pollution and disease (the latter being the direct cause of many mass mortalities) is impairment of the immune function. Studies with harbor seals (*Phoca vitulina*) in Europe (de Swart et al. 1994, Ross et al. 1996) demonstrated contaminant-induced immunosuppression. One group of harbor seals was fed herring from the polluted Baltic Sea, whereas a control group was fed herring from the relatively unpolluted Atlantic Ocean. The test period was almost 2 yr, during which time immunological factors (natural killer cell activity and T-cell–mediated immune responses) were monitored. Seals fed the contaminated Baltic-caught fish (in which PCBs, PCDDs, and PCDFs were three to six times higher than in Atlantic-caught fish) had significantly lower levels of the immunological factors studied, as well as reduced thyroid hormone and vitamin A levels — findings that, in the words of the investigators, "show a functional impairment of cells of both the innate and adaptive immune systems of harbor seals after chronic exposure to environmental contaminants at concentrations occurring in their natural habitat" (De Swart et al., p. 158). (The authors did, however, recognize the impossibility of identifying any single contaminant as responsible for the impaired immune responses found in the study.)

A related examination of blubber biopsies from the same groups of experimental harbor seals, did, however, implicate PCBs rather than dioxins or furans in the observed immunosuppression (Ross et al. 1995). But the study provided much more information about immunosuppression by pollutants. Seals fed contaminated fish were less able to mount a specific response in skin tests with ovalbumin, and responses correlated well with *in vitro* tests of 1-lymphocyte function — all suggesting pollutant-induced immunosuppression. On the basis of on these results, the authors went on to speculate that "anthropogenic contaminants, in particular PCBs, played a role in the 1988 phocine distemper virus (PDV) epizootic in Europe and other recent mass mortalities of marine mammals caused by virus infections" (p. 166).

This European research provides good data on an inverse relationship between pollution and disease resistance in marine mammals. It has been augmented by other lines of evidence (as assembled and described by Ross et al. 1995):

- Mortalities of marine mammals (harbor seals in northern Europe and Baikal seals in Lake Baikal) in 1988, bottlenose dolphins in the Gulf of Mexico in 1987–1988, and striped dolphins in the Mediterranean Sea in 1990–1991 all occurred in areas where pollution could have impaired immunocompetence, rendering members of those populations more susceptible to viral infections.

- Harbor seals that survived the disastrous European PDV epizootic in 1988 had lower tissue organochlorine burdens than did seals that died (Hall et al. 1992).
- Phocine distemper virus or a similar one had infected Canadian harbor seals before the European epizootic, with no apparent mortalities (Henderson et al. 1992, Ross et al. 1992, Barrett et al. 1993).
- Harbor seals living in less contaminated waters around Britain had lower mortality rates during the viral outbreak than did those from polluted areas (Simmonds, Johnston, & French 1993).*

Two very recent mass mortality episodes on the west coast of North America suggest that environmental elements other than morbilliviruses may on occasion kill large numbers of marine mammals. Sea lions and dolphins died off the coast of Mexico in 1997, possibly due to ingesting fish containing domoic acid toxin from a bloom of the diatom *Pseudo-nitschia australis* (Ochoa et al. 1998). Then a larger incident of toxin-related mortalities (estimated at about 500 deaths) of sea lions occurred in May and June of 1998 off the central California coast (Gulland et al. 2000, Scholin et al. 2000). Domoic acid toxins were demonstrated in forage fish and in sea lion body fluids, and a bloom of *P. australis* coincided with mortalities. It should be noted, though, that none of the cited published reports on sea lion mortalities mentioned examinations for microbiological pathogens. Histopathological studies did, however, disclose brain lesions similar to those seen in humans and laboratory mammals exposed to domoic acid.

Although less dramatic than mass mortalities due to viral pathogens or environmental biotoxins, reproductive failure due to pollution can affect population abundance of marine mammals. Some of the effects of pollutants such as PCBs on reproduction in fish were discussed in Chapter 9, and there are counterpart effects on marine mammals. One study began with a search for causes of collapse of the harbor seal population of the western Dutch Wadden Sea between 1950 and 1975 — dropping from more than 3000 individuals to fewer than 500. Increasing industrial pollution, especially of PCBs from the Rhine River, was suspected (Reijnders 1986). Reijnders set up a field study in which captive seals were placed into two experimental groups and fed differentially for 2 yr — one with mixed fish from the PCB-polluted western Wadden Sea and the other with fish from the northeast Atlantic. Reproductive success was significantly lower in seals fed with fish from the polluted Wadden Sea. Estradiol levels were lower, and the reproductive process in females was disrupted in the postovulation/implantation stage. A simultaneous study substituting mink (*Mustela vison*) for seals and using very low levels of PCBs (25 mg/d in an artificial diet) produced very similar physiological changes. Earlier studies of PCB effects on mink had found impaired reproduction in the form of effects on ovulation, mating, and implantation, followed by early abortion or resorp-

---

* Parenthetically, it should be pointed out (and it was, by the investigators) that the Baltic herring used in the harbor seal feeding experiment that demonstrated impaired immunocompetence were originally destined for human consumption, because the species is highly prized as food in northern Europe. This should raise concern about comparable effects on other fish-eating mammals — including an erect featherless biped called *Homo sapiens.*

tion (Jensen et al. 1977). The author of the seal report (Reijnders 1986) concluded that reproductive failure in seals from the Dutch Wadden Sea was related to consuming fish from that polluted area, and he suggested that the organochlorines were the main cause of the failure.

CONCLUSIONS

The killing of many marine mammals by human predators remains a principal environmental cause of mortality, despite restrictive legislation and the existence of regulations on a global scale. Beyond this reality, though, other causes of mass mortalities have been identified over a period of almost 2 centuries. Toxic algal blooms were suspects in earlier incidents (see Dietz, Heide-Jorgensen, & Härkönen 1989), but more recently disease and debilitating effects of chemical pollutants have assumed dominant roles. Beginning in the 1980s, viruses, especially the morbilliviruses, have been implicated in mass deaths of cetaceans and pinnipeds in Europe and North America. In some instances, a contributory role for elevated levels of industrial pollutants such as organochlorines, butyltins, and mercury has been seen in producing negative physiological effects and reducing disease resistance (Brouwer, Reijnders, & Koeman 1989; Becker et al. 2000). Organochlorines — particularly the PCBs — have also been associated with reproductive failure as well as physiological debilitation and immunosuppression.

As an example of effects of pollutants, in an analysis of reasons for mass deaths of harbor seals in northern Europe in 1988, Harwood and Grenfell (1990) pointed out that despite findings of high organochlorine contaminants in blubber, "none of the involved investigators suggested that contaminants were a cause of the mass mortality, only that they may have been a contributory factor in some areas" (p. 286). Harwood and Grenfell also reminded us of two other facts — that "The distemper viruses are in themselves immunosuppressive" (p. 286) and that "In the early stages of the epizootic, death was most frequently caused by secondary bacterial infections, contracted as a result of immunosuppression associated with PDV, rather than the direct effect of the virus itself" (p. 286).

Investigations of mass mortalities — of any marine animal, whether invertebrate, fish, or mammal — are never simple, and conclusions must always be to some degree tentative. Harwood and Grenfell have again said it well: "[A]lthough the quantitative features of the PDV epizootic may be explicable simply in terms of the known epidemiology of morbilliviruses, we cannot rule out the influences of environmental factors" (p. 287). These factors should clearly include pollution.

REFERENCES

Aguilar, A. and J.A. Raga. 1993. The striped dolphin epizootic in the Mediterranean Sea. *Ambio 22:* 524–528.

Anderson, D.M. and A.W. White, editors. 1989. Toxic dinoflagellates and marine mammal mortalities. Tech. Rep. No. CRC-89-6, Woods Hole Oceanographic Institution, Woods Hole, MA. 65 pp.

Barrett, T., I.K.G. Visser, L. Mamaev, L. Goatley, M.F. Van Bressem, and A.D.M.E. Osterhaus. 1993. Dolphin and porpoise morbilliviruses are genetically distinct from phocine distemper virus. *Virology 193:* 1010–1012.

Becker, P.R., M.M. Krahn, E.A. Mackey, R. Demiralp, M.M. Schantz, M.S. Epstein, M.K. Donais, B.J. Porter, D.C.G. Muir, and S.A. Wise. 2000. Concentrations of polychlorinated biphenyls (PCBs), chlorinated pesticides, and heavy metals and other elements in tissues of belugas, *Delphinapterus leucas*, from Cook Inlet, Alaska. *Mar. Fish. Rev. 62:* 81–98.

Brouwer, A., P.J.H. Reijnders, and J.H. Koeman. 1989. Polychlorinated biphenyl (PCB)–contaminated fish induces vitamin A and thyroid hormone deficiency in the common seal (*Phoca vitulina*). *Aquat. Toxicol. 15:* 99–106.

Dietz, R., M.-P. Heide-Jorgensen, and T. Härkönen. 1989. Mass deaths of harbor seals (*Phoca vitulina*) in Europe. *Ambio 18:* 258–264.

Duignan, P.J., C. House, J.R. Geraci, N. Duffy, B.K. Rima, M.T. Walsh, G. Early, D.J. St. Aubin, S. Sadov, H. Koopman, and H. Rhinehart. 1995a. Morbillivirus infection in cetaceans of the western Atlantic. *Vet. Microbiol. 44:* 241–249.

Duignan, P.J., C. House, J.R. Geraci, G. Early, H. Copland, M.T. Walsh, G.D. Bossart, C. Cray, S. Sadove, D.J. St. Aubin, and M. Moore. 1995b. Morbillivirus infection in two species of pilot whales (*Globicephala* sp.) from the western Atlantic. *Mar. Mamm. Sci. 11:* 150–162.

Forcada, J., A. Aguilar, P.S. Hammond, X. Pastor, and R. Aguilar. 1994. Distribution and numbers of striped dolphins in the western Mediterranean Sea after the 1990 epizootic outbreak. *Mar. Mamm. Sci. 10:* 137–150.

Geraci, J.R. 1989. Clinical investigation of the 1987–88 mass mortality of bottlenose dolphins along the U.S. central and south Atlantic coast. Final Rep. to National Marine Fisheries Serv., U.S. Navy, Office of Naval Research, and Marine Mammal Commission. 63 pp.

Geraci, J.R., D.J. Aubin, I.K. Barker, R.G. Webster, V.S. Hinshaw, W.J. Bean, H.L. Ruhnke, J.H. Prescott, G. Early, A.S. Baker, S. Madoff, and R.T. Schooley. 1982. Mass mortality of harbor seals: Pneumonia associated with influenza A virus. *Science 215:* 1129–1131.

Grachev, M.A., V.P. Kumarev, L.V. Mamaev, V.L. Zorin, L.V. Baranova, N.N. Denikina, S.I. Belikov, E.A. Petrov, V.S. Kolesnik, R.S. Kolesnik, V.M. Dorofeev, A.M. Beim, V.N. Kudelin, F.G. Nagieva, and V.N. Sidorov. 1989. Distemper virus in Baikal seals. *Nature (London) 338: 209.*

Gulland, F.M., M. Haulena, T. Rowles, L.J. Lowenstine, T. Spraker, T. Lipscomb, V. Trainer, F. Van Dolah, and C. Sholin. 2000. Unusual marine mammal mortality event — Domoic acid toxicity in California sea lions (*Zalophus californianus*) stranded along the central California coast, May–October 1998. NOAA Tech. Memo. NMFS-OPR-8, U.S. Dept. Commerce, National Marine Fisheries Service, Silver Spring, MD.

Hall, A.J., R.J. Law, J. Harwood, H.M. Ross, S. Kennedy, C.R. Allchin, L.A. Campbell, and P.P. Pomeroy. 1992. Organochlorine levels in common seals (*Phoca vitulina*) which were victims and survivors of the 1988 phocine distemper epizootic. *Sci. Tot. Environ. 115:* 145–162.

Harkönen, T. and M.-P. Heide-Jorgensen. 1990. Short-term effects of the mass dying of harbour seals in the Kattegat-Skagerrak area during 1988. *Z. Säugetierkde 55:* 233–238.

Harwood, J., and A. Hall. 1990. Mass mortality in marine mammals: Its implications for population dynamics and genetics. *TREE 5:* 254–257.

Harwood, J. and B. Grenfell. 1990. Long term risks of recurrent seal plagues. *Mar. Pollut. Bull. 21:* 284–287.

Henderson, G., A. Trudgett, C. Lyons, and K. Ronald. 1992. Demonstration of antibodies in archival sera from Canadian seals reactive with a European isolate of phocine distemper virus. *Sci. Total Environ. 115:* 93–98.

Jensen, S., J.E. Kilström, M. Olsson, C. Lundberg, and J. Orberg. 1977. Effects of PCBs on mink reproduction. *Ambio 6:* 239–246.

Krafft, A., J.H. Lichy, T.P. Lipscomb, B.A. Klaunberg, S. Kennedy, and J.K. Taunbenberger. 1995. Postmortem diagnosis of morbillivirus infection in bottlenose dolphins (*Tursiops truncatus*) in the Atlantic and Gulf of Mexico epizootics by polymerase chain reaction–based assay. *J. Wildl. Dis. 31:* 410–415.

Laws, R.M. and R.J.F. Taylor. 1957. A mass dying of crabeater seal, *Lobodon carcinophagus* (Gray). *Proc. Zool. Soc. London,* 315–324.

Lipscomb, T.P., F.Y. Schulman, D. Moffett, and S. Kennedy. 1994a. Morbilliviral disease in Atlantic bottlenose dolphins (*Tursiops truncatus*) from the 1987–1988 epizootic. *J. Wildl. Dis. 30:* 567–571.

Lipscomb, T.P., S. Kennedy, D. Moffett, and D.K. Ford. 1994b. Morbilliviral disease in an Atlantic bottlenose dolphin (*Tursiops truncatus*) from the Gulf of Mexico. *J. Wildl. Dis. 30:* 572–576.

Ochoa, J.I., A.P. Sierra-Beltrán, G. Oláiz-Fernández, and L.M. Del Villar Ponce. 1998. Should mollusk toxicity in Mexico be considered a public health issue? *J. Shellfish Res. 17:* 1671–1673.

O'Shea, T.J. 1999. Environmental contaminants and marine mammals, pp. 485–536. In: J.E. Reynolds and S.A. Rommel (eds.), *Biology of Marine Mammals.* Smithsonian Institution Press, Washington, DC.

Osterhaus, A.D.M.E., J. Groen, F.G.C.M. Uytdehaag, I.K.G. Visser, M.W.G. van de Bildt, A. Bergman, and B. Klingeborn. 1989. Distemper virus in Baikal seals. *Nature 338:* 209–210.

Osterhaus, A.D.M.E., J. Groen, H.E.M. Spijkers, H.W.J. Broeders, F.G.C.M. UytdeHaag, P. de Vries, J.S. Teppema, I.K.G. Visser, M.W.G. van de Bildt, and E.J. Vedder. 1990. Mass mortality in seals caused by a newly discovered morbillivirus. *Vet. Microbiol. 23:* 343–350.

Reijnders. P.J.H. 1986. Reproductive failure in common seals feeding on fish from polluted coastal waters. *Nature 324:* 456–457.

Reynolds, J.E. and S.A. Rommel, editors. 1999. *Biology of Marine Mammals.* Smithsonian Institution Press, Washington, DC. 578 pp.

Rosenberg R. O. Lindahl, and H. Blanck. 1988. Silent spring in the sea. *Ambio 17:* 289–290.

Ross, P.S., I.K.G. Visser, H.W.J. Broeders, M.W.G. van de Bildt, W.D. Bowen, and A.D.M.E. Osterhaus. 1992. Antibodies to phocine distemper virus in Canadian seals. *Vet. Rec. 130:* 514–516.

Ross, P.S., R.L. De Swart, P.J.H. Reijnders, H. Van Loveren, J.G. Vos, and A.D.M.E. Osterhaus. 1995. Contaminant-related suppression of delayed-type hypersensitivity and antibody responses in harbor seals fed herring from the Baltic Sea. *Environ. Health Perspect. 103:* 162–167.

Ross, P., R. De Swart, R. Addison, H. Van Loveren, J. Vos, and A.D.M.E. Osterhaus. 1996. Contaminant-induced immunotoxicity in harbour seals: Wildlife at risk? *Toxicology 112:* 157–169.

Scholin, C.A., F. Gulland, G.J. Doucette, S. Benson, M. Busman, F.P. Chavez, J. Cordaro, R. DeLong, A. De Vogelaere, J. Harvey, M. Haulena, K. Lefebvre, T. Lipscomb, S. Loscutoff, L.J. Lowenstine, R. Marin III, P.E. Miller, W.A. McLellan, P.D.R. Moeller, C.L. Powell, T. Rowles, P. Silvagni, M. Silver, T. Spraker, V. Trainer, and F.M. Van Dolah. 2000. Mortality of sea lions along the central California coast linked to a toxic diatom bloom. *Nature 403:* 80–85.

Schulman, F.Y., T.P. Lipscomb, D. Moffett, A.E. Krafft, J.H. Lichy, M.M. Tsai, J.K. Taubenberger, and S. Kennedy. 1997. Histologic, immunohistochemical, and polymerase chain reaction studies of bottlenose dolphins from the 1987–88 United States Atlantic coast epizootic. *Vet. Pathol. 34:* 288–295.

Simmonds, M.P., P.A. Johnston, and M.C. French. 1993. Organochlorine and mercury contamination in United Kingdom seals. *Vet. Rec. 132:* 291–295.

Swart, R.L. de, P.S. Ross, L.J. Vedder, H.H. Timmerman, S. Heistercamp, H. Van Loveren, J.G. Vos, P.J.H. Reijnders, and A.D.M.E. Osterhaus. 1994. Impairment of immune function in harbor seals (*Phoca vitulina*) feeding on fish from polluted waters. *Ambio 23:* 155–159.

Tougaard, S. 1989. Monitoring harbour seal (*Phoca vitulina*) in the Danish Wadden Sea. *Helgol. Meeresunters. 43:* 347–356.

Van Bressem, M.F., I.K.G. Visser, M.W.G. Van De Bildt, J.S. Teppema, J.A. Raga, and A.D.M.E. Osterhaus. 1991. Morbillivirus infection in Mediterranean striped dolphins (*Stenella coeruleoalba*). *Vet. Rec. 129:* 471–472.

Van Bressem, M.-F., I.K.G. Visser, R.L. de Swart, C. Örvell, L. Stanzani, E. Androukaki, K. Siakavara, and A.D.M.E. Osterhaus. 1993. Dolphin morbillivirus infection in different parts of the Mediterranean Sea. *Arch. Virol. 129:* 235–242.

Van Bressem, M.F., P. Jepson, and T. Barrett. 1998. Further insight on the epidemiology of cetacean morbillivirus in the northeastern Atlantic. *Mar. Mamm. Sci. 14:* 605–613.

Van Bressem, M.-F., K.V. Waerebeek, and J.A. Raga. 1999. A review of virus infections of cetaceans and the potential impact of morbilliviruses, poxviruses, and papillomaviruses on host population dynamics. *Dis. Aquat. Org. 38:* 53–65.

Wyatt, T. 1980. Morell's seals. *J. Cons. CIEM 39:* 1–6.

Section III

Effects of Coastal Pollution on Humans

I began this book with eight chapters (actually nine, counting the sludge monster in the prologue) describing some man-made undersea horrors. In case you have already forgotten, they are cholera, mercury poisoning, PCB contamination, polluted ocean beaches, toxic algae, anoxia, oil pollution, and invasions by alien species. That was the frosting. Then, in Part 2, I examined the effects of coastal pollution on marine animals, with special consideration of effects on fish and shellfish population abundance, and a short chapter on pollution effects on marine mammals. We are now as ready as we will ever be to consider in Part 3 the effects of coastal pollution on humans. This key area of concern — impacts on humans — the area's further review here, even though it has been given some cursory attention in several earlier chapters.

The human species is fortunate to have survived thus far in its short and brutish existence with only a few known episodes of mass disabilities and deaths caused by industrial pollution of coastal/estuarine waters. Of those, the one that has received greatest attention occurred in and around the city of Minamata in southern Japan, almost half a century ago. We examined that dreadful period of human suffering in Chapter 2 — mercury poisoning resulting from eating contaminated seafood.

The genuine horror story of the effects of mercury contamination in Minamata Bay illustrates, better than most examples can, the multiple consequences of coastal pollution to humans. Three principal kinds of impacts are apparent in this historical tale of pollution-associated disease — effects that can be discerned to varying degrees in other pollution events wherever they occur — as:

- Effects on human health
- Economic impacts
- Effects on the quality of human life

Effects of pollution on public health are the most visible consequences, and the ones given greatest coverage in the news media, but the costs of pollution extend beyond health considerations, to include economic losses to producers and consumers, and degradation of the quality of life for all of us.

Any realistic treatment of the topic of effects of pollution on humans must contain heavy emphasis on public health aspects, for several reasons — especially because this is where most of the quantitative information can be found, and also because it is our natural tendency, as members of the species, to give high priority to human health matters. Economic impacts of pollution are important to us but are more difficult to quantify, even though some attempts have been made. Quality of life considerations are mostly nonquantifiable but are still important consequences of environmental degradation.

All three kinds of effects — on public health, economics, and quality of life — should be closely integrated in our thinking and acting, when confronted with coastal pollution problems and the need for decisions about their solutions, but each kind of effect will be treated separately in Section III.

12 Effects of Coastal Pollution on Public Health

INTRODUCTION

Contamination of coastal waters can result in risks to human health through three principal routes:

1. Illnesses caused by microbial contamination of seafood
2. Illnesses caused by chemical contamination of seafood
3. Illnesses caused by environmental exposure to toxic chemicals and microbial contaminants in coastal waters

Microbial and chemical pollutants may affect the health of humans, either when they consume contaminated fish and shellfish or when they are exposed to waterborne pollutants in the coastal environment.

In an effort to make the subject of health effects of pollution a little easier to handle, I have subdivided it into the three very unequal segments just listed, with those having to do with microbial and chemical contamination of seafood given much greater status because of their relatively larger impacts on public health. Thus, in my opinion, the two seafood contaminant categories contain the principal problems, whereas the environmental exposure category is of much smaller stature but is still significant — especially the recreational exposures of the kinds discussed in Chapter 4.

Illnesses resulting from microbial contamination of seafood — especially contamination of shellfish — have emerged as significant problems as more and more people crowd the shorelines of industrialized countries like ours, as international transport of raw or frozen seafood products (often from countries with poor or nonexistent sanitary controls) expands, and as illogical practices of eating raw or inadequately cooked seafood persist and even prosper among the lunatic fringes of society. (The appearance of a quivering freshly opened raw oyster is repulsive enough, but when the visual turn-off is accompanied by the almost certain knowledge that potentially pathogenic microorganisms are lurking within that slimy mass, sensible people will practice total abstinence.)

Accompanying the risks from microbial contamination of shellfish and fish, although on a lesser level, is chemical contamination of seafood — either with toxicants of industrial origin or with toxins from marine microalgae (already considered in Chapter 5). Then, much further down the list of sources of risk to human health is environmental exposure to microbial contaminants and toxic chemicals, usually by swimming and diving, but also including some occupational activities in polluted waters (such as fishing, aquarium supply, and commercial diving).

Scientific studies of microbial and chemical contamination of shellfish and fish have provided much information about the disease risks involved, as well as the methodology to assess risks and to reduce them through regulation. I consider the material in this chapter on human health effects of coastal pollution to be of critical importance to my story — long, maybe, but vital to an appreciation of that part of coastal pollution effects that transcends the marine environment itself.

ILLNESSES CAUSED BY MICROBIAL CONTAMINATION OF SEAFOOD

Cholera epidemics, described in Chapter 1 of this book, constitute only one example of the serious human diseases with microbial etiology that can be transmitted by contaminated seafood. Infection may be acquired by eating raw or improperly processed shellfish and fish that have ingested and accumulated (or have had their flesh or external surfaces contaminated by) microorganisms infective to humans. Included here would be microbial pathogens that cause typhoid fever, hepatitis, and several types of gasteroenteritis.

As we dump more and more untreated or inadequately treated domestic sewage into rivers, estuaries, and coastal waters, the populations in those waters of microorganisms of human origin — bacteria and viruses in particular — will be increased. Dilution occurs as a result of river outflow, tidal flushing, and inshore currents, but this may not take place fast enough to remove the risk of infection soon enough. Many bacteria that cause human disease neither reproduce nor survive very long in more saline ocean waters. However, they may not be killed instantaneously and so can constitute a threat to human health. Of particular concern are the microorganisms that cause cholera, typhoid, dysentery, skin infections, hepatitis, botulism, and eye and ear infections. Disease-causing viruses and bacteria of human origin, present in domestic sewage, may persist for days, weeks, or months in the intestines of fish, on the body surfaces or gills of fish and shellfish, and within the digestive tracts of shellfish, or on their gills, as well as in bottom sediments. Swimmers, skin divers, and fishermen obviously expose themselves to infection by venturing too close to ocean outfalls, sludge dumpsites, or badly degraded estuarine waters. Frequently, though, pollutants may be carried for miles by currents, so that it is difficult to determine which waters are safe and which are not, except by more or less continuous monitoring.

An added element of danger results from handling or eating uncooked fish and shellfish from polluted areas. Marine animals can and do ingest contaminated material, and certain shellfish may accumulate viruses and bacteria. Public health problems related to microbial contamination can be a major deterrent to full utilization of coastal resource species. Diseases such as typhoid and hepatitis have been transmitted by ingestion of raw shellfish from polluted waters (Mason & McLean 1962); hepatitis is an especially persistent problem.

Viral Diseases of Humans Transmitted by Shellfish

Half a century of epidemiological studies have indicated a causal relationship between viral hepatitis and consumption of raw, fecally contaminated molluscan shellfish (Ross 1956). However, of the total number of cases of infectious hepatitis reported annually from all causes, the percentage transmitted by consumption of raw contaminated shellfish is a small, persistent, and controllable segment (Liu et al. 1966). Despite the availability of information about disease risks, each year brings new reports of hepatitis outbreaks that can be traced to consumption of raw shellfish. As an early example, outbreaks of hepatitis A affecting almost 300 people, traced to eating raw oysters, occurred in Texas and Georgia in 1973 (Hughes 1979). The oysters were from Louisiana growing areas approved for harvesting under guidelines of the National Shellfish Sanitation Program. The source of contamination seemed to be floodwaters that occurred several months earlier (Portnoy et al. 1975). During the period 1961 to 1990, some 1400 cases of oyster- and clam-associated hepatitis A were documented in the United States (NOAA 1991).

Until 1974, all outbreaks of hepatitis associated with raw shellfish were thought to be caused by hepatitis A virus. In that year, hepatitis B virus was reported in repeated samples of clams (*Mya arenaria*) from one location on the Maine coast (Mahoney et al. 1974). The site (one of 24 closed shellfish areas sampled) received untreated sewage from a coastal hospital in which two individuals with type B hepatitis were patients during the 3 months preceding the study. Transmission of the pathogen to previously unexposed clams in closed aquaria was achieved experimentally. The investigators concluded that clams must be considered potential vectors for hepatitis and that under special circumstances they could serve as reservoirs for type B hepatitis virus as well as type A.

Viruses have been found experimentally to have variable, but in some instances surprisingly long survival time in saline waters — often under what would appear to be adverse conditions (Metcalf & Stiles 1966). Rates of inactivation of enteric viruses in seawater increase with increasing temperature. For example, in one study (Gerba & Schaiberger 1975a, 1975b) 90% of poliovirus 2 was inactivated in sterile seawater in 48 d at 1°C, whereas 99.9% was inactivated in 30 at 40 d at 22°C. The virus survived four times longer in filter-sterilized seawater than in natural seawater, indicating that microorganisms in seawater (or their metabolites) are factors responsible for inactivation of the viruses. Important survival factors for viruses in seawater seem to be aggregation and adsorption onto particulates (Schaiberger, Gerba, & Estevez 1976).

There is much research interest in procedures to inactivate or remove viruses from sewage treatment wastewater and sludge. Methods are mechanical, biological, and chemical, but none seems to be completely effective, and the number of complicating factors (for example, temperature, pH, particle size, electrical charge, flocculation, organic content) is daunting (Cooper 1975).

During the past 3 decades, outbreaks of shellfish-associated viral diseases not only have continued, but they seem to have intensified. Hepatitis and acute gastroenteritis have been dominant problems, with noroviruses and rotaviruses mentioned most frequently as being involved in gastroenteritis outbreaks (Richards 1985, 1987) — so that by the end of the century, Norwalk-like viruses (NLVs; now called noroviruses) were considered a leading cause of foodborne illnesses in the United States (Mead et al. 1999). Most adults have antibodies for that group of viruses, suggesting widespread exposure of the U.S. population. Earlier, Norwalk virus was determined to be the cause of a widespread acute gastroenteritis epidemic in Australia. More than 2000 people became ill in 1978, presumably after eating oysters (*Crassostrea commercialis*) from the Georges River estuary in New South Wales (Murphy et al. 1979).

Occurrences and outbreaks of liver disease caused by another foodborne viral group, the hepatitis A viruses (HAVs), have been described by Halliday et al. (1991); Tang et al. (1991); and Kingsley, Meade, & Richards (2002). An estimated 80,000 cases of hepatitis A occur each year in the United States, according to Mead et al. (1999), and, even more significantly, an epidemic of hepatitis A, with an estimated 290,000 cases (about 5% of the city's population), occurred in Shanghai, China, in 1988. The cause was attributed to consumption of raw contaminated clams.

Imported raw Manila clams (*Ruditapes philliparum*) from China were fingered as culprits in a recent (2000) small outbreak of gastroenteritis in Cortland Manor, New York (Kingsley, Meade, & Richards 2002). Hepatitis A viruses were detected by reverse transcription polymerase chain reaction methodology, as were noroviruses. The clams obviously came from a highly polluted source in China, for the fecal coliform level averaged a most probable number (MPN) of 93,000/100 µg meats — which is about 300 times higher than the U.S. standard for shellfish meats. In this instance of gross contamination, the fecal coliform standard alone would have resulted in rejection of the clams for human consumption, but, as pointed out by Kingsley and colleagues, even

> Low fecal coliform levels in shellfish do not always indicate that the shellfish are free of viral contamination, since viruses may persist within shellfish for relatively long periods after bacterial levels have been reduced in surrounding waters. (p. 3917)

Viruses of human fecal origin in coastal waters and in shellfish have been examined with ever greater intensity during the past half century, and, with the recent availability of PCR tests for their environmental occurrence, knowledge about distribution and abundance has increased significantly. Some relevant characteristics of these viruses of enteric origin are presented in Table 12.1. The global prevalence of shellfish-associated viral gastroenteritis was addressed by Le Guyader et al. (1994) as follows:

> One of the most important consequences of the contamination of coastal areas is the concentration of viruses by shellfish through filter feeding. Standards based on coliform bacteria and established to protect shellfish consumers are known not to be correlated with the presence of viruses, and little about viral depuration is known. Outbreaks of shellfish-transmitted viral disease occur periodically, causing problems for public health and resulting in economic losses for the seafood industry.

> The development of molecular technology has provided sensitive, specific, and rapid tools for viral detection, and the applicability of these methods to environmental samples is beginning to be demonstrated.

TABLE 12.1
Viral Groups of Human Enteric Origin in Coastal Waters and Shellfish

Viral Group	Defining Features
Noroviruses	Noroviruses have been divided into two distinct genogroups, both with broad genetic diversity: Norwalk virus type, and Snow Mountain virus type.
	Noroviruses cause acute gastroenteritis. Globally, up to 42% of gastroenteritis cases are estimated to be caused by noroviruses. In Japan, in 2001, noroviruses accounted for 28% of all food poisoning cases and 99% of purely viral cases.
	Water and foodborne transmissions can occur, but large epidemics have resulted from consumption of contaminated molluscan shellfish.
	No conventional cell culture method has been developed for propagation of noroviruses; detection now depends on reverse transcription polymerase chain reaction methods (RT-PCR), enzyme-linked immunosorbent assays (ELISA); and electron microscopy (EM).
Enteroviruses	Enteroviruses are important environmental contaminants of fecal origin; the group includes polioviruses, cocksakievirus groups A and B, and echoviruses (Gantzer et al. 1998).
	RT-PCR techniques have been developed for detection of the enterovirus genome (Kopecka et al. 1993), but cell culture is the method of choice to determine the infectious nature of specific viral isolates.
Adenoviruses	Many types of adenoviruses exist, of which Adenovirus type 2 (prototype) and type 12 (prototype-like) are the most common enteric viruses in coastal/estuarine waters.
	Adenoviruses are difficult to isolate in cell culture.
	Adenoviruses and hepatitis A viruses (among the enteroviruses) are relatively stable in seawater.
	In a recent comparative study in Spain (Pina et al. 1998), human adenoviruses were the viruses most frequently detected throughout the year, and all samples that were positive for enteroviruses or hepatitis A viruses were also positive for human adenoviruses.
	It has been suggested that the detection of adenoviruses by PCR could be used as an index of the presence of human viruses in the environment, where a molecular index is acceptable [that is, where verification of the infectiousness of the isolate is not required] (Pina et al. 1998).
Hepatitis A viruses (HAVs)	This group includes three genotypes: Genotype I contains about 80% of all HAV isolates, Genotype II is rare, and Genotype III contains almost 20% of all human isolates (Robertson et al. 1992).
	Hepatitis A viruses, like noroviruses and rotaviruses, grow poorly or not at all in cell culture. Use of molecular methods such as PCR, which do not require cell cultivation, for detection of viruses in environmental samples has enhanced understanding of distribution and abundance.
	Hepatitis A viruses and noroviruses share the questionable distinction of being the causes of viral illnesses most frequently associated with shellfish consumption in Europe and United States. An estimated 1.4 million cases of HAV-mediated illnesses occur annually worldwide, with about 85,000 cases annually in the United States alone (Kingsley, Meade, & Richards 2002).

TABLE 12.1 (Continued)
Viral Groups of Human Enteric Origin in Coastal Waters and Shellfish

Viral Group	Defining Features
Rotaviruses	Group A rotaviruses have 14 serotypes (serotyping is viral classification based on neutralization of viral infectivity). Of these serotypes, type 1 is most prevalent throughout the world, followed by types 3 and 2 (Woods et al. 1992). Assays of environmental samples with RT-PCR have been developed and applied to detection of types 1 to 4 Group A rotaviruses in sewage samples (Gajardo et al. 1995). Human rotaviruses (HRVs) are a principal cause of viral gastroenteritis in children (Cubitt 1991). On the basis of recent research, Gajardo et al. (1995) reached the following conclusions: "Although serotyping is a classification based on neutralization of virus infectivity, the available information on gene 9 sequences of rotavirus strains allows the prediction of the serotype of a given strain by PCR with type-specific primers. This powerful technique could permit the acquisition of actual epidemiological data on the prevalent rotavirus serotypes in the environment and at the same time provide information on the occurrence of asymptomatic rotavirus infections in the community" (p. 3462).

Viruses affecting humans, then, constitute a critical problem for fishing or aquaculture operations in coastal/estuarine areas where even marginal domestic pollution exists — and because of non–point source runoff, this includes most of the areas now used or planned for use in marine aquaculture. Additionally, viral contamination is and will be an important issue where treated sludges or other fecal degradation products are used for enrichment of growing areas until large-scale, inexpensive techniques are available that will ensure total viral destruction. Shellfish purification (depuration) procedures must also take viral survival into account.

BACTERIAL DISEASES OF HUMANS TRANSMITTED BY FISH AND SHELLFISH

Although viruses constitute a definite public health problem in utilizing inshore species as food, pathogenic bacteria also form a continuing threat when raw or partially processed products are consumed by humans. Much attention has been paid during the past 40 yr to the role of the vibrios, *Vibrio parahaemolyticus* and *Vibrio cholerae*, in outbreaks of gastroenteritis and cholera, respectively, that have been associated with consumption of raw or improperly processed seafood. Although the vibrios are normal constituents of the inshore flora, their abundance may be increased facultatively by organic enrichment of coastal and estuarine areas, marine animals may carry or be infected by members of the genus, and seafood may be contaminated by improper handling. Most marine bacteria are not harmful to humans, but some of the vibrios can cause acute digestive disturbances, particularly when fish and shellfish carrying those bacteria are consumed raw or undercooked. One species in particular, *Vibrio vulnificus*, can also cause fatal wound infections.

Vibrio parahaemolyticus

Beginning in the 1950s, summer bacterial gastroenteritis outbreaks in Japan have been traced to human ingestion of raw marine fish and invertebrates (Iida et al. 1957). The largest outbreak, affecting 20,000 people, occurred in Niigata Prefecture in 1955 and was traced to eating cuttlefish from the Sea of Japan. Examples of the involvement of marine products in gastroenteritis outbreaks can be seen often in the statistics of the Japanese Ministry of Health and Welfare. The causative organism in many outbreaks was identified as the halophilic bacterium *Vibrio parahaemolyticus*. Numerous pathogenic and nonpathogenic strains have been isolated from coastal seawater, plankton organisms, bottom mud, and the body surfaces and intestines of marine fish and shellfish. Many strains have been recognized, and an extensive body of Japanese literature on *V. parahaemolyticus* has accumulated.

The Oriental custom of eating raw fish and shellfish (i.e., sushi and sashimi) has undoubtedly contributed to the severity of the vibrio problem there; 70% of all reported gastroenteritis outbreaks have been associated with *V. parahaemolyticus*. The organism was first recognized in Japan in 1951 as the cause of "shirasu food poisoning" (Fujino et al. 1953). During the 20 yr after recognition of the problem (1951 to 1971), more than 1200 technical papers on *V. parahaemolyticus* as well as several books were published. The natural habitat of the organism seems to be in estuaries rather than in the open sea. The infective dose for humans is 1 million to 1 billion organisms. *Vibrio parahaemolyticus* has a short generation time (9 to 11 min) — twice as fast as the common fecal bacterium *Escherichia coli* (at about 20 min) — which means that infective dose levels can be reached from an original population of only *10 organisms* in 3 to 4 h — a remarkably short time.

An important observation that emerged from investigations conducted during the 1970s is that *V. parahaemolyticus* could cause outbreaks even when fish and shellfish were cooked. Improper processing procedures — undercooking; use of raw seawater to wash work surfaces; allowing raw seafood to drain onto cooked products; or placing cooked seafood on surfaces where raw marine animals have been shucked, cleaned, or sliced — can lead to ingestion of the pathogens by humans, with resultant gastrointestinal infections (Colwell et al. 1975).

In addition to gastrointestinal disturbances, there have been earlier reports of injury-induced tissue infections caused by marine vibrios, including *V. parahaemolyticus*. Case histories of such marine vibrio-related infections — some of them fatal and some requiring amputation — have been described in the literature before 1980 (Craun 1975), and other lesser cases, in which *V. parahaemolyticus* was isolated from infected wounds, have also been reported (Poores & Fuchs 1975). Questions arose as to whether *V. parahaemolyticus* isolated from localized tissue infections acquired from coastal/estuarine waters were enteric pathogens with an altered route of entry, or whether they were "nonpathogenic" vibrios with previously unsuspected virulence. One extensive study indicated that isolates from wound infections were clearly similar to enteric forms isolated from cases of gastrointestinal illnesses in Japan and were unlike isolates from estuarine waters (Twedt, Spaulding, & Hall 1969). (A more recent question about these earlier reports of wound infections is whether the pathogens were actually *V. parahaemolyticus* or members of a species unrecognized before 1976, *Vibrio vulnificus*, to be discussed later.)

Vibrio parahaemolyticus has been described as a leading cause of seafood-associated bacterial enteritis in the United States and a major cause of foodborne illness in the world (Joseph, Colwell, & Kaper 1983; Mead et al. 1999; DePaola et al. 2003b). The species has been subdivided into a number of strains or serotypes. A virulent clone of Serotype 03:K6 emerged in India in 1996 and spread quickly throughout Asia. The new clone caused large outbreaks with a high attack rate (Matsumoto et al. 2000). Recent research has indicated that pathogenic strains of *V. parahaemolyticus* generally produce a thermostable direct hemolysin (TDH) — a virulence factor coded for by the gene labeled (tdh) (DePaola et al. 1990, Honda & Iida 1993). Japanese investigators, whose professional predecessors had called attention almost half a century earlier to the role of *V. parahaemolyticus* in summer gastroenteritis outbreaks, have recently published the genome of the organism.

Vibrio parahaemolyticus has enjoyed a resurgence of research interest in the United States in the late 1990s as a consequence of outbreaks in the states of Washington, Texas, and New York. The first outbreak, with more than 200 confirmed oyster-associated cases, occurred in Washington in 1997, followed in 1998 by an outbreak of more than 400 cases linked to consumption of raw oysters from Galveston, Texas, and, also in 1998, much smaller outbreaks (43 cases in Washington and 8 cases associated with shellfish from Oyster Bay, Long Island, New York; DePaola et al. 2000). The 03:K6 strain was the causative agent, and concern has been expressed about the apparent increase in *V. parahaemolyticus* infections from consumption of shellfish.

Vibrio cholerae

To return to our slightly frayed historical thread, research and publication on *V. parahaemolyticus* by marine microbiologists were diverted in the late 1970s and the early 1980s to a surge of research activity with *Vibrio cholerae* from coastal/estuarine sources (DePaola 1981). Microorganisms with characteristics of *V. cholerae* were isolated from many estuaries in many countries (see, for example, Kaysner et al. 1987). Extensive studies by the noted marine microbiologist Dr. Rita Colwell and her associates led to the conclusion that *V. cholerae* is a normal component of the flora of brackish waters, estuaries, and salt marshes of the temperate zone (Colwell et al. 1981). Other conclusions were that *V. cholerae* can occur in the absence of fecal contamination and that outbreaks can be expected in humans when proper food-handling techniques are not used. Sporadic outbreaks have occurred in a number of temperate zone countries — in Italy in 1973 and 1980, in Portugal in 1974, and in the United States (Louisiana) in 1978 (this was the first reported outbreak in the United States since 1911). Contaminated shellfish were implicated in each outbreak — mussels in Italy, cockles in Portugal, and crabs in Louisiana. Whereas *V. cholerae* may be a normal part of the brackish water microflora, its potential for causing human disease seems to be enhanced in heavily polluted shellfish-growing areas, especially if raw or improperly processed products are consumed or if confirmed cases of cholera have been reported in the adjacent towns.

As seen with certain other pathogens, whenever even one case of cholera occurs in a local human population, the danger of shellfish contamination will exist in surrounding waters. As an example, a study in Portugal (Ferreira & Cachola 1975) disclosed the presence of *V. cholerae* in 38% of 166 samples of molluscan shellfish taken in 1974 from the vicinity of Tavira, where a single case of cholera had been reported. This report was a sequel to an earlier paper (Cachola & Nunes 1974) pointing out extensive pollution of shellfish growing areas on the southern (Algarve) coast of Portugal. The *New York Times* (November 2, 1975) reported more than 200 cases of cholera with three deaths in Coimbra, Portugal. Health authorities attributed the outbreaks to contaminated cockles from the Mondego River estuary.

The recent history of cholera in the United States needs to be placed in the perspective of a longer time span. During much of the 20th century, cholera was not reported in the United States — until 1973, when a case was reported from Texas. Since then, sporadic small outbreaks have occurred: 11 cases in Louisiana in 1978, 2 cases in Texas in 1981, 17 cases on a Texas oil rig in 1982, and 13 cases in Louisiana and Florida in 1986. Most of the cases were associated with eating contaminated shellfish.

The most recent major outbreak of cholera in the Western Hemisphere — as discussed briefly in Chapter 1 — began in 1991 in the port city of Chimbote, Peru. The first case, caused by a virulent Asian strain of the vibrio, was diagnosed in January. Early spread of the disease was attributed to eating fecally contaminated uncooked fish and shellfish in a popular dish called ceviche. Further spread was aided by ingestion of fecally contaminated drinking water as well as food, including raw vegetables. A little over a year later (March 1992), more than 3000 Peruvians had died from the disease, and the epidemic had spread and continued to spread erratically through much of Central and South America. In early 1993, Brazil had become one of the foci of the epidemic, with 32,313 cases and 389 deaths reported, principally along the Atlantic coast of that country. By the end of that year, the grand total of cholera cases in Latin America and the Caribbean had reached 700,000, with an estimated 6400 deaths.

Cases were reported in Mexican cities near the U.S. border, and isolates of a *V. cholerae* strain identical to that found in Peru were recovered from oyster reefs in Alabama as early as September 1991, resulting in closure of the beds. The source of the pathogens was not determined, but human carriers from South America were suspected. In another incident, 65 (of 336) passengers on an Argentine airplane bound for Los Angeles were stricken with cholera in February 1992. One person died, and the outbreak was blamed (arguably) on eating contaminated seafood salad brought on board during a stop in Lima, Peru. Other isolated cases in the United States (totaling 24) have been associated with the South American epidemic — mostly travelers who ate contaminated seafood while in Central or South America or family members who ate contaminated seafood transported home by the travelers.

The likelihood of a major cholera outbreak in the United States is considered to be slight, because the disease is associated with primitive hygienic conditions not often found in this country. One exception might be among inhabitants of poorer districts along the Mexican border, who lack public water or sewage disposal systems.

So much then for the anguish and death caused by the most notorious of the vibrios, *V. cholerae.*

Vibrio vulnificus

The most recent vibrio on the scene is one that can be severely pathogenic to some humans — *V. vulnificus*. The species was first recognized as a human pathogen in 1976, and the taxonomic group now includes some organisms formerly identified as *V. parahaemolyticus* (Hollis et al. 1976, Farmer 1980). Gastroenteritis, and, in some cases, primary septicemias caused by *V. vulnificus* may result from ingestion of contaminated raw oysters or clams; wound infections with *V. vulnificus* result from contact with marine animals (lacerations from barnacles, shark bites) or exposure of pre-existing wounds to seawater containing the pathogens (Oliver 1981). Septicemias can cause death (61% of patients), especially among individuals with pre-existing liver damage or immunodeficiencies, in a matter of hours or days (Blake, Weaver, & Hollis 1980). Wound infections have a lower death rate (22%) but sometimes require amputations.

Three subgroups (biotypes) of the species *V. vulnificus* have been described: Biotype 1, associated with human infections; Biotype 2, pathogenic for eels and an opportunistic pathogen of humans; and Biotype 3, recently described by Bisharat et al. (1999) and isolated thus far only in Israel as a cause of infections in humans who handled *Tilapia* spp. (all infections were acquired while cleaning fish). Biotypes 1 and 2 have been isolated from humans, shellfish, sediments, and seawater; Biotype 2 was originally isolated from eels (Biosca et al. 1991) and subsequently from humans (Dalsgaard et al. 1996).

Vibrio vulnificus has been the object of substantial research effort, especially during the 1980s and 1990s — probably because of its virulence in human infections, acquired by consumption of shellfish or contamination of wounds. Its pathogenicity for eels is, of course, a concern for Scandinavia and other countries that culture the animals as food (Høi et al. 1998b).

Relevant findings from recent research include these:

- *Vibrio vulnificus* has been found on all coasts of the United States, as well as in coastal waters of Europe, Asia, Africa, and South America (Oliver 1989).
- Whereas *V. vulnificus* is most prominently implicated in warmer environments (such as the Gulf of Mexico) as a pathogen acquired by eating raw shellfish, its major route of entry in colder climates (Denmark, for example) seems to be through wound infections in summer. Fishermen and fish processors with lesions on their hands contributed most of the human cases in one Scandinavian study (Høi et al. 1998b).
- *Vibrio vulnificus* is the leading cause of death in the United States associated with consumption of seafood, and consumption of raw Gulf coast oysters from April to November of each year is responsible for nearly all the cases (Shapiro et al. 1998, Oliver & Kaper 2001, DePaola et al. 2003a).
- Water temperature is an important determinant of risk of human infections by *V. vulnificus*, with 15°C a critical point (Kelly 1982, Høi et al. 1998a). Isolation of the organism is most prevalent when water temperatures exceed that point. Existence of a viable but nonculturable state has been demonstrated.

- Correlations between occurrences of *V. vulnificus* and coliform bacteria in seawater have been reported by some investigators (Tamplin et al. 1982; Oliver, Warner, & Cleland 1983; Høi et al. 1998b) but were not found in other studies (Koh, Huyn, & LaRock 1994; Pfeffer, Hite, & Oliver 2003).
- Infections of humans by *V. vulnificus* are predominantly found in males (82%), possibly because estrogen promotes a protective response against induced toxic shock (Merkel et al. 2001).
- One of the principal virulence factors in *V. vulnificus* infections is the presence of a polysaccharide capsule; encapsulated cells are highly virulent, with a 50% lethal dose of <10 colony forming units (CFU; Strom & Paranjpye 2000; Pfeffer, Hite, & Oliver 2003).

A concise description of infection risks from *V. vulnificus* has been published recently by Danish scientists:

Shellfish are often implicated in the transmission of *V. vulnificus* infections in the United States, especially in states bordering the Gulf of Mexico. Concentrations of *V. vulnificus* in raw oysters from this region are reported to be as high as 10^3 to 10^6 organisms per g of oyster during the summer, when more than 90% of raw oyster-associated *V. vulnificus* infections, mainly septicemia, occur. Wound infections due to occupational activities around seawater have been reported to show a similar seasonal pattern, with the highest number of cases occurring from April to October. In Denmark, infections due to *V. vulnificus*, mainly wound infections, occurred only in warm summers. To date, no *V. vulnificus* infections have been associated with consumption of raw shellfish in Denmark or elsewhere in Europe. (Høi et al. 1998a, p. 12)

In summary, it is important to emphasize that vibrios, including *V. parahaemolyticus*, *V. cholerae*, and *V. vulnificus*, are present in and on shellfish and in seawater, not as contaminants but as part of the normal microflora. The abundance of these organisms, however, may be enhanced by organic loading of coastal/estuarine waters from human sources or by augmentation, via sewage contamination, with pathogens from infected individuals.

Because some misguided humans (fortunately in diminishing numbers) persist in eating raw bivalve molluscs — especially oysters — outbreaks of seafood-borne gastrointestinal disease are grimly summarized every year in the aptly named "Morbidity and Mortality Report" of the federal CDC in Atlanta. Recent epidemics picked up from a CDC report by the news services in late January 1995 were of acute gastroenteritis in more than 100 people who ate sewage-contaminated raw oysters from Apalachicola Bay, Florida, and Galveston Bay, Texas, during and after the Christmas holiday period.

OTHER MICROBIAL DISEASES OF HUMANS THAT MAY HAVE SOME ASSOCIATION WITH MARINE POLLUTION

Other bacterial genera, such as *Clostridium*, *Salmonella*, and *Shigella*, that are more directly pollution related should not be ignored in this discussion, because a single outbreak of disease related to any marine species can have a drastic impact on markets for *all* marine products.

Most studies of the relationship of fish to *Salmonella* infections in humans conclude that fish can serve as passive vectors of waterborne pathogens and that the bacteria disappear from body surfaces and gut when the fish leave contaminated areas. An investigation in 1970 found that *Salmonella paratyphi A* survived for 2 weeks in filtered sterilized estuarine water from Chesapeake Bay and for 2 months in filtered sterilized seawater from the Delaware coast (Janssen 1970). Such a survival time, whether in sediments, in the water column, or in or on fish, could provide a passive mechanism for possible infection of humans, even without active infection of the fish.

Experimental infections with *Salmonella typhimurium* were obtained in mullet (*Mugil cephalus*) and pompano (*Trachinotus carolinus*) by 2-h exposure to 10^7 cells/ml in static aquarium systems (Lewis 1975). Infections, in the form of hemorrhagic areas of the intestine from which pure cultures of *S. typhimurium* were recovered, were seen 10 to 14 d after exposure in some of the experimental fish. The organism was recovered from the alimentary tracts of the two fish species up to 30 d after exposure. It may be important to note, though, that the original isolates of *S. typhimurium* on which the experimental exposures were based were from the digestive tract of a mullet and not from an active mammalian infection or type culture collection. These results with *Salmonella* species indicate that fish can harbor the pathogens for appreciable periods after exposure and that at least some exposed animals may actually become infected.

A government publication summarizing information on seafood-poisoning microorganisms listed nine bacterial genera as having been isolated from raw or processed seafood and, in some instances, having caused human disease (Cockey & Chai 1988). Present as contaminants in raw fish and shellfish, or as contaminants introduced during processing, were representatives of the genera *Vibrio, Salmonella, Shigella, Staphylococcus, Clostridium, Yersinia, Listeria, Campylobacter*, and *Escherichia* (*E. coli*). Of these, the vibrios are clearly the most significant from a public health perspective, with three species — *Vibrio parahaemolyticus, V. cholerae,* and *V. vulnificus* definitely implicated as human pathogens acquired from consumption of raw or inadequately cooked seafood. The other genera are contaminants introduced during processing and can be acquired from other kinds of animal products as well as from seafood.

Outbreaks of shellfish-associated typhoid fever in the United States had diminished by the 1960s to be replaced by outbreaks of shellfish-associated viral diseases — hepatitis A and nonspecific gastroenteritis in particular. Norwalk and rotaviruses have been mentioned most frequently as being involved in gastroenteritis outbreaks (Richards 1985, 1987).

ILLNESSES CAUSED BY CHEMICAL CONTAMINATION OF SEAFOOD

We humans have a remarkably stupid approach to the use of living marine resources: we either kill too many individuals for stocks to be maintained, or we poison their habitats so that the flesh of survivors is inedible anyway. Some naturally occurring chemicals (an example would be mercury in large predators such as swordfish) can reach toxic levels in fish because of biomagnification as marine food chains are ascended. But humans have often added to existing natural levels by industrial contamination or have invented new toxic chemicals (such as PCBs and pesticides) that have become widely dispersed and bioaccumulated in marine animals.

We have created, in our irresponsible industrial practices, especially in coastal species such as striped bass, subpopulations that have to be described collectively as "those incredible inedible fish." Consumers can no longer be sure that *any* fish, regardless of its geographic origin, will not provide them with a ration of assorted toxicants along with the sought-for (and increasingly costly) fish protein. This uncertainty is part of the price paid for past and continuing use of estuarine/coastal waters as convenient dumps for wastes from inefficient technological processes. Uneasiness about eating seafood is and should be especially prevalent among pregnant women and mothers of small children — and to some extent among all of us. When we face the stunning fact that 46 of the 50 states have published advisories or bans limiting or prohibiting fish or shellfish consumption (although usually for selected species and locations), we can begin to appreciate, at least dimly, the problems we have generated for ourselves.

Of course, many of these warnings and advisories are against consumption of contaminated *freshwater* fish — mostly sportfish — where problems have been demonstrated to be real. In the Great Lakes, for example, some advisories date back to the 1970s and are concerned with long-lasting pollutants such as DDT and PCBs. Associations have been found in a number of studies between consumption (by women) of sportfish from heavily polluted Lake Michigan and abnormal reproductive and developmental effects in offspring. Regulatory actions have reduced levels of contamination, but risks to human health still exist.

Problems also occur in coastal waters, usually near industrial or municipal outfalls. A study published in 1994 of recreationally caught fish and shellfish from the southern California coast (Cross 1994) disclosed persistently high contaminant levels in organisms taken near known outfalls when compared with data collected 2 decades earlier, despite regulatory efforts in the intervening years. Chlorinated hydrocarbons and metals were foci of the investigations. Highest levels of DDT and PCBs were recorded in a popular sportfish, the white croaker, and decreases in average tissue levels of contaminants since the early 1970s were variable; some pollutants were higher than in previous samples from the same stations.

The advisories limiting consumption, the occasional reports of human illnesses from eating raw shellfish, and the rare admonitions about mercury in larger ocean-caught fish all have effects on perceptive seafood consumers. One such effect is greater resistance to buying seafood of any kind; another is limiting selection to a few species that are trawled in offshore waters or reared in aquaculture. Such practices indicate the distressing reality — that many consumers distrust the safety level of food from the sea. They do not have acceptable access to information about amounts of contaminants that they are eating or about the effects of various quantities of those polluting chemicals on their health. They know that inshore habitats for fish are to some degree polluted, especially near cities or industrial facilities. They read reports of the inadequacies of government seafood inspection, especially insofar as any chemical analysis is concerned. They may then reach the conclusion that the special flavor and texture of seafood is not worth the risk to them or to their families. They slump back to chicken and red meat, which offer at least the facade (but certainly not the actuality) of fewer safety problems.

Three general categories of industrial chemical pollutants command most of the attention in human health matters related to coastal waters:

1. PCBs, pesticides, and related chlorinated hydrocarbons
2. Metals
3. Carcinogens

Representatives of the first two classes of pollutants — chlorinated hydrocarbons and metals — have wide-ranging effects on human health, not the least of which is their activity as carcinogens, which will be considered briefly in this section. Additionally, natural biotoxins, whose origins and intensities may be affected by levels of anthropogenic nutrients, are of increasing importance in the scientific literature and have already been discussed at length in Chapter 5.

PCBs and Related Chlorinated Hydrocarbons as Pollutants

Since their initial use in the 1930s, PCBs have permeated waters of all the world's oceans, principally by riverine transport to coastal waters in association with sediment particles and by airborne transport, followed by deposition and subsequent movement through aquatic food chains. Effects of PCBs on humans were demonstrated in Japan in 1968 and in Taiwan in 1979 by unfortunate outbreaks of accidental poisonings through contamination of cooking oil by PCBs and polychlorinated dibenzofurans (PCDFs). More direct demonstrations of effects on human health were disclosed by studies in the Great Lakes states of the effects on offspring of a maternal diet that included significant intake of PCB-contaminated fish. These studies were considered at length in Chapter 3; they included maternal health effects as well as effects on offspring.

Possible toxic effects of chlorinated organic pesticides in food have been discussed for at least the past 3 decades, but some conclusions are still controversial. Even though indiscriminate use of persistent pesticides is coming under some measure of control in the United States and in some other industrialized countries, their use in other parts of the world is expanding, and contamination of the world's oceans is continuing and may even be increasing. Because of their persistence in the environment and their accumulation by successive levels of food chains, pesticides continue to be threats in nearshore ocean areas, including those devoted to marine aquaculture and those important as nursery areas for fish and shellfish. The sublethal effects of long-term exposure to low levels of pesticides in the diets of most marine animals are incompletely understood.

Another aspect of chemical pollution of seafood that has not been fully appreciated until recently is the possible long-term effect on humans of consumption of low levels of contaminants in food. Some contaminants are readily metabolized and excreted; others may accumulate in storage tissues as a result of continued ingestion. Certain of the heavy metals may accumulate, if the ingestion rate exceeds rates of detoxification and excretion. Several of the fat-soluble contaminants — especially the chlorinated hydrocarbons — can build up in humans as well as other animals if the diet provides continuing low-level dosages.

Extensive studies have been made of the effects of chlorinated hydrocarbons — especially PCBs — on reproduction in humans. Beginning in the early 1990s, many pollutant chemicals have received new attention as potential "endocrine disrupters," principally because of their ability to mimic or block activity of hormones such as estrogens or testosterone. Exposure to estrogens or estrogen mimics during early sexual differentiation can induce abnormalities in duct development and intersexuality, whereas exposure during sexual maturation may inhibit gonadal growth and development (Jobling et al. 1996, 1998).

METALS AS POLLUTANTS

Fish and shellfish may accumulate dangerously high levels of pesticides, heavy metals, and other potentially toxic chemicals in grossly polluted waters. Of the chemicals that could occur in seafood at levels harmful to humans, mercury has justly received the most attention. The horrors of Minamata disease, caused by mercury contamination of fish and cultivated shellfish in a bay in southern Japan and described briefly in Chapter 2 of this book, were publicized over 30 yr ago on television, in news magazines, and in several books (Harada 1972, Huddle & Reich 1975, Smith & Smith 1975). Severe permanent neural damage characterized those individuals most seriously afflicted. Partly as a consequence of this mass poisoning, increased surveillance of mercury and other heavy metals in all kinds of seafood lessens the likelihood of another Minamata incident, although whenever marine products are grown near industrial operations there is always a risk of chemical contamination through negligence, deliberate dumping, or accidental spills. It is not feasible to provide adequate continuous chemical surveillance of every localized area where fish and shellfish are produced — especially because some of the analytical methods are very time-consuming and costly, and the toxic action levels for some contaminants are not fully understood.

Although mercury has achieved the most notoriety among heavy metal contaminants of food, public health problems have also been created by toxic levels of other metals. Ingestion of cadmium-contaminated water resulted in a disease called itai-itai in Japan during the decade after World War II (Kobayashi 1969). The problem developed from the use of cadmium-contaminated river water in two towns near metal mines; the contaminant caused severe disturbance of calcium metabolism, characterized by neurologic symptoms and extreme skeletal fragility.

A detailed evaluation of potentially harmful metals in fish and other seafood was made recently by an international joint "Group of Experts on the Scientific Aspects of Marine Pollution" (GESAMP). This prestigious U.N.-sponsored group reached a number of major conclusions about an array of metals (GESAMP 1985a, 1985b, 1986a, 1986b, 1986c; Friberg 1988):

Mercury — Populations with high fish intake or intake of fish with a high methylmercury content can easily exceed the World Health Organization/Food and

Agriculture Organization (WHO/FAO) provisional tolerable intake level. Pregnant women constitute a special risk group.

Cadmium — Only under exceptional circumstances will cadmium intake from fish constitute an important part of the total daily intake via food. High consumption of certain shellfish may increase considerably the intake, and, over many years, may increase cadmium concentrations in the kidney to toxic levels.

Arsenic — Exposure to arsenic via seafood may be substantial. Most of this arsenic is in the form of arsenobetaine, which is considered relatively nontoxic. Extreme seafood consumption may give rise to an intake of several hundred micrograms of inorganic arsenic per day, an exposure level which over a lifetime may be related to a significant increase in skin cancer.

Lead — Lead in seafood does not greatly contribute to the daily intake of lead, but other sources of lead (such as paint) will be additive.

Tin — The contribution of seafood to the daily intake of tin is low. However, more data are needed for trimethyltin, which is synthesized by marine organisms and which may produce neural pathology in humans.

Selenium — Selenium does not pose a toxicological problem, but its interaction with mercury compounds may be biochemically significant.

The GESAMP report on metals in seafood (as summarized by Friberg 1988) indicated greatest current concern with arsenic and mercury.

The effects of arsenic were summarized as follows:

Seafood is the predominant source of human arsenic intake. From the toxicological point of view, there are two forms of arsenic in marine organisms which should be considered, namely arsenobetaine, which is the dominant form in most seafood, and inorganic arsenic, which constitutes 2 to 10% of the total arsenic content in seafood. Inorganic arsenic is by far the most toxic form and has given rise to skin lesions, such as hyperkeratosis, hyperpigmentation and skin cancer, peripheral blood vessel pathologies, effects on the central nervous system, and chromosome damage. In cases of extreme consumption of seafood, the intake of inorganic arsenic would reach levels at which the increased risk for skin cancer is definitely no longer negligible. (Friberg 1988, p. 383)

Statements from the summary about the effects of mercury are:

Mercury, in the form of methylmercury (MeHg), is still considered a prime pollutant in fish, including marine fish. Its possible implications for human health are important, and more and more emphasis is being put on the study of developmental effects, as observed in young children prenatally exposed to low concentrations of MeHg. Population groups consuming one normal fish meal/day (150 gm fish) will reach the provisionally tolerable weekly intake (PTWI) of 200 μg mercury even when MeHg concentrations in the fish consumed are very low. For people who eat only one seafood meal per week (about 20 gm fish/day), the PTWI will not be exceeded, even when the average MeHg concentration (in the fish) is very high. (Friberg 1988, p. 381)

Mercury in fish has persisted as a sporadic problem for more than 3 decades. Beginning with early concerns about high levels in swordfish, tuna, and halibut, more recent examples are found in elevated human blood levels from consuming "sea bass" (not further identified) imported from Chile. Some general background information about mercury in fish is:

- Current public health standards for methylmercury allow seafood to contain up to 0.1 µg/kg of body weight (ATSDR/FDA 1994; EPA reference level is 0.01 µg/kg of body weight, and the FDA action level is 0.05 µg/kg). The ATSDR (Agency for Toxic Substances and Disease Registry) has recently proposed raising its standard to 0.5 µg/kg body weight.
- Normal human blood mercury levels range from below detection to 5.0 µg/l.
- Blood levels from 10 to 20 µg/l can be associated with tremors, impaired coordination, and memory disturbances (ATSDR/FDA 1994).
- Mercury levels above 200 µg/l occurred in Minamata victims; no overt symptoms of toxicity occurred in adults with levels below that figure (200 µg/l would be attained by a 70-kg adult having a steady diet of 0.3 µg/d of methylmercury).

Impacts of methylmercury exposure in children of the Faroe Islands in the North Atlantic were reported in 1997. Statistical relationships of neurological dysfunction and maternal methylmercury exposure were found among mothers who frequently ate pilot whale meat and other seafood during pregnancy. Effects associated with exposure included impaired attention spans and memory and language functions (Grandjean et al. 1997).

The most recent flap about mercury in seafood received extensive media coverage in April 2004, after public release of a CDC (Communicable Disease Centers) study indicating that 8% of American women of childbearing age have blood mercury levels that would be potentially hazardous to unborn fetuses. Another study, by other scientists, with results released a month later, found no correlation between consumption of fish and fetal neurological problems. The concern was and is about mercury in large tuna, and it was reflected in a subsequent joint FDA and EPA guidance to potentially vulnerable consumers (especially children and pregnant women) to eat no more than one meal (6 oz) of albacore tuna per week and for women of childbearing age to avoid eating the flesh of other large predators such as shark, swordfish, king mackerel, and tilefish.

CARCINOGENS IN THE AQUATIC ENVIRONMENT

It might be well at this point to discuss the matter of carcinogens (cancer-causing agents) in the marine environment, because among the sublethal effects of chemical contaminants in estuarine and coastal waters are those that involve carcinogenic properties of the chemicals. Marine animals themselves may be affected, or, more significantly, the carcinogens may be accumulated in fish and shellfish that are then consumed by humans. The public health risks from ingestion of carcinogen-contaminated marine products can easily be appreciated (intuitively), but the extent of the present contamination of seafood is poorly documented, and the long-term effects of eating such contaminated products are unknown, except in the negative sense that no reported cases of human cancer have been traced directly thus far to ingestion of contaminated fish or shellfish.

Roughly 40 chemicals or groups of chemicals are considered to be carcinogenic for humans. Best known are arsenic and certain other metals, PCBs, DDT derivatives, benzo[a]pyrene (BaP) and other PAHs, dioxin, and toxaphene. Some of them accumulate, occasionally at high levels, in fish and shellfish in areas of local pollution or in larger enclosed bays and estuaries. Risk assessments with pollutants such as PCBs have suggested increased risk of cancer as a consequence of consumption of fish containing high levels of the contaminant (Cordle, Locke, & Springer 1982), but direct relationships between seafood consumption and cancer have not yet been demonstrated.

There are, however, disquieting pieces of information that emphasize the importance of determining the levels and effects of chemical carcinogens from the contaminated marine environment. Some investigations have focused on the PAHs, particularly BaP, which is highly carcinogenic. In one study, BaP levels as high as 121 ppm of dry sediment were found in the immediate vicinity of a Pacific coast sewage treatment plant, with diminishing concentrations at increasing distances from the outfall (GESAMP 1983, Friberg 1988). In a related study, the authors examined BaP levels in mussels (*Mytilus edulis*) from stations near Vancouver, British Columbia, and found some values as high as 215 ppm wet weight of tissue (Dunn & Stich 1976).

Arsenic has been implicated in recent studies as a carcinogen. Studies in Sweden indicated an increased risk of skin cancers (squamous carcinomas) associated with consumption of fish with high levels of inorganic arsenic (GESAMP 1985a, 1985b). Higher incidences were found in fishermen when compared with other occupational groups. An earlier paper also reported frequent occurrences of skin cancers in deep-sea fishermen and fishing industry wharf workers (Cabre & Lasanta 1968). Other factors, such as exposure to ultraviolet (UV) radiation, may of course be involved in the genesis of skin tumors, so the conclusion must be that available epidemiological data do not support or refute an association between cancer and arsenic intake via fish.

In addition to the apparent global increase in the frequency and duration of toxic algal blooms (discussed at length in Chapter 4), it seems that other kinds of pollution-associated chemical events with direct or potential public health significance are increasing in frequency in coastal/estuarine waters. Although some of this change may be due to greater public awareness and some improvement in surveillance, it is probable that the remarkable expansion in synthetic chemical production and use in the past 3 decades has contributed substantially. Synthetic chemicals that simply did not exist even a decade or two ago are now being viewed with some alarm as environmental contaminants. Many such chemicals have been dumped indiscriminately into rivers and estuaries, and some are still being released. Disclosure of most pollution problems is accidental, and even after such disclosure, regulatory agencies frequently encounter strong resistance from the contaminating industries.

Public alarm about contaminants in food and water that may affect human health can lead to closure of industries or modifications of production methods, but usually only after clear evidence of danger to humans has appeared and has been widely publicized in news media. There have been several incidents in North America during the past 3 decades that illustrate a common sequence of events:

1. Release of toxic chemicals in the absence of surveillance and control, and with little knowledge of or concern about their public health effects
2. Some preliminary accidental or fortuitous indication of danger to humans or to resources
3. Vigorous denial of danger or responsibility by the polluting industry involved
4. Preliminary investigation and half-hearted action by regulatory agencies
5. Legal delaying tactics by the polluting industry
6. Reluctant compliance by the offending industry in the presence of mounting data, advisory legal opinions, and a rising crescendo of expressed public concern
7. Grudging and usually minuscule payments (mostly absorbed by lawyers) to settle damage claims

Examples of this sequence include the release of PCBs in the Hudson River and Great Lakes; the release of Kepone in the James River in Virginia; and the release of phosphorous in Placentia Bay, Newfoundland.

The PCB story in the Hudson River is still unfolding. It includes deliberate long-term release of PCBs into the upper river by two units of the General Electric Company, the finding of dangerously high levels of PCBs in fish, the closing of the river to fishing (except for certain anadromous species), the reluctant reduction of contamination and token cleanup efforts by the offending industry after much legal foot-dragging, and (the ultimate insult) the successful attempt by that industry to shift most of the financial burden of adequate cleanup to the U.S. taxpayers.

Kepone, a highly toxic and persistent insecticide, was deliberately discharged into the upper James River (an important oyster-producing area) over a period of 16 months in 1974–1975 by a subsidiary of Allied Chemical Company. Only after obvious toxic effects on chemical plant workers were disclosed was there any concern about pollution of waters and shellfish by the plant effluents. The producing facility was closed, and the river was also closed to fishing for an extended period.

The Placentia Bay (Newfoundland) phosphorous contamination event was first disclosed by observations of "red herring" and extensive herring mortalities in that bay in 1969. A new industrial operation had begun in 1968, and an investigation revealed that it had been releasing phosphorous into the bay (which is also a very important fishing area). Again, as with the Kepone event, the plant was closed temporarily, as was the fishery, in the presence of a clear danger to public health.

So, in summary, although there is little evidence of significant direct danger to human life in existing chemical contaminant levels in marine resource species (except for very localized incidents of gross pollution such as those just described), there are instances (such as mercury in black marlin, large halibut, and swordfish and PCBs in striped bass) of contaminant levels high enough to warrant attention, further study, and possibly controlled consumption. There is also a great need for much more scientific examination of possible long-term sublethal effects on humans caused by contaminants in food. Sufficient evidence now exists about carcinogenic, mutagenic, and other long-term toxic effects of many industrial chemicals to warrant more attention and reasoned action, even in the absence of incontrovertible proof of risk to public health.

ILLNESSES CAUSED BY ENVIRONMENTAL EXPOSURE TO TOXIC CHEMICALS AND MICROBIAL CONTAMINANTS IN COASTAL WATERS

Until the late 1980s, any discussion of illnesses caused by environmental exposure to toxins and industrial toxicants would have been severely circumscribed — pretty much limited to anecdotal accounts of algal toxins in sea spray during blooms, causing bronchitis or eye irritation in local residents. But the story is changing. Laboratory exposures to toxins from cultures of the dinoflagellate *Pfiesteria piscicida* — the so-called microbe from hell — in the late 1980s led to association of a sequence of human ills — asthmatic bronchitis, skin lesions, eye irritation, short-term memory loss — resulting from unprotected contact with cultured life history stages of the organism. Then, in 1997, the effects noted in laboratory workers were observed among fishermen, field technicians, and a water skier, all of whom were in the vicinity of a *Pfiesteria* outbreak with accompanying fish mortalities and who had contact with the water in a tiny tidal river on Maryland's Eastern Shore.

The toxic organism and its possible relatives exists in other mid-Atlantic estuaries, especially in Pamlico Sound, North Carolina, but the most noteworthy effect of its presence to date is production of sudden fish kills and skin lesions in survivors. There is some recent indication that a number of related species occur in those waters and those as far south as Florida.

The role of anthropogenic nutrient loading of coastal/estuarine waters in increasing the risks of biotoxin-induced human illnesses may be greater than present data will support, although there is some suggestive information available. I see at least three possible environmental situations in which nutrient loading could have an indirect effect on human health:

- Proliferation of known or unknown toxin-producing microalgae — including forms such as *Pfiesteria piscicida* that have neurotoxic capabilities — may be encouraged. A cause-and-effect relationship of proliferation of such forms with nutrient loading from agricultural sources has been proposed but needs further substantiation.
- It is possible that some microalgal species not known as toxin producers may become toxic if environmental nutrient concentrations are augmented from human sources (Smayda 1989, Burkholder 1998).
- Proliferation of salinity-tolerant or salinity-requiring potentially pathogenic bacterial populations (*Vibrio* and *Aeromonas* in particular) may occur in brackish-water habitats in which nutrient concentrations have been increased from anthropogenic causes (Cabelli 1978).

In addition to the indirect effects of nutrient loading of coastal waters, there are always the direct effects of exposure to microbial contaminants in seawater on swimmers, divers, and others who enter polluted waters for recreation or any other purpose. This topic was explored in Chapter 4.

In the technologically advanced countries of North America, Europe, and Asia, attitudes toward coastal pollution seem to be shifting gradually from emphasis on reducing industrial point sources of toxic chemicals to that of reducing inputs of nutrient chemicals from all sources. This shift in focus is promoted by observed ecological and economic effects of increasing occurrences of harmful algal blooms and related oxygen depletion in estuarine/coastal waters of many countries. The industrialized nations have acted (however sluggishly and inadequately) to reduce their contributions of toxic chemicals to the marine environment and have thereby exposed a problem of comparable dimensions — effects of nutrient loading from anthropogenic activities. Such a shift in perspective is less discernible in third world and developing countries, which are still producing and releasing toxic contaminants such as DDT and other persistent pesticides, heavy metals, and microbial pathogens.

CONCLUSIONS

In this chapter on effects of coastal pollution on public health, we have scrutinized some of the available information from three perspectives:

1. Illnesses caused by microbial contamination of seafood
2. Illnesses caused by chemical contamination of seafood
3. Illnesses caused by environmental exposure to toxic chemicals and microbial contaminants in coastal waters

After coastal pollution has taken its toll on resource and food chain organisms — in the form of disabilities, reduced fecundity, and death — the survivors may be dangerous to predators, including humans, who consume them in quantity. Microbial contaminants in fish and shellfish may cause life-threatening diseases in human consumers; viral and bacterial pathogens can be transmitted to humans who ingest raw or improperly processed molluscan shellfish. Chemical contaminants in the flesh of fish and shellfish can be bioaccumulated and can reach toxic levels at the upper ends of food chains. People are not exempt from these effects. To compound the toxicity problem, many marine species that manage to survive in abnormal chemical environments do so by developing degrees of tolerance to otherwise damaging effects of pollutants, often by sequestering toxicants in specific body tissues. The higher body burdens may then be available — sometimes at lethal levels — to predators, including humans.

Public health matters are by definition important to all of us, so it would be logical in concluding this chapter on coastal pollution and public health to look briefly beyond present data, and even beyond present operational concepts, at the potential for future harmful effects of contamination of coastal living resources. We have in this chapter cited several instances, such as Minamata disease in a coastal area of Japan and cholera in a coastal area of Peru, of resource-related damage to the human population caused by pollution. These may be dramatic illustrations of insidious long-term damage that can only be speculated about now. We know, for example, that some heavy metals and many hydrocarbons can be carcinogenic and mutagenic, or can produce physiological and biochemical changes in test animals. Much of the information has been derived from acute exposures of laboratory animals to specific toxicants, but information about effects of long-term chronic exposures is already appreciable, and it suggests that continuous exposure to low levels of contaminants can be dangerous to human health. Of course, contamination of fishery products from estuaries and coastal waters is only a small part of the total problem of chemical and microbial contamination of food, but because coastal waters are the recipients of many chemicals of terrestrial origin, because marine organisms can selectively accumulate some contaminants (especially at higher trophic levels), and because seafood constitutes a significant part of the diet in a number of countries and peoples, the long-term effects of seafood contamination on human health cannot be ignored.

Humans may serve as bioindicators of the extent of evil that they have inflicted on coastal waters. Effluents from aggregations of people can contain toxic and nutrient chemicals, as well as microbial pathogens, which may have direct or indirect effects on public health. Exposure during recreational activities (swimming, diving) or eating seafood from contaminated zones can result in illnesses of varying severity, depending on the nature and extent of pollution in the coastal area of contact.

Emerging diseases is a topic that has received recent attention in scientific publications and in the news media (see, for example, Colwell 1996 and Harvell et al. 1999, or read any copy of the new scientific journal *Emerging Infectious Diseases*). As Dr. Colwell pointed out early in her review of current understanding of cholera, "Emerging diseases are considered to be those infections that are either newly appearing in the population or are rapidly increasing in incidence or expanding in geographic range" (p. 2025). She goes on to make the point that is most relevant to this book: "*Human activities* [emphasis mine] drive emergence of disease, and a variety of social, economic, political, climatic, technological, and environmental factors can shape the pattern of a disease and influence its emergence into populations" (p. 2025). Coastal pollution is of course one environmental change that has earned its place in the emergence of those diseases of humans and marine animals considered in several chapters of this book, beginning with Chapter 1 and index.

REFERENCES

ATSDR/FDA (Agency for Toxic Substances and Disease Registry/Food and Drug Administration). 1994. Physiological responses to 10–20 μg/L methyl mercury, pp. 170–171.

Biosca, E.G., C. Amaro, C. Esteve, E. Alcaide, and E. Garay. 1991. First record of *Vibrio vulnificus* biotype 2 from diseased European eel, *Anguilla anguilla* L. *J. Fish Dis.* 14: 103–109.

Bisharat, N., V. Agmon, R. Finkelstein, R. Raz, G. Ben-Dror, L. Lerner, S. Soboh, R. Colodner, D.N. Cameron, D.L. Wykstra, D.L. Swerdlow, and J.J. Farmer for the Israel *Vibrio* Study Group. 1999. Clinical, epidemiological, and microbiological features of *V. vulnificus* biogroup 3 causing outbreaks of wound infection and bacteraemia in Israel. *Lancet 354:* 1421–1424.

Blake, P.A., R.E. Weaver, and D.G. Hollis. 1980. Diseases of humans (other than cholera) caused by vibrios. *Annu. Rev. Microbiol. 34:* 341–367.

Burkholder, J.M. 1998. Implications of harmful microalgae and heterotrophic dinoflagellates in management of sustainable marine fisheries. *Ecol. Appl. 8*(Suppl.): S37–S62.

Cabelli, V.J. 1978. New standards for enteric bacteria, pp. 23–33. In: R. Mitchell (ed.), *Water Pollution Microbiology.* Wiley, New York.

Cabre, J. and J. Lasanta. 1968. Malignant epithelial tumors in deep-sea fishermen. *Actas Dermo-Sifiliogr. 59:* 361–364.

Cachola, R. and M.C. Nunes. 1974. Quelques aspects de la pollution bacteriologique des centres producteurs de mollusques de l'Algarve (1963–1972). Bol. Inf., Inst. Biol. Marit. 13, 12 pp.

Cockey, R.R. and T. Chai. 1988. An update on seafood-poisoning microorganisms. *Seafood Inf. Tips 14*(1): 1–6.

Colwell, R.R. 1996. Global climate and infectious disease: The cholera paradigm. *Science 274:* 2025–2031.

Colwell, R.R., T.E. Lovelace, L. Wan, T. Kaneko, T. Staley, P.K. Chen, and H. Tubiash. 1973. *Vibrio parahaemolyticus* — Isolation, identification, classification and ecology. *J. Milk Food Technol. 36:* 202–213.

Colwell, R.R., R.J. Seidler, J. Kaper, S.W. Joseph, S. Garges, H. Lockman, D. Maneval, H. Bradford, N. Roberts, E. Remmers, I. Huq, and A. Huq. 1981. Occurrence of *Vibrio cholerae* serotype 01 in Maryland and Louisiana estuaries. *Appl. Environ. Microbiol. 41:* 555–558.

Cooper, R.C. 1975. Waste water management and infectious disease. II. Impact of waste water treatment. *J. Environ. Health 37:* 342–353.

Cordle, F., R. Locke, and J. Springer. 1982. Risk assessment in a federal regulatory agency: An assessment of risk associated with the human consumption of some species of fish contaminated with polychlorinated biphenyls (PCBs). *Environ. Health Perspect. 45:* 171–182.

Craun, G.F. 1975. Microbiology — Waterborne outbreaks. *J. Water Pollut. Control Fed. 47:* 1566–1569.

Cross, J.N., editor. 1994. Contamination of recreational seafood organisms off southern California. Southern California Coastal Water Research Program (SCCWRP) Annu. Rep. 1992–1993, pp. 100–110.

Cubitt, W.D. 1991. A review of the epidemiology and diagnosis of waterborne viral infections. *Water Sci. Technol. 24:* 193–203.

Dalsgaard, A., N. Frimodt-Møller, B. Bruun, L. Høi, and J.L. Larsen. 1996. Clinical manifestations and epidemiology of *Vibrio vulnificus* infections in Denmark. *Eur. J. Clin. Microbiol. Infect. Dis. 15:* 227–231.

DePaola, A. 1981. *Vibrio cholerae* in marine foods and environmental waters: A literature review. *J. Food Sci. 46:* 66–70.

DePaola, A., L.H. Hopkins, J.T. Peeler, B. Wentz, and R.M. McPhearson. 1990. Incidence of *Vibrio parahaemolyticus* in U.S. coastal waters and oysters. *Appl. Environ. Microbiol. 56:* 2299–2302.

DePaola, A., C.A. Kaysner, J.C. Bowers, and D.W. Cook. 2000. Environmental investigations of *Vibrio parahaemolyticus* in oysters following outbreaks in Washington, Texas, and New York (1997, 1998). *Appl. Environ. Microbiol. 66:* 4649–4654.

DePaola, A., J.L. Nordstrom, J.C. Bowers, J.G. Wells, and D.W. Cook. 2003a. Seasonal abundance of total and pathogenic *Vibrio parahaemolyticus* in Alabama oysters. *Appl. Environ. Microbiol. 69:* 1521–1526.

DePaola, A., J. Ulaszek, C.A. Kaysner, B.J. Tenge, J.L. Nordstrom, J. Wells, N. Puhr, and S.M. Gendel. 2003b. Molecular, serological, and virulence characteristics of *Vibrio parahaemolyticus* isolated from environmental, food, and clinical sources in North America and Asia. *Appl. Environ. Microbiol.* 69: 3999–4005.

Dunn, B.P. and H.F. Stich. 1976. Release of the carcinogen benzo(a)pyrene from environmentally contaminated mussels. *Bull. Environ. Contam. Toxicol. 15:* 398–399.

Farmer, J.J. 1980. Revival of the name *Vibrio vulnificus*. *Int. J. Syst. Bacteriol. 30:* 656.

Ferreira, P.S. and R.A. Cachola. 1975. *Vibrio cholerae* El Tor in shellfish beds of the south coast of Portugal. Int. Counc. Explor. Sea, Doc. C.M. 1975/K:18, 7 pp.

Friberg, L. 1988. The GESAMP evaluation of potentially harmful substances in fish and other seafood with special reference to carcinogenic substances. *Aquat. Toxicol. 11:* 379–393.

Fujino, T., Y. Okuno, D. Nakada, A. Aoymama, K. Fukai, T. Mukai, and T. Ucho. 1953. On the bacteriological examination of Shirasu-food poisoning. *Med. J. Osaka Univ. 4:* 299–304.

Gajardo, R., N. Bouchriti, R.M. Pinto, and A. Bosch. 1995. Genotyping of rotaviruses isolated from sewage. *Appl. Environ. Microbiol. 61:* 3460–3462.

Gantzer, C., A. Maul, J.M. Audic, and L. Schwartzbrod. 1998. Detection of infectious enteroviruses, enterovirus genomes, somatic coliphages, and *Bacteroides fragilis* phages in treated wastewater. *Appl. Environ. Microbiol. 64:* 4307–4312.

Gerba, C.P. and G.E. Schaiberger. 1975a. Effect of particulates on virus survival in seawater. *J. Water Pollut. Control Fed. 47:* 93.

Gerba, C.P. and G.E. Schaiberger. 1975b. Aggregation as a factor in loss of viral titer in seawater. *Water Res. (G.B.) 9:* 567.

GESAMP. 1983. Impact of oil and related chemicals on the marine environment. Rep. Stud. GESAMP 50, 180 pp.

GESAMP. 1985a. The impact of carcinogenic substances on marine organisms and implications concerning public health. GESAMP XV/2/4, WHO Int. Rep., Geneva.

GESAMP. 1985b. Review of potentially harmful substances — Cadmium, lead and tin. WHO Rep. Stud. 22, Geneva.

GESAMP. 1986a. Review of potentially harmful substances — Arsenic, mercury and selenium. WHO Rep. Stud. 18, Geneva.

GESAMP. 1986b. Impact of carcinogenic substances on marine organisms and implications concerning public health. GESAMP SVI/2/3, WHO Int. Rep., Geneva.

GESAMP. 1986c. Occurrence of potential carcinogens and mutagens in marine organisms. GESAMP SVI/2/4, WHO Int. Rep., Geneva.

Grandjean, P., P. Weihe, R.F. White, F. Debes, S. Araki, K. Yokoyama, K. Murata, N. Sorensen, R. Dahl, and P. Jorgensen. 1997. Cognitive deficit in 7-year-old children with prenatal exposure to methylmercury. *Neurotoxicol. Teratol. 19:* 417–428.

Halliday, M.L., L.Y. Kang, T.K. Zhou, M.D. Hu, Q.C. Pan, T.Y. Fu, Y.S. Huang, and S.L. Hu. 1991. An epidemic of hepatitis A attributable to the ingestion of raw clams in Shanghai, China. *J. Infect. Dis. 164:* 852–859.

Harada, M. 1972. *Minamata Disease.* [In Japanese.] Iwanami Shoten, Tokyo, 274 pp.

Harvell, C.D., K. Kim, J.M. Burkholder, R.R. Colwell, P.R. Epstein, D.J. Grimes, E.E. Hofmann, E.K. Lipp, A.D.M.E. Osterhaus, R.M. Overstreet, J.W. Porter, G.W. Smith, and G.R. Vasta. 1999. Emerging diseases: Climate links and anthropogenic factors. *Science 285:* 1505–1510.

Høi, L., J.L. Larsen, I. Dalsgaard, and A. Dalsgaard. 1998a. Occurrence of *Vibrio vulnificus* biotypes in Danish marine environments. *Appl. Environ. Microbiol. 64:* 7–13.

Høi, L., I. Dalsgaard, A. DePaola, R.J. Siebeling, and A. Dalsgaard. 1998b. Heterogeneity among isolates of *Vibrio vulnificus* recovered from eels (*Anguilla anguilla*) in Denmark. *Appl. Environ. Microbiol. 64:* 4676–4682.

Hollis, D.G., R.E. Weaver, C. Baker, and C. Thornberry. 1976. Halophilic *Vibrio* species isolated from blood culture. *J. Clin. Microbiol. 3:* 425–432.

Honda, T. and T. Iida. 1993. The pathogenicity of *Vibrio parahaemolyticus* and the role of the thermostable direct haemolysin and related haemolysins. *Rev. Med. Microbiol. 4:* 106–113.

Huddle, N. and M. Reich. 1975. *Island of Dreams: Environmental Crisis in Japan.* Autumn Press, New York. 225 pp.

Hughes, J.M. 1979. Epidemiology of shellfish poisoning in the United States, 1971–1977, pp. 23–28. In: D.L. Taylor and H.H. Seliger (eds.), *Developments in Marine Biology. Vol. 1. Toxic Dinoflagellate Blooms.* Elsevier/North Holland, New York.

Iida, H., T. Iwamoto, T. Karashimada, and M. Kumagai. 1957. Studies on the pathogenesis of fish-borne food-poisoning in summer. II. Studies on cholinesterase inhibition by culture filtrates of various bacteria. *Jpn. J. Med. Sci. Biol. 10:* 177–185.

Janssen, W.A. 1970. Fish as vectors of human bacterial diseases. *Am. Fish. Soc. Spec. Publ. 5:* 284–290.

Jobling, S., D. Sheahan, J.A. Osborne, P. Matthiessen, and J.P. Sumpter. 1996. Inhibition of testicular growth in rainbow trout (*Oncorhynchus mykiss*) exposed to estrogenic alkyl-phenolic chemicals. *Environ. Toxicol. Chem. 15:* 194–202.

Jobling, S., M. Nolan, C.R. Tyler, G. Brighty, and J.P. Sumpter. 1998. Widespread sexual disruption in wild fish. *Environ. Sci. Technol. 32:* 2498–2506.

Joseph, S.W., R.R. Colwell, and J.B. Kaper. 1983. *Vibrio parahaemolyticus* and related halophilic vibrios. *Crit. Rev. Microbiol. 10:* 77–123.

Kaysner, C.A., C. Abeyta, Jr., M.M. Wekell, A. DePaola, Jr., R.F. Scott, and J.M. Leitch. 1987. Incidence of *Vibrio cholerae* from estuaries of the United States west coast. *Appl. Environ. Microbiol. 53:* 1344–1348.

Kelly, M.T. 1982. Effect of temperature and salinity on *Vibrio* (*Beneckea*) *vulnificus* occurrence in a Gulf coast environment. *Appl. Environ. Microbiol. 44:* 820–824.

Kingsley, D.H., G.K. Meade, and G.P. Richards. 2002. Detection of both hepatitis A virus and Norwalk-like virus in imported clams associated with food-borne illness. *Appl. Environ. Microbiol. 68:* 3914–3918.

Kobayashi, J. 1969. Investigations of the cause of itai-itai disease. I–III. [In Japanese.] *Kagaku (Science) 39:* 286, 369, 424.

Koh, E.G.L., J.H. Huyn, and P.A. LaRock. 1994. Pertinence of indicator organisms and sampling variables to *Vibrio* concentrations. *Appl. Environ. Microbiol. 60:* 3897–3900.

Kopecka, H., S. Dubrou, J. Prevot, J. Marechal, and J.M. Lopez-Pila. 1993. Detection of naturally occurring enteroviruses in waters by reverse transcription-polymerase chain reaction and hybridization. *Appl. Environ. Microbiol. 59:* 1213–1219.

Le Guyader, F., E. Dubois, D. Menard, and M. Pommepuy. 1994. Detection of hepatitis A virus, rotavirus, and enterovirus in naturally contaminated shellfish and sediment by reverse transcription-seminested PCR. *Appl. Environ. Microbiol. 60:* 3665–3671.

Lewis, D.H. 1975. Retention of *Salmonella typhimurium* by certain species of fish and shrimp. *J. Am. Vet. Med. Assoc. 167:* 551–552.

Liu, O.C., H.R. Seraichekas, and B.L. Murphy. 1966. Viral pollution and self-cleansing mechanism of hard clams, pp. 419–437. In Berg, G. (ed.), *Transmission of Viruses by the Water Route.* Wiley-Interscience, New York.

Mahoney, P., G. Fleischner, I. Millman, W.T. London, B.S. Blumberg, and I.M. Arias. 1974. Australia antigen: Detection and transmission in shellfish. *Science 183:* 80–81.

Mason, J.O. and W.R. McLean. 1962. Infectious hepatitis traced to the consumption of raw oysters. An epidemiologic study. *Am. J. Hyg. 75:* 90.

Matsumoto, C., J. Okuda, M. Ishibashi, M. Iwanaga, P. Garg, T. Rammamurthy, H. Wong, A. DePaola, Y.B. Kim, M.J. Albert, and M. Nishibuchi. 2000. Pandemic spread of an O3:K6 clone of *Vibrio parahaemolyticus* and emergence of related strains evidenced by arbitrarily primed PCR and *toxRS* sequence analysis. *J. Clin. Microbiol. 38:* 578–585.

Mead, P.S., L. Slutsker, V. Dietz, L.F. McGaig, J.S. Bresee, C. Shapiro, P.M. Griffin, and R.V. Tauxe. 1999. Food-related illness and death in the United States. *Emerg. Infect. Dis. 5:* 607–625.

Merkel, S.M., S. Alexander, E. Zufall, J.D. Oliver, and Y.M. Huet-Hudson. 2001. Essential role for estrogen in protection against *Vibrio vulnificus*-induced endotoxic shock. *Infect. Immunol. 69:* 6119–6122.

Metcalf, T.F. and W.C. Stiles. 1966. Survival of enteric viruses in estuary waters and shellfish, pp. 439–447. In: G. Berg (ed.), *Transmission of Viruses by the Water Route.* Wiley-Interscience, New York.

Murphy, A.M., G.S. Grohmann, P.J. Christopher, W.A. Lopez, G.R. Davey, and R.H. Millsom. 1979. An Australia-wide outbreak of gastroenteritis from oysters caused by Norwalk virus. *Med. J. Austr. 2:* 329.

NOAA (National Oceanic and Atmospheric Administration). 1991. *The 1990 National Shellfish Register of Classified Estuarine Waters.* NOAA, Office of Oceanographic and Marine Assessment, Rockville, MD. 99 pp.

Oliver, J.D. 1981. The pathogenicity and ecology of *Vibrio vulnificus:* A particularly virulent microorganism. *Mar. Technol. Soc. J. 15:* 45–52.

Oliver, J.D. 1989. *Vibrio vulnificus*, pp. 569–600. In: M.P. Doyle (ed.), *Foodborne Bacterial Pathogens.* Marcel Dekker, New York.

Oliver, J.D. and J.B. Kaper. 2001. *Vibrio* species, pp. 263–300. In: M.P. Doyle, L.R. Beuchat, and T.J. Montville (ed.), *Food Microbiology: Fundamentals and Frontiers.* ASM Press, Washington, DC.

Oliver, J.D., R.A. Warner, and D.R. Cleland. 1983. Distribution of *Vibrio vulnificus* and other lactose-fermenting vibrios in the marine environment. *Appl. Environ. Microbiol. 45:* 985–998.

Pfeffer, C.S., M.F. Hite, and J.D. Oliver. 2003. Ecology of *Vibrio vulnificus* in estuarine waters of eastern North Carolina. *Appl. Environ. Microbiol. 69:* 3526–3531.

Pina, S., M. Puig, F. Lucena, J. Jofre, and R. Girones. 1998. Viral pollution in the environment and in shellfish: Human adenovirus detection by PCR as an index of human viruses. *Appl. Environ. Microbiol. 64:* 3376–3382.

Poores, J.M. and L.A. Fuchs. 1975. Isolation of *Vibrio parahaemolyticus* from a knee wound. *Clin. Orthop. 106:* 245–252.

Portnoy, B.L., P.A. Mackowiak, C.T. Caraway, J.A. Walker, T.W. McKinley, and C.A. Klein. 1975. Oyster-associated hepatitis: Failure of shellfish certification programs to prevent outbreaks. *JAMA 233:* 1065–1068.

Richards, G.P. 1985. Outbreaks of shellfish-associated enteric virus illness in the United States: Requisite for development of viral guidelines. *J. Food Prot. 48:* 815–823.

Richards, G.P. 1987. Shellfish-associated enteric virus illness in the United States, 1934–1984. *Estuaries 10:* 84–85.

Robertson, B.H., R.W. Jansen, B. Khanna, A. Totsuka, O.V. Nainan, G. Siegl, A. Widell, H.S. Margolis, S. Isomura, K. Ito, T. Ishizu, Y. Mortisugu, and S.M. Lemon. 1992. Genetic relatedness of hepatitis A virus strains recovered from different geographical regions. *J. Gen. Virol. 73:* 1365–1377.

Ross, B. 1956. Hepatitis epidemic transmitted by oysters. *Sven. Laekartidn. 53:* 989–1003.

Schaiberger, G.E., C.P. Gerba, and E.G. Estevez. 1976. Survival of viruses in the marine environment, pp. 97–109. In: S.P. Meyers (ed.), *Proceedings of the International Symposium on Marine Pollution Research.* EPA-600/9-76-032. U.S. Environmental Protection Agency, Environmental Research Lab., Gulf Breeze, FL. 171 pp. (Avail. from NTIS, Springfield, VA: PB-267 601)

Shapiro, R.L., S. Altekruse, L. Hutwagner, R. Bishop, R. Hammond, S. Wilson, B. Ray, S. Thompson, R.V. Tauxe, P.M. Griffin, and the *Vibrio* Working Group. 1998. The role of Gulf coast oysters harvested in warmer months in *Vibrio vulnificus* infections in the United States, 1988–1996. *J. Infect. Dis. 178:* 752–759.

Smayda, T.J. 1989. Homage to the International Symposium on Red Tides: The scientific coming of age of research on Akashiwo; algal blooms; flos-aquae; tsvetenie vody; wasserblute, pp. 23–30. In: *Proceedings of the First International Symposium on Red Tides.* Appleton-Lange, Norwalk, CT.

Smith, W.E. and A.M. Smith. 1975. *Minamata.* Holt Rinehart Winston, New York. 220 pp.

Strom, M.S. and R.N. Paranjpye. 2000. Epidemiology and pathogenesis of *Vibrio vulnificus. Microbes Infect. 2:* 177–188.

Tamplin, M.L., G.E. Rodrick, N.J. Blake, and T. Cuba. 1982. Isolation and characterization of *Vibrio vulnificus* from two Florida estuaries. *Appl. Environ. Microbiol. 44:* 1466–1470.

Tang, Y.W., J.X. Wang, Z.Y. Zu, Y.F. Guo, W.H. Quian, and J.X. Xu. 1991. A seriologically confirmed, case-control study, of a large outbreak of hepatitis A in China, associated with consumption of clams. *Epidemiol. Infect. 107:* 651–657.

Twedt, R.M., P.L. Spaulding, and H.E. Hall. 1969. Morphological, cultural, biochemical, and serological comparison of Japanese strains of *Vibrio parahaemolyticus* with related cultures isolated in the United States. *J. Bacteriol. 98:* 511–518.

Woods, P.A., J. Gentsch, V. Gouvea, L. Mata, A. Simhon, M. Santosham, Z.S. Bai, S. Urasawa, and R.I. Glass. 1992. Distribution of serotypes of human rotavirus in different populations. *J. Clin. Microbiol. 30:* 781–785.

13 Economic Effects of Coastal Pollution: A Resource Perspective

INTRODUCTION

Thus far in this book, we have examined the effects of coastal pollution on individual marine animals, on fisheries resource populations, on aquaculture production, and on humans. With all this as background, it seems logical now to put a price tag on pollution of coastal/estuarine waters — to assess economic effects, especially from the perspective of fish and shellfish resources and human well-being. That kind of assessment can be performed with some degree of adequacy for short-term immediate impacts, such as those associated with oil spills, but longer-term effects, such as possible reductions in available living resources or rejection of seafood products because of fear of contamination, are more difficult to quantify. An attempt at economic evaluation is almost obligatory, though, because, to many people, translation of biological or environmental findings into dollars and cents terms is a necessary prelude to any conclusions that might be reached about pollution effects — and to some of those same people, unless an economic effect can be clearly demonstrated, the issue is unimportant.

The topic of economic impacts of coastal pollution can be introduced with two back-to-back vignettes about historical aspects of the problem. The first and most recent is called "Plight of the Hudson River Fisherman," and the second "The Great Contaminated Fish Scare in Japan."

Plight of the Hudson River Fisherman

The lower reaches of the Hudson River, just above the compressed insanity that is New York City, but still within the range of the tides, have for entire lifetimes been favorite hunting grounds for George Kundera and his friends — all members of a special breed of small-boat commercial fishermen. His town is near the shores of that much-abused waterway, and his prey varies with the changing seasons: shad, eels, striped bass, menhaden, clams, and assorted other species. He catches whatever is available (and marketable) at the moment, then moves on to other species, whether fish or shellfish.

George is a present-day representative of an independent race of survivors, clinging to a way of life that is older than the city that now sprawls around the *ocean access to his river. But his livelihood is threatened now, more than ever, by multiple pressures of an expanding human population, and by continued degradation of the river's waters. That threat became very real to George, and his world collapsed, in early spring of 1976, just as the azaleas were blooming in his front yard. New regulations were announced by the government; some of the species of fish that he depended on — like striped bass — were declared "unfit for human consumption" by an anonymous bureaucrat in the state public health agency, and the taking of those fish was prohibited. According to reports in the <u>New York Times</u>, two manufacturing plants far up the Hudson — units of the General Electric Company — had been dumping toxic chemicals — especially polychlorinated biphenyls [PCBs] — into the river for decades, and enough of the obnoxious stuff had been carried downstream to contaminate the fish that inhabited the lower reaches. So-called "action levels" of the pollutants were found in the flesh of the fish, and the "action" was to close the fishery, as a protection for consumers.*

George, and others like him, spent a long time alternating between frustration and rage — a repetition of their reactions during the previous summer, when the clam beds outside the town were closed to fishing because of increasing levels of fecal contamination from the expanding human population. He and his dispossessed cohorts felt increasingly mistreated and rejected by a changing society that had no respect for their way of life. They felt control of their existence slipping away, almost capriciously and certainly without their consent.

But the response of that changing society to George's questions would be that there is no answer, except that he is an anachronism — a living relic of an earlier, gentler time when the human species had not yet achieved sufficient numbers to despoil the land and then foul the coastal waters. So-called "environmental awareness" had appeared too late to save him from a land-based service job, or maybe one on a still-surviving industrial assembly line — far from the pristine edges of the sea that exist only in his memory.

From Field Notes of a Pollution Watcher
(C.J. Sindermann, 1978)

The Great Contaminated Fish Scare in Japan

Post-World War II industrial development in Japan was a source of astonishment and envy for most of the rest of the world — but it had been achieved at a staggering environmental cost. Some elements of the cost became apparent on a local level during the Minamata Bay mercury poisoning incident of the 1950s and 1960s, but the true extent of damage to coastal waters was first impressed on the Japanese national consciousness in 1973, during what has been described as "the great fish panic." The episode was initiated by two reports that appeared almost simultaneously in the spring of that year. One, by a university medical research group, disclosed the discovery of new cases of mercury poisoning (Minamata Disease) in inhabitants of an area near the Ariaki Sea, 40 km north of Minamata Bay. The other disclosure, by the Japanese Fisheries Agency, was that PCBs were found to be above established human safety standards in more than 80% of fish sampled from six major coastal fishing areas.

Fear of the consequences of eating contaminated seafood, in a country whose animal protein source consists to a major extent of products from the sea, grew quickly and precipitated a violent and unusual response by the people. Within a few weeks, fish sales had plummeted by as much as 50%, and widespread mass demonstrations against polluting industries and unresponsive government agencies occurred. Dead fish were heaped in front of factory gates and government offices by angry fishermen, and thousands of placard-carrying consumers picketed the headquarters of the largest polluting industries. Media treatment was inflammatory, and inadequate solutions (government-sponsored surveys and industry promises to reduce contamination) were proposed to reassure an alarmed public.

But the event faded all too quickly — within two months. Newspaper coverage declined in favor of new crises of the moment (in this instance the hijacking of a Japanese commercial plane), and fish sales returned to normal levels. There were, however, a few persistent residues. One was heightened sensitivity of the average Japanese citizen to some of the real costs of industrial expansion — the consequences of the policy of economic growth at any price. Another was the realization that industrial pollution problems in coastal waters were national in scope, and required vigorous government intervention instead of the former laissez faire attitude of national regulatory agencies.

From Field Notes of a Pollution Watcher
(C.J. Sindermann, 1981)

These two vignettes illustrate, in different ways, the economic impacts of coastal pollution. Important from that perspective are four principal problem areas:

1. Consumer resistance: rejection of fish and shellfish as food
2. Reduced yields from commercial fisheries
3. Reduced revenues from recreational fishing
4. Economic effects of "nutrient pollution"

Each category will get some limited attention in the pages ahead.

CONSUMER RESISTANCE: REJECTION OF FISH AND SHELLFISH AS FOOD

Media attention to technical reports about effects of coastal pollution on the quality of fish and shellfish has increased consumer concerns about the wisdom of buying and eating certain kinds of fish or shellfish — or of eating any kind of seafood, cooked or uncooked. The following vignette illustrates the nature of the problem.

Query from a Pregnant Editor

During the summer of 1988, at the height of public and media unhappiness about fouled ocean beaches and contaminated fish, I made a business trip to the New York office of a book publisher. One of the senior editors, a remarkably perceptive woman in her 30s, asked me to have lunch. Knowing my background in marine biology and coastal pollution, she wanted to discuss the possible risks from eating chemically contaminated seafood. Her concern was not academic; she was in the early months of pregnancy; she was a seafood aficionado; and she wanted to follow a diet that would minimize the potential dangers to her developing child. Disturbing news stories had appeared during the year about PCBs in Hudson River striped bass, eels, and other fish, and mercury in larger oceanic species like swordfish and tuna, and she was obviously upset by the reports and uncertain about what, if any, seafood she should eat.

Since the book publisher was picking up the lunch tab, and since I had a small amount of relevant secondhand information, we spent an enjoyable two hours constructing a partial diet, with a list of high-risk seafood and a much shorter one of reasonably safe seafood (and with an extensive accompanying list of species for which no information was available).* I wondered, after our conference, about all the other pregnant women in New York or elsewhere — about what their concerns about contaminated dietary choices might be, if any. Were they aware of developmental problems, especially in early pregnancy, that might be consequences of eating chemically contaminated food? How frequently did such problems occur in early embryonic development, and how often could dietary contaminants be implicated?

I left that meeting feeling uneasy about my responses to her concerns; they seemed less than adequate, except for my assurances that the problem was real.

From Field Notes of a Pollution Watcher
(C.J. Sindermann, 1989)

[Addendum, 2004:

Preparation of a list of "safe foods" is an exercise that is fraught with uncertainty. As an example, my earlier advice about the relative safety of products from aquaculture was wrong! To demonstrate this fact, several papers, the most recent being that of Hites et al. (2004), indicate that flesh from cultivated salmon (annual global production now over one million tons) contains higher levels of certain bioaccumulative contaminants (especially PCBs, dioxins, and dieldrin) than do wild-caught salmon. This probably results from feeding cultured fish on formulations (pellets) made from small low-cost forage fish species with high contaminant burdens. Although still below FDA action

---

* The luncheon special that day was broiled cod — a moderately low-risk choice, depending on how far from shore the fish had been caught. Unfortunately, overfishing in the Western Atlantic has caused the virtual disappearance of cod from most menus.

levels, according to Hites et al., the amounts of these chemicals in salmon fillets would put such seafood products in a "minimal consumption" category (one-half or one meal per month). Of the three principal producing areas, concentrations of contaminants in farmed salmon from Europe and North America were higher than those from Chile, and all were higher than wild-caught salmon.]

This vignette exposes one facet of the general uncertainty that surrounds consumption of seafood that may contain chemical contaminants (especially mercury or PCBs) or (in shellfish) microbial pathogens (*Vibrios* and other bacteria as well as enteric viruses). Uncertainty about a food product can easily be translated into total rejection, or at least heightened sensitivity to any abnormality, such as the presence of external tumors, worms, discolorations, eye damage, or ulcers. Wary seafood customers can and will refuse products that they perceive as possibly injurious to family health, as many did during the *Pfiesteria* toxin scare of 1997 in the Middle Atlantic states (discussed in Chapter 5). During that grim period, seafood brokers refused to handle seafood from affected areas, such as the Pocomoke River on the Eastern Shore of Chesapeake Bay, and some retailers actually posted signs proclaiming that none of their products came from those suspect waters.

Despite such attempts at reassurance, a "halo effect" or "ripple effect" could be seen clearly, in which consumers seemed to avoid seafood of any kind, whether it was taken from affected areas or not. A similar effect has been noted repeatedly in other "media science events," such as periodic scares about mercury or arsenic or PCBs or endocrine disrupters in fish.

Wary seafood retailers can and do reject shipments of shellfish from marginally safe producing areas and shipments of fish with obvious abnormalities in the flesh. For example, I learned in 1999, on good authority, about a Florida fish market that refused to accept a 6000-lb shipment of swordfish because of obvious ulcers and tumors on a number of fish — an act that undoubtedly resulted in a substantial economic loss somewhere in the marketing chain. As another example, in 1991 brokers in Denmark refused to accept a shipload of live eels from the east coast of the United States because an inspector there reported the presence of nematode worms in the fish. (At the time, much of Europe was suffering from a severe outbreak of eel nematodes of a different species.)

Reduced consumption of seafood because of fear of contaminated market products is not amenable to the same kind of quantification that is possible with reduced production due to closure of shellfish beds, but effects can be dramatic, even though they are episodic and transient. The news media follow such contamination events carefully. They report with great enthusiasm the closing of the New England clam fishery in summer due to high levels of a natural toxin (paralytic shellfish poisoning toxin) derived from algal blooms, the closing of the Hudson River striped bass fishery because of high tissue levels of PCBs in the fish, and warnings to Great Lakes sportfishermen not to eat their catches of introduced salmon, warnings against eating large swordfish because of high mercury levels in the flesh. All receive extensive media coverage, leading to justifiable unease on the part of consumers about the advisability of eating *any* seafood. This unease is expressed during crisis periods with any species by an accompanying sharp drop in sales of other seafood products (the "halo effect") — so that, for example, when the Maine clam fishery is closed because of an outbreak of paralytic shellfish poisoning, people stop buying lobsters, crabs, and even fish, although these species rarely carry enough of the toxin to be dangerous. The unease is expressed more subtly in daily decisions by food shoppers, who may select from a great array of protein sources other than fish and shellfish — especially after having been exposed to a half-hour documentary on pollution in the New York Bight or Puget Sound or after reading a newspaper story about tar balls or condoms fouling bathing beaches on Long Island.

We can conclude this discussion of consumer resistance by pointing out that, partly because of the consequences of coastal pollution, those consumers (and other elements in the marketing chain) view any purchase of fish and shellfish with suspicion, compared with buying most other animal products. News media help by playing a legitimate public information role, creating a smarter consumer population, with some persistent sensitivity to the ways in which environmental abuses in coastal waters impinge on everyday life.

REDUCED YIELDS FROM COMMERCIAL FISHERIES

Reduced yields as consequences of regulatory restrictions constitute an identifiable and most visible component of coastal pollution effects on fish and shellfish production. Fish from specific contaminated habitats, such as the Hudson River, Great Lakes, and New Bedford Harbor, are or have been subject to catch prohibitions or limitations, consumer advisories, and negative media attention that result in lower production.

It is important to note, though, that direct effects of coastal pollution on abundance of fish species are not easy to detect with any certainty. Only in extreme instances of gross contamination — usually in enclosed seas or in localized bays and estuaries with restricted circulation — can a case be made for a causal relationship between pollution and fish abundance. In many instances pollution is too easily proposed as a cause of decline or scarcity, when the controlling factor may well be, and usually is, overfishing and closely related mechanical destruction of habitats. At present, with exceptions in restricted bodies of water (such as the Black Sea), decline in abundance or disappearance of commercial fish species cannot be attributed clearly and specifically to effects of pollution.

The greatest impact of pollution on coastal seafood production can be seen in molluscan shellfish species — clams, oysters, and mussels. Public health risks from microbial contamination have resulted in extensive closures of some clam and oyster producing areas, with resulting and often permanent economic loss unless the animals are relaid in cleaner waters and then sold. Recent estimates have been made by coastal states and the federal government (National Marine Fisheries Service) that about one-third of existing U.S. oyster-producing beds are closed to harvesting because of high coliform counts in sediments.

But closure of growing areas because of microbial contamination is not the only reason for reduced shellfish production. Most molluscan shellfish are essentially immobile and hence very vulnerable to man-induced environmental excesses, such as petroleum or other chemical spills, and increases in toxic/anoxic zones resulting from nutrient loading of estuarine waters. Closures and restrictions may also result from longer-term chemical contamination, as, for example, in PCB-contaminated parts of New Bedford Harbor in Massachusetts and Escambia Bay in Florida.

In addition to removal of fish and shellfish from commerce because of regulations designed primarily to protect humans from foodborne diseases or to protect fish stocks from overexploitation, other causes of lost production exist. Some of them are:

- Losses from trawled catches can occur because tumors and ulcers must be removed in the filleting process, and areas of heavy worm parasitization (even though not related to pollution) must be avoided by fishing vessels.
- Losses in aquaculture production can occur when increasing chemical pollution causes abandonment of formerly productive areas or causes decreased survival and growth of fish and shellfish.
- Losses in aquaculture may occur because of harmful algal blooms, caused in part by self-pollution of growout areas and in part from anthropogenic nutrient loading of those areas. Examples can be found in the mussel-producing areas of Portugal; in the yellowtail growout pens in restricted bays and inlets of the island of Shikoku, Japan; and wherever intensive aquaculture is practiced in waters with limited circulation.
- Toxins from algal blooms may cause mortalities of larval, juvenile, and adult fish and shellfish (White 1988). Blooms may also result in permanent toxicity in some commercial species in some areas, resulting in closure of fisheries and in their permanent exclusion from exploitation.

It is apparent to me that two pollution-associated problems account for principal economic losses to the fishing industry. They are:

1. Closure of shellfish beds because of high coliform counts or because of proximity to waste disposal sites
2. Harvest restrictions because of chemical pollution, such as PCBs in lobsters from New Bedford Harbor and in striped bass from the Hudson River

Other sources of lost production exist, of course, but direct effects of coastal pollution on abundance of marine commercial species remain difficult to isolate and quantify.* Probably the most quantifiable effects of coastal pollution are the gradually increasing closures of shellfish beds because environmental fecal coliform levels exceed regulatory tolerances. Most shellfish-producing states, as well as the federal government, now have detailed maps indicating the location and extent of closures. Shellfish production by state and area is carefully monitored by several agencies, state and federal, so the loss in potential landed product due to closures can be calculated directly. Each shellfish-producing state (and the federal government) maintains statistics on total productive acreage, acreage lost by closures, total

---

* Such a statement should not imply that quantitative biologists have not approached the problem of comparing the relative effects of pollution and exploitation on fish populations. I referred to a number of studies with this objective in Chapter 10 on population effects, and I recommend a discussion by Boreman (1997) of methods that have been or could be used to compare population-level effects of pollution and fishing.

value of the marketed crop, and estimated loss due to closure (assuming that closed beds would be exploited at the same level as open beds if they were not polluted).

Scrutiny of those statistics permits certain conclusions:

1. The total annual loss due to pollution of shellfish production areas is substantial.
2. If total potential production figures are used instead of actual production figures, the shellfish industry could increase gross income by over 30% — even without utilizing the potential increase due to expansion of aquaculture.

But of course these conclusions represent oversimplifications of a much more elaborate story. Some closed beds have been reopened because of antipollution measures taken (but, simultaneously, other beds have been closed as population pressures in some coastal areas increase). Some beds are closed seasonally or conditionally and are therefore only partly out of production. Environmental factors other than pollution — shellfish diseases, dredge and fill operations, overharvesting — have also caused major losses in shellfish production, so the stage is crowded with villains, of which pollution is only one.

What we can do much less efficiently, however, is to estimate losses in *fish* production due to sublethal impacts of pollution — factors such as poor growth resulting from exposure to contaminants in the diet or reduced fecundity resulting from body burdens of contaminants such as PCBs or chlorinated pesticides. Estimates of partial economic losses due to these and other pollution-related influences can be combined to provide the total percentage of a resource population that is lost, which, when combined with market value estimates, could provide a rough and probably misleading guess of the costs of coastal pollution. The imprecision would be staggering; economists would undoubtedly retch at the dimensions of the oversimplification.

And yet, knowledge of biological and population characteristics is critical to any economic valuation of pollution effects, beyond the somewhat tenuous estimates of short-term losses due to mortality alone. A small cadre of quantitative fisheries biologists, including Dr. John Boreman of the Northeast Fisheries Science Center at Woods Hole, Massachusetts, exists and can communicate with resource economists in the preparation of more realistic appraisals of the totality of pollution effects on fish populations. Encouraging indicators have appeared, such as joint symposia on economic consequences of environmental stress and on the economic effects of hypoxia. The remarkably opaque jargon of economists is a persistent roadblock, but progress can be made with dogged determination by both sides, economists and quantitative biologists. A critical problem that biologists and economists both confront is the extreme variability in all the environmental factors that determine abundance of resource populations, as well as the amount of statistical noise that surrounds any influences of coastal pollution.

Relevant to this problem is an interesting quotation from two well-known resource economists (Lipton & Strand 1997): "Assumptions will often depend on biological models of fish populations on which the economic models must build, which creates two layers of measurement and modeling error" (p. 518).

Keep trying.

REDUCED REVENUES FROM
RECREATIONAL FISHING

"Recreational fishing" is an inclusive term that enfolds the "land-based fisherman" standing in the surf at Cape Hatteras; the owner of a trailered outboard boat in Chesapeake Bay; a day-trip fisherman on a "headboat" out of Cape May, New Jersey — or almost anyone with a rod, reel, and license (where required). Many of the fish caught by so-called sportfishermen are either eaten at home, forced on reluctant friends, or sold to dealers. A substantial part of that fish-catching population is therefore better described as "subsistence" fishermen. Furthermore, a significant segment of the subsistence population catches fish in contaminated coastal waters near their homes and sells fish with high body burdens of pollutants to local dealers (or consumes them at home).

Advisories and area closures have only modest effects on these subsistence fishermen. Some do not hear about the advisories, and many others choose to ignore them. The advisories can thus have bad and good effects: they can reduce the food base and the meager profits of subsistence fishermen, but they also reduce consumption of contaminated seafood.

An American Chemical Society publication reported that, in the mid-1990s, state-issued advisories restricting or banning fishing for certain species or in specific areas were increasing annually, with 95% of them resulting from fish contamination by mercury; dioxins; and now-banned PCBs, chlordane, and DDT. A major contributor to the continued increase was thought to be long-range atmospheric transport of toxic pollutants (Anonymous 1997). The total number of advisories (in the form of recommendations to limit or avoid consumption of contaminated fish, with more stringent consumption limits for pregnant women and children) reached an amazing total of 2193 for the country as a whole in that reference year.

Those advisories, when combined with visible degradation of some coastal/estuarine waters and observed abnormalities in occasional sport-caught specimens, result in reduced incentives to go fishing and encourage trips to local bowling alleys instead. Actions or inactions by recreational fishermen set in motion a whole chain of reductive economic consequences:

- Fewer fishermen fish, so they do not buy equipment or boats.
- Fewer fishermen buy licenses (which support state natural resources management programs).
- Existence of fewer fishermen translates into less pressure on those management/regulatory entities to maintain staff numbers or conservation programs.
- Fewer fishermen patronize marinas, bait shops, and other shore facilities.
- Fewer fishermen travel by car to fishing spots and do not, therefore, buy gasoline or require restaurants or motels.

As a consequence of these trends, economists in a number of universities, often with grants from the NOAA Sea Grant Program, have tried to quantify costs of pollution insofar as impacts on recreational fishing are involved. The prologue of this book ("Menace of the Sludge Monster") described the peak (or nadir) of pollution events on the New Jersey and Long Island (NY) coasts in 1987 and 1988 — sewage and medical wastes washed up on bathing beaches, and the menace of the sludge monster just offshore (Swanson et al. 1991). A recent economic study of that period (Ofiara & Brown 1999) estimated that annual losses to the sport-fisheries were 4 to 20%, ranging from $35 million to $178 million. It should be pointed out that these figures represent only a fraction of the total economic loss from the 1987–1988 pollution event; a much larger estimated annual loss ($379 million to $1597 million) resulted from impacts on beach use as well as on marine recreational fishing.

An earlier economic study of impacts of chemical pollution (PCBs) in New Bedford Harbor (MA) on the local recreational fishery estimated aggregate damages of $67 thousand for 1986 alone (McConnell 1986). Bottom fishing in the harbor had been banned because of PCB pollution (as discussed in Chapter 3).

Comparable economic studies have been conducted in many other states with coastal pollution problems, and damage estimates for recreational fisheries are invari-ably substantial. It is difficult, however, for noneconomists to grasp the often-con-voluted terminology and the degree of robustness of numerical estimates given. Despite such limitations, however, it seems clear from these and other studies (see, for example, Gold & van Ravensway 1984 and Wegge, Hanemann, & Strand 1986) that pollution events, with subsequent media coverage, beach closures, and fish advisories, can have significant impacts on recreational fishing and its contribution to coastal economies. Much (but not all) of the eventual economic impact results from unease about health effects, as summarized recently by Kimbrough (1991):

> Concern has been raised that subsistence fishermen and sportfishermen ... may be at a greater risk of developing chronic health effects. This concern is based on the assumption that these populations would be exposed to higher levels of persistent organic chemicals such as DDT, dieldrin, chlordane, hexachlorobenzene, polychlori-nated biphenyls (PCBs), polychlorinated dibenzodioxins (PCDDs), and dibenzofurans (PCDFs) through the consumption of fish from polluted water ways. (p. 82)

ECONOMIC EFFECTS OF "NUTRIENT POLLUTION"

The term "nutrient pollution" has been applied to overenrichment of coastal/estuarine waters by chemicals of anthropogenic origin (especially nitrogen and phosphorus) that enhance excessive growth of microalgae. This overproduction leads to blooms and anoxia, resulting in exclusion or mortality of animal populations.

On the one hand, it would seem that supplying nutrient chemicals should be a positive factor in increasing the carrying capacity of coastal waters for fish and shellfish production, and there is substantial evidence that this is so. A recent study of yields from European estuaries and semi-enclosed seas (DeLeiva Moreno et al. 2000) indicated higher landings of fish from those waters with higher nutrient levels and seasonal oxygen depletion. Similar findings were reported earlier by Nixon et al. (1986), who found a positive correlation between nitrogen loading of coastal waters and fish production. Breitburg (2002) pointed out that "Nutrient enrichment

... typically increases prey abundance in more highly oxygenated surface waters and beyond the boundaries of the hypoxic zone" (p. 769).

It is, on the other hand, the uncontrolled and excessive additions of nutrient chemicals that create problems and lead to negative economic effects. Those effects include:

- Harmful algal blooms (Hoagland et al. 2002) with negative effects on tourism, bottom vegetation, fish and shellfish growth and survival, and — in the case of toxin-producing algal species — temporary or even permanent closure of fisheries, as well as human foodborne illnesses, as discussed earlier in this chapter.
- Development of expanded zones of hypoxia or anoxia, which can kill sedentary forms such as molluscan shellfish, reduce growth rates of shellfish, and change migration pathways and distributions of finfish.

From an economic perspective, the negative effects of nutrient overenrichment seem to have garnered most of the attention, as indicated by a recent report by the National Research Council (NRC 2000) and by a recent economic viewpoint paper by Segerson and Walker (2002). The present environmentally driven policy is to reduce nutrient pollution by voluntary or mandatory actions, but inclusion of economic considerations that specify some of the positive aspects of nutrient enrichment (not overenrichment) should be part of every discussion.

CONCLUSIONS

Quantifying the impacts of coastal pollution on living resources and on humans — especially the costs to society — is a legitimate and necessary objective of scientific research. It requires the concerted and integrated efforts of many disciplines (population dynamics, ecosystem monitoring, physiology, pathology, economics — even sociology). One important yield from research should be accurate determination of the costs of pollution, followed by ensuring public awareness of those costs. Because many of those costs (incurred by loss of seafood production, rejection by consumers of seafood products, loss of recreational fishing opportunities, and deleterious changes in coastal ecosystems) relate directly or indirectly to living resources, a close interaction must exist among quantitative fisheries biologists, pollution scientists, and economists to ensure that defensible estimates of abundance, models of yields, and even predictions are generated.

Subsequent policy-level actions would include regulations to limit any continuing environmental degradation and the "internalizing" of the costs of pollution (getting the polluters to pay and not someone else; Whitmarsh 1993). Assignment of financial responsibility is a difficult regulatory task that requires the best available synthesis of data — biological, environmental, and economic — if it is to be acceptable to society. As Whitmarsh pointed out, though, regulatory instruments tend to be inflexible and inefficient and to become more complex and restrictive with the passage of time. In his opinion, regulatory measures with an adequate economic base (such as tradeable emission permits for industries or transferable catch quotas for fishermen) could replace legal coercion with commercial incentives. I think this would be feasible in a more perfect world, where causality (and costs) in pollution-related issues could be assigned with greater confidence than is possible now.

REFERENCES

Anonymous (American Chemical Society). 1997. Fish consumption advisories are up 26%. *Environ. Sci. Technol. 31:* 451A–451B.

Boreman, J. 1997. Methods for comparing the impacts of pollution and fishing on fish populations. *Trans. Am. Fish. Soc. 126:* 506–513.

Breitburg, D. 2002. Effects of hypoxia, and the balance between hypoxia and enrichment, on coastal fishes and fisheries. *Estuaries 25:* 767–781.

DeLeiva Moreno, J.I., V.N. Agostini, J.F. Caddy, and F. Carocci. 2000. Is the pelagic-demersal ratio from fishery landings a useful proxy for nutrient availability? A preliminary data exploration for the semi-enclosed seas around Europe. *ICES J. Mar. Sci. 57:* 1091–1102.

Gold, M.S. and E.O. van Ravensway. 1984. Methods for assessing the economic benefits of food safety regulations: A case study of PCBs in fish. Agric. Econ. Rep. No. 460, Michigan State Univ., East Lansing.

Hites, R.A., J.A. Foran, D.O. Carpenter, M.C. Hamilton, B.A. Knuth, and S.J. Schwager. 2004. Global assessment of organic contaminants in farmed salmon. *Science 303:* 226–229.

Hoagland, P., D.M. Anderson, Y. Kaoru, and A.W. White. 2002. The economic effects of harmful algal blooms in the United States: Estimates, assessment issues, and information needs. *Estuaries 25:* 819–837.

Kimbrough, R.D. 1991. Consumption of fish: Benefits and perceived risk. *J. Toxicol. Environ. Health 33:* 81–91.

Lipton, D.W. and I.E. Strand. 1997. Economic effects of pollution in fish habitats. *Trans. Am. Fish. Soc. 126:* 514–518.

McConnell, K.E. 1986. The damage to recreational activities from PCBs in New Bedford Harbor. Prepared for Ocean Assessment Div., NOAA (National Oceanic and Atmospheric Admin.), Rockville, MD.

Nixon, S.W., C.A. Oviatt, J. Frithsen, and B. Sullivan. 1986. Nutrients and the productivity of estuarine and coastal marine ecosystems. *Limnol. Soc. South. Afr. 12:* 43–71.

NRC (National Research Council). 2000. *Clean Coastal Waters: Understanding and Reducing the Effects of Nutrient Pollution.* National Academy Press, Washington, DC.

Ofiara, D.D. and B. Brown. 1999. Assessment of economic losses to recreational activities from 1988 marine pollution events and assessment of economic losses from long-term contamination of fish within the New York Bight to New Jersey. *Mar. Pollut. Bull. 38:* 990–1004.

Segerson, K. and D. Walker. 2002. Nutrient pollution: An economic perspective. *Estuaries 25:* 797–808.

Swanson, R.L., T.M. Bell, J. Kahn, and J. Olha. 1991. Use impairments and ecosystem impacts of the New York Bight. *Chem. Ecol. 5:* 99–127.

Wegge, T.C., W.M. Hanemann, and I. Strand. 1986. An economic assessment of marine recreational fishing in southern California. NOAA-TM-NMFS-SWR 015. Southwest Region, NMFS (National Marine Fisheries Serv.), Terminal Island, CA.

White, A.W. 1988. Blooms of toxic algae worldwide: their effects on fish farming and shellfish resources, pp. 9–14. In *Proceedings of International Conference on Impact of Toxic Algae on Mariculture*. Aqua-Nor 87 International Fish Farming Exhibition, August 1987, Trondheim, Norway.

Whitmarsh, D. 1993. Economics and the marine environment. *Mar. Pollut. Bull. 26:* 588–589.

14 Effects of Coastal Pollution on the Quality of Human Life

INTRODUCTION

Perspectives on coastal pollution vary with the observer, but the average newspaper-reading, television-watching citizen has concerns about the subject that can be aggregated into three general categories:

1. Fear of consuming seafood products that may be contaminated
2. Disgust with environmental encounters in degraded coastal areas
3. Unease about the world's food base — about global losses in abundance of many living marine resources

I have a true story (the next vignette) about a pollution-associated incident that illustrates one person's awakening to the realities of degraded environments.

A Small Incident on the Wharf in East Boston

The winter flounder, technically labeled <u>Pleuronectes americanus</u>, lives and thrives in the coastal waters of the northeastern states, where it is a favorite catch of hook and line fishermen, and is also taken by inshore trawlers sailing from such historic ports as Gloucester, Rockland, New Bedford, and even Boston. On a cloudy morning in early June, Jim Driscoll, fishing off pier 41 in East Boston, brought up a specimen that ruined his day. It was clearly a winter flounder, but most of the fins had been eroded away, and near the tail were several large, bloody sores. Jim was tempted at first to throw the repulsive thing back, but the thought occurred to him that something must be very wrong down there to do such damage to any fish. He knew that there was a government statistical agent stationed down the street at the Custom House, so he wrapped the animal in a newspaper and delivered it, asking for some kind of explanation, but not really expecting much.

Jim wasn't aware of it, of course, but the statistical agent was part of a network of people with similar jobs deployed in all the major northeastern fishing ports, to collect data on catches and environmental conditions, and to transmit the information back to a research center in Woods Hole. One of the research staff members in that center actually looked at the specimen and even got excited about its condition — to the point where she called Jim a few weeks later with a report. The fish, she said, was suffering from three abnormalities, all associated with polluted habitats: the eroded fins and the skin ulcers were obvious, but additionally, microscopic examination of the liver disclosed numerous tumors — again characteristic of badly degraded habitats like Boston Harbor. Furthermore, chemical analyses of the flesh of the fish showed much higher than normal levels of PCBs, aromatic hydrocarbons, and cadmium. She then gently warned him not to eat fish from the harbor too often.

Now Jim is no dummy. He knew that the water below the pier couldn't be quite the quality of a mountain spring, but he had not previously been confronted with visual evidence of just how dreadful it really must be. That single experience really upset him — to the extent that he gave up fishing altogether, and now doesn't even eat fish anymore, regardless of the source. He has concluded, not without cause, that the coastal pollution problem near big cities like Boston is probably intractable, and that seafood from those areas might even be dangerous to his health and that of his family.

From Field Notes of a Pollution Watcher
(C.J. Sindermann, 1988)

Jim Driscoll has, from this seemingly minor episode of the diseased flounder, suffered a blow to the quality of his existence. He has unwittingly become an illustration of the truism that, in addition to its impacts on public health and economics, coastal pollution can degrade the quality of human life in many obvious and not so obvious ways. Strong negative emotions can be aroused by confrontations with evidence of pollution. Fear is a common response — fear of contaminated food products, of diseases such as hepatitis or cancer, or of abnormalities in early development of precious children. This emotion is often accompanied by disgust with grossly polluted coastal/estuarine environments and personal experiences resulting from contacts with them. These feelings of disgust can quickly turn to rage — rage at polluting industries and inadequate public sewage authorities; at local, state, and federal regulators for allowing abuses to continue; at legislators for being reactive or inactive instead of proactive; and at politicians and administrators for using environmental issues as political bargaining chips. Permeating all of these feelings are vague but persistent concerns — about future generations and their environments, about a decline in abundance or the disappearance of some commercial species from the markets, and about permanent loss of species from the biosphere.

IMPACTS OF COASTAL POLLUTION ON THE HUMAN PSYCHE

To make loss in quality of life (admittedly a nebulous concept) a little more tangible, we can take some examples from the human population that is being affected by coastal pollution. These are of course highly selected, but what better way to demonstrate a point, for they are responses to the inevitable question, "Whose quality of life is being impacted?"

THE BATHER/SCUBA DIVER/WATER SKIER

People searching for outdoor recreation may journey to the seashore only to find beaches fouled by noxious debris, and harbor or nearshore waters sometimes visibly contaminated by human excrement. If not quite so grossly polluted, then coastal waters are almost certain to contain microorganisms from human wastes, whether or not levels of imprecise indicator organisms are high enough to close nearby beaches. And then there are the rumors and reports of outbreaks of warts and digestive disturbances, as well as eye and ear infections among those who take risks by entering yon murky waters. Bathers may also encounter algal blooms, local or widespread, some with associated fish kills. Very few people enjoy swimming in dense algal blooms, even if the causative organism is a nontoxic form — although the beaches may be closed anyway.

Urban beachgoers are also some of our best readers of newspapers and watchers of televised news. Inserted in their daily media dosages of murder, rape, and robbery, they may find small squibs about beach closures, breakdowns in coastal sewage plants, small oil spills fouling shorelines, industries cited for violating EPA clean water regulations, and chemical spills from vessels on coastal routes — all of which fail to improve attitudes toward continued participation in today's society. This book was originally titled *The Sludge Monster and Other Undersea Horrors.* Those horrors can seem very close at hand to someone standing on a barely marginal urban beach at Sandy Hook, New Jersey, or Brooklyn, New York.

THE RECREATIONAL FISHERMAN

A typical recreational fisherman often begins an outing by going to the marina where his boat is moored. There he can be immediately affronted and dismayed by conditions in and near the facility — floating plastic and other debris, oil slicks, sanitary napkins, condoms, and even raw feces (the last from small boats that pump the holding tanks directly into crowded waters). On his way to the boat slip, he may see posted advisories against eating fish that are caught or recommending that they should not be eaten too often (it is always reassuring to be told in print by a natural resources agency not to consume too large a dose of poison). In the absence of such warnings, the fear is always present that the fish are contaminated and will make family members sick if eaten. The final insult, on this day of fun and relaxation, is catching fish that are in some way visibly abnormal, with ulcers, tumors, frayed fins, bulging or missing eyes, or skin hemorrhages. Such abnormalities seem to be occurring more commonly now in inshore fish than was the case even a few decades ago.

THE PREGNANT WOMAN

Today's expectant mother is inundated with advice and warnings about chemicals that may harm the developing fetus. She has heard about the awful tragedy of "thalidomide babies" in an earlier era, born with incomplete appendages; about the consequences in female offspring of maternal use of diethylstilbesterol to prevent miscarriages; and about the still-developing information from the Midwest about developmental abnormalities in children born of mothers who consumed PCB-contaminated fish from the Great Lakes. Decisions about taking or not taking medications during pregnancy are controllable, but the danger of toxic chemicals in foods — often at undetermined levels or not yet known to be toxic to the unborn — can give rise to fear and uncertainty about every item of food the prospective mother eats. Seafood is immediately high on the list of suspects because of chemical contamination of fish and shellfish habitats and because of known high levels of pollutants, such as PCBs and mercury, in certain species and locations.

What is that prospective mother to do — at least insofar as eating fish and shellfish from coastal waters is concerned? Adequate geographic and species-specific information is not readily available to her (or to anyone) about contaminant levels in all the choices of seafood available — and even if it were, the dietary intake and retention of persistent toxicants before pregnancy occurred could also have an effect on fetal health. Avoiding *any* seafood from chemically polluted waters would be a logical defensive measure, but even this step is not a guarantee, because some high-seas predators, such as swordfish and tuna, can sequester in their flesh potentially dangerous levels of mercury (just as an example).

THE ENVIRONMENTALLY CONSCIOUS CONCERNED CITIZEN

Environmental awareness has shown a long-term growth in the United States, despite periods of stasis and even regression during political regimes in which environmental issues were not of high priority. Average citizens are today far more sensitive to and knowledgeable about environmental abuses than they were, for example, at the time of the first Earth Day in 1970. A substantial cadre of participants in the struggle for environmental protection, conservation, and environmentally sustainable development exists today, supported by a broadening base of public concern.

The people who make up this base come to recognize that participation has its costs and that life can't be the same once a decision to join the fight has been made. Some of the costs that quickly emerge include, but are certainly not limited to:

- The physiological consequences of stress, frustration, and rage against intransigent polluting industries
- Anger at well-financed developers and others who buy, disfigure, and degrade coastal lands, despite public protests
- More anger at clever amoral lawyers who are the willing servants of such destructive financial interests
- Still more anger at the inaction, wrong-headedness, arbitrariness, and stonewalling of government agencies at all levels — local, state, and federal — but especially of growth-oriented local agencies and commissions
- Time spent in committee meetings, public meetings, and other forms of active protest against environmentally degrading practices by privately financed groups and governmental agencies

- Expenses of supporting environmental activist organizations at every level, from local to international

An important point, though, is that although participation in environmental affairs has its price in time, money, and energy, it does serve as a source of satisfaction when victories are won, sound practices and policies prevail, or the despoilers are halted momentarily.

THE ENVIRONMENTAL SCIENTIST

Scientists are among the most fortunate people on this earth, being committed to a stimulating, demanding, and internally rewarding career. Some aspects of their existence may, however, be slightly less than perfect at times — especially if they become involved in coastal pollution matters. There are instances when the quality of life of an environmental scientist can be severely impacted by outside forces. These are some of them:

- An offer of environmental advice and recommendations by a scientist, acting in good faith but naively as an expert witness in legal proceedings, may be accepted, but then he or she may suffer career damage at the hands of clever lawyers, hostile expert witnesses, or hearing examiners.
- The environmental scientist may, if a government employee, be silenced by politician/administrators if his or her findings and conclusions are not in accord with agency positions.
- The environmental scientist may, if a private consultant, be pushed beyond proper ethical limits by clients who expect findings favorable to their preconceptions and justifying environmental damage.
- The environmental scientist may, if an academic, find strong (and frequently public) opposition from other scientists with conflicting points of view and different sources of funding — or may find that the university administration has unwritten guidelines about expressions of opinion in environmental disputes.

But it should be admitted that some of these strictures could apply to almost any subdiscipline of science, even though this admission does little to improve a specific situation.

THE KIDS ON THE BEACH

The joy of going to the beach — from a kid's perspective — is severely diminished if:

- The sand is fouled with debris, or the water is murky and apt to cause sickness if it is swallowed or if it gets into eyes and ears.
- Parents are uneasy enough to discourage or even eliminate ocean swimming as a family activity because of pollution and the risk of illness.

- Beaches are closed periodically because of toxic algal blooms or malfunction of nearby sewage treatment plants — so the day's adventures
  may be canceled abruptly.

But the really insidious element here is that city kids adapt to degraded conditions
in their environment, and then may never experience the pure pleasure and freedom
of a completely natural beach, far removed from the impacts of people and their
wastes. That, to me, represents a severe loss in quality of life.

CONCLUSIONS

My examples of groups of people affected by coastal pollution — swimmers,
fishermen, pregnant women, concerned citizens, environmental scientists, kids on
the beach — are only a few choices from a list that could be expanded indefinitely.
The actuality is that we are *all* affected in some way by coastal pollution, whether
we live close to the ocean or not. Our health can be affected by eating contaminated
seafood, the cost and availability of seafood can also be affected, and the way we
view our lives on this planet can be influenced by the totality and the wholesomeness
of our surroundings. Factors (like coastal pollution) that detract from the quality of
human life can often be best defined in nonnumerical terms. Unlike economic costs,
they may not be subject to precise quantification because they include intangibles
such as feelings of loss, or a vague sense of malaise, or disgust with existing
environmental conditions, or a realization that all is not right with the world. I am
convinced that the quality of human existence — whatever its essence may be —
is an important part of any consideration of the topic "Effects of Coastal Pollution
on Humans."

The human species has treated the edges of the seas abominably, but until recent
millennia the impacts were apparent only locally. Now, however, people are numerous enough and their technology wasteful enough to exert negative influences on
the entire biosphere. The three concluding chapters in this part of this book have
considered some of the consequences of coastal pollution in terms most relevant to
today's society — public health, economics, and quality of life. Signs of awakening
environmental concerns have begun to appear, beginning in the 1970s, but responses
are generally too slow or too limited in scope to address emerging problems. We
can do better!

15 Summary and Conclusions

It seems so long ago that we began the review of coastal pollution sketched briefly in these pages. This, the last chapter, should logically be a summing up of information covered and (for the author) the growing realization that so little of the total story has been included. Every chapter could be and should be of book length. Because this is not likely, an alternative might be to attempt the reverse — a *précis* of the topics covered to see if the distillation process releases additional juices of insight and wisdom.

I feel at this point that some chapters are strong and others not so strong. I particularly like the long chapters on quantitative effects of coastal pollution on fish, on effects of alien species, on effects of coastal pollution on public health, and on sublethal effects of pollutants on fish — but these pets are probably just reflections of my own research interests. I am less pleased with the chapters on effects of environmental exposures of humans to pollutants and on economic effects of coastal pollution, but it is too late and I am too tired now to begin patching. What you see is what you get.

However, in this concluding chapter I do want to revisit, fleetingly, most of the principal topics covered in the chapters and in rough order of their appearance in the book.

The introductory chapter considered the *health of coastal waters* with the general finding that progress has been made, but environmental problems still exist, especially in enclosed or semienclosed seas and in estuaries with restricted exchanges. The toxic waste disposal problem has been addressed and is being addressed in western industrialized nations, but it is still acute in former communist bloc eastern European countries and in developing countries.

Cholera is still a global threat, but significant advances have been made in understanding the ecology and genetics of the disease agent and of vibrio bacteria as a group. Relationships of epidemics with coastal environmental cycles have been elucidated, but universal access to uncontaminated water is an elusive goal and a major impediment to reducing effects of this ancient pestilence.

Minamata disease was an important consciousness-raising episode in the history of human industrial development, and its effects can be seen in today's society as extreme public sensitivity and reactivity to reports of metal contamination of food and water (mercury, lead, and arsenic would be good examples). Even today, there is little that can be done to atone for the absolute horrors of that experience for the Japanese — except for commitment to increased vigilance and improvement of industrial practices.

Research and publicity about potential damage to humans and their offspring caused by the global presence of *PCBs and chlorinated pesticides* have helped to raise public awareness, particularly about contaminants in seafood. Furthermore, public suspicions have been raised about any new industrial chemicals, including pharmaceuticals, that may enter environments and food.

Pollution of coastal *recreational waters* is a continuing public health threat as more and more humans crowd into shoreline areas. Investment by governments in improved sewage treatment facilities has not moved rapidly enough to do any more than maintain a fragile status quo; public use of urban beaches is still a high-risk activity, even though most (but not all) of the human health dangers are less severe than they were early in the past century.

Harmful algal blooms have increased in frequency, extent, and toxicity in recent decades, undoubtedly due in part to organic loading of coastal/estuarine waters from human sources (farm and other industrial runoff, sewage treatment plant effluents, and so forth). Some of the algal toxins can cause serious human illnesses, and blooms — toxic or not — create hypoxic/anoxic zones that can be lethal to marine animals.

Anoxia, especially in estuaries and seas with limited water exchanges, is an increasing problem related to eutrophication from organic loading. Anoxic zones, such as those in Chesapeake Bay, Long Island Sound, and the Louisiana coast, appear to be expanding, with consequent reduction in fish and shellfish habitats in affected areas.

Petroleum in quantity in coastal waters — from the many minor and the few major oil spills — creates severe local economic problems and local environmental impacts (most, but not all of which, are transient). Media attention is usually confined to the major events. That attention has resulted in a substantial flow of government and industry research that has provided good, but not complete, understanding of environmental and resource animal interactions with oil. The number of major events (involving pollution in excess of 10^7 gal of oil) is surprisingly small, considering the amount of oil transported daily throughout the world. Improved tanker design (double hulls) has undoubtedly helped to keep the number low.

Introductions of alien marine species — deliberate or accidental — have occurred throughout recent human history, but the pace has increased since World War II, with more and faster ships in a worldwide network and with the phenomenal expansion of marine aquaculture. The introduced organisms may live and die without reproducing, or they may produce offspring but not dominate any habitat, or they may multiply to a point where they become a dominant part of the flora or fauna. The aliens (and their progeny) may produce drastic changes in ecosystems, they may interbreed with indigenous species, or they may carry pathogens that infect and kill members of the indigenous species. The aliens may become economically important in aquaculture (oysters, salmon), or they may become unwanted predators (green crabs, oyster drills) on wild resource species.

Coastal/estuarine fish can be subject to localized mass mortalities caused by extremely high levels of pollutants (near industrial chemical spills or industrial effluent discharges), but much of the damage to individual fish is caused by exposure to *sublethal levels of chemical pollutants*. Effects can be especially severe on embryos, larvae, and early juveniles of species that use coastal/estuarine waters as spawning and nursery areas. Effects on early development may be morphological or physiological, and they often result in later mortality. Pollutants such as PCBs and chlorinated pesticides may be present in the tissues of adult fish and may be transferred to eggs of females, causing developmental abnormalities and early death of offspring.

After reviewing the many ways that pollutants may cause sublethal impacts on individual fish, it is logical to ask *whether pollution affects abundance* of coastal/estuarine commercial fish species. The answer to that question is not satisfying. It is possible to identify quantitative effects of severe pollution on local populations of fish and shellfish, but pollution has not been demonstrated conclusively to cause significantly reduced abundance of an entire species. Too many other factors, such as overfishing, predation, starvation, or abnormal environmental conditions, can exert major influences on abundance, so it is difficult to isolate and quantify the effects of pollution, even if those effects are occurring. Intuitively, we would expect that the impacts of contaminants that have been demonstrated experimentally in individual fish should be translated ultimately into population effects, but this is difficult to prove in wild stocks of fish.

Whenever *marine mammals* — dolphins, whales, seals, sea lions, and their relatives — die in large numbers (or even in small numbers), a likely suspect is almost always coastal pollution. Many of us feel some affinity for these evolutionary "kissing cousins." We are concerned that the coastal pollution for which we are responsible may be involved in their demise, and we must be sensitive to the observation that mass deaths have become more frequent during the past 3 decades. From information that is available, coastal pollution may be a culprit, but other players are in the game, too. Many investigations have concluded that a wave of mortalities of dolphins and seals in Europe and North America beginning in the mid-1980s was caused by viral diseases, possibly abetted by reduced disease resistance and physiological debilitation as consequences of pollution. A few other mortality events — ancient and recent — were possibly caused by algal toxins present in forage fish eaten by the mammals. No mass mortalities of marine mammals have been attributed directly and exclusively to coastal pollution, to my knowledge (with the exception of animals engulfed in spilled oil).

Although effects of pollution on fish and marine mammals are interesting subjects of research and make good occasional headlines for the news media, it is the *public health* aspect of coastal pollution that properly gets most of the respect and attention. Microbial pathogens in shellfish, toxic chemicals (mercury, PCBs, algal toxins) in fish and shellfish, and disease agents lurking in the surf of bathing beaches all contribute to massive regulatory problems and cause illnesses in humans. Undoubtedly the largest public health problem associated with coastal pollution is *microbial contamination of molluscan shellfish*, which are still, in the 21st century, consumed raw and quivering by hordes of stupid humans and which carry a panoply of viral and bacterial pathogens. Coming in as a poor second is *chemical contamination of fish and shellfish* — shellfish that are grown in polluted waters or large predatory fish that feed on contaminated animals lower in food chains.

Economic losses due to coastal pollution have been examined in detail by every coastal state and are always found to be substantial. Such losses may result from reduced catches and sales of seafood, loss of revenue from recreational fishing, or reduced use of recreational beaches. Indirect costs (decreases in transportation of products, sale of equipment, user fees, and travel and accommodations) have been calculated for most coastal states, especially those with large commitments to exploitation of coastal/estuarine resources. Published analyses of problem areas related to pollution usually lead to increased regulatory activity and grudging improvements to sewage treatment facilities — all aimed at the most visible effects of urbanization of the coastal zone.

Although not amenable to precise quantification, the effects of coastal/estuarine pollution are important factors in *degrading the quality of life* of every person who lives at or near the edges of the oceans and of many who live inland as well. Some of the nation's coastal waters and harbors (Boston Harbor, the New York Bight, Biscayne Bay, San Francisco Bay, and Puget Sound, to name only a few) have been abused in the past by industrial effluents, ocean dumping, contamination from ships, dredge spoil deposition, and a host of other degrading practices. Enforcement of old and new regulations has caused the discontinuation of many evil activities (such as sludge dumping in the waters only 12 mi from New York City). So a new ethic has developed — one emphasizing care of coastal/estuarine waters as a priceless asset, replacing the previous ethic that condoned using the ocean and its watery margins as dumping grounds for society's wastes.

These *précis* or summary statements may have some utility as very brief reminders of the book's contents, but they lack the stature of good generalizations about effects of coastal pollution. I have tried another approach to a graceful exit — a list of principal conclusions about effects of pollution on marine organisms, ecosystems, and humans:

First, *the technical literature is replete with examples of effects of pollutants on individual fish and shellfish or on local populations, but no specific evidence exists to indicate widespread damage to major fishery resources that can be attributed directly to pollution.* Other factors, such as repeated year-class successes or failures, shifts in geographic distribution of fish populations, or overfishing, may cause pronounced changes in fisheries — changes that could obscure any effects of habitat degradation.

Additionally, the magnitude of fluctuations in fish population abundance, due to incompletely known natural environmental causes, makes it extremely difficult to isolate, quantify, and demonstrate the possible role of environmental contaminants in causing such fluctuations.

It may be, of course, that coastal pollution *is* exerting some overall influence on certain resource species, but that this may be masked by increased fishing effort or by favorable changes in other environmental factors that create a positive effect on abundance outweighing any negative effects of pollutants. Many experimental studies, particularly those concerned with long-term exposures of fish and shellfish to low levels of contaminants, suggest that some long-term quantitative effects should be felt, but that our statistics, our monitoring, and our population assessments are not adequate to detect them.

It may also be that present levels of contaminants in coastal environments are not high enough to exert significant or observable impacts on fish populations except in extremely localized, grossly degraded habitats. Local reductions in abundance could easily be masked in large, often migratory populations that are being influenced simultaneously by a host of natural environmental factors such as predation, temperature, ocean currents, and food availability.

One additional reason that the effects of pollutants on marine resource species are not easy to demonstrate may be the resiliency of estuarine and coastal populations and ecosystems. Human activities (other than overfishing) may not have damaged entire species enough to show significant effects — even though local populations may have been reduced or eliminated, even though eggs and larvae may have been killed, and even though forage organisms may have been affected. Counterbalances acting for the species or the ecosystem, such as the great reproductive potential (fecundity) of most marine species, the great species diversity in most ecosystems, and the adaptive capabilities of many coastal-estuarine species to drastic changes in their physical-chemical environment; all may offset the negative influence of pollutants.

Second, *proof of biological damage to resource populations caused by pollutants is very difficult to acquire in the midst of many other factors that influence the abundance of coastal species.* Effects may be long term, with average annual increments so small as to escape detection. Experimental information, particularly from long-term chronic exposures to contaminants, suggests that some deleterious effects can be expected and that they probably occurring, but they escape our methods of detection. Other evidence suggests that natural phenomena are still overriding influences in the survival of coastal populations, but that the dynamic and precarious equilibrium that permits survival may be overturned or distorted by human interference. Still other evidence indicates that overexploitation of resource populations is of overriding importance.

Third, *investigations of effects of chemical contaminants on resource species have produced much useful information, including the following*:

- Contaminant levels in tissues of fish have been examined in a number of geographic areas. In many instances, the widespread occurrence of detectable levels of selected contaminants has been demonstrated (PCBs, DDT, mercury, petroleum components).
- Environmental contaminant levels in certain restricted coastal/estuarine areas are sufficiently high to produce the acute or chronic effects seen in experimental systems.
- In certain localized areas, tissue levels of contaminants have been recognized in fish that are well above the few legal action levels that exist. Toxic effects on the fish at such levels are largely unknown, as are toxic effects of exposure prior to the buildup of measurable tissue levels. Beyond this, the relationship between toxicity and any tissue level has been inadequately explored. High body burdens of contaminants may be sequestered and not affect health until a level is reached where accommodation is no longer feasible and spillover may occur.
- Inadequate data are available about the synergistic effects on fish of extremely complex mixtures of contaminant chemicals and dissolved or particulate organics in polluted waters, but enough experimental data exist

to state that antagonisms and synergisms in complex mixtures of contaminant chemicals may be important factors in producing any net effect.

- A much broader perspective focuses on cumulative effects, which include chemical synergisms, as well as interactions among fisheries, physical degradation, and chemical contamination. It is possible to consider each item separately, but in the end it is the totality of effects that is significant. As an example, fish do not encounter just dioxin or mercury but a whole range of contaminants, and they can do so within highly degraded environments. Most fish are highly mobile and tend to go from one estuary or coastal area to another. Temporal effects may become apparent when fish visit a highly degraded area, are exposed to contaminants, but then move on to another area where the effects of the contaminants may be manifested in what appears to be a pristine environment.
- Surveys have indicated that some pathological conditions in fish (fin erosion, ulcerations, liver lesions, certain epidermal papillomas, chromosomal damage) can be associated statistically with severe environmental contamination. The statistical relationships have varying degrees of "robustness," and cause-and-effect relationships are difficult to demonstrate for specific contaminants.

Fourth, *the effects of pollutants on marine animals may be expressed in a wide variety of forms*, including diminished reproductive activities, damage to genetic material of the egg or embryo with resulting mortality or abnormal development, direct chemical damage to cell membranes or tissues, modification of physiological and biochemical reactions, changes in behavior (often due to chemical damage to sensory equipment), increased infection pressure from facultative microbial pathogens, and reduced resistance to infection. *Effects are often expressed as disease* either infectious or noninfectious. Infectious diseases may result from lowered resistance to primary pathogens, from invasion of damaged tissues by facultative (secondary) pathogens, or from proliferation of latent microbial infections. Noninfectious diseases may result from early genetic damage or from chemical modification of bone and soft tissues, leading to skeletal anomalies and tumors (Valentine & Bridges 1909, Valentine 1975, Longwell 1976).

Fifth, *the capacity of many coastal/estuarine species to accommodate to environmental extremes, including pollutants, has probably been underestimated.* Looking specifically at pollutant chemicals, survival may be enhanced by:

- Induction of cytochrome P450 enzymes to metabolize hydrocarbons
- Protein binding of heavy metals
- Increased mucus secretion to mitigate chemical stress
- Selection for high immunological competence
- Selection of resistant individuals from the population at risk, leading to establishment of survivor populations more tolerant at some life stages to high levels of contaminant chemicals

Physiological and population responses such as these can lead to prolonged survival and even reproduction in environments that might be considered hostile to aquatic life.

Sixth, *early life history stages of marine animals are usually (but not always) more susceptible to chemical stressors than are later stages.* The effects may be obvious, resulting in death, or less obvious, such as genetic damage, which may not be expressed until later developmental stages as abnormalities in structure or function. Genetic damage to sex products, embryos, or larvae may be an important and as yet poorly understood consequence of industrial contamination of coastal/estuarine waters.

Seventh, *pollutants may transform marine resource species into potential health hazards to humans.* If we persist in consuming contaminated seafood, we may reach a point where it may have low appeal because of odor or flavor, and it may poison us or cause infection. In a sense, humans may be the ultimate bioassay organisms — detecting (by the appearance of abnormalities and disease) over the long term the buildup of carcinogens, mutagens, and other toxic chemicals in the marine environment and marine resource populations. The human species may become (in the presence of continued contamination of the coastal environment) "subjects of a vast experiment in chronic toxicology" (Huddle & Reich 1975, p. 181).

Eighth, *given the present rate of human population growth and the likelihood of its increasing impact on coastal/estuarine waters, it is easy to be pessimistic.* We may be facing a future in which:

- Currently isolated zones of severe environmental contamination may gradually expand, intensify, and coalesce as human population density and industrialization in adjacent land areas increase.
- Sublethal effects of toxic industrial contaminants may, in presently unknown ways, affect the survival and well-being of resource species, and, indirectly, humans as well.
- Wild fish and shellfish stocks that survive in the presence of industrial or municipal pollution may be increasingly excluded from markets for public health reasons because they may carry dangerous levels of toxic chemical contaminants and microbial pathogens.
- Growing consumer unease and periodic panics about toxic contaminants in seafood may act to reduce the acceptability and hence the market value of products from natural aquatic sources — and even from aquaculture sources.
- Fish stocks have collapsed (some spectacularly) in many parts of the world, and no relief seems to be in sight. Although fisheries have reached low points in the past, it is, for example, difficult to find a parallel to the present extremely low stock abundance of cod and certain other species in the Northwest Atlantic. Overfishing and habitat degradation are usually identified as principal culprits; some investigators have suggested, however, that overfishing is a problem for the short run, but habitat degradation is a problem for the long run.

That such a dismal future is close at hand has been reinforced by emerging environmental news from Europe. We have learned of the extent to which humans can and will degrade their surroundings, as the horrors of aquatic and terrestrial pollution in former communist bloc eastern European countries have been revealed. We discovered that the totalitarian regimes of those nations, during the period from 1945 to 1991, tolerated and even encouraged:

- Ocean dumping (including radioactive material) on a scale that trivializes that carried on in North America and western Europe
- The treatment of great rivers as open sewers, without even the pretense of abatement measures
- The almost total disregard for dangers to public health from effluents and other emissions of industrial complexes

What has unfolded is evidence for almost half a century of deliberate despoliation and degradation of the environment, fostered by an overwhelming indifference of government functionaries concerned only with meeting quotas and private gain. The history of that half-century may eventually supply an epic example of the cumulative effects of humans on aquatic resources and their habitats — where rivers have been diverted throughout entire river-basin systems or have been used primarily for effluent discharge, and where fish have been removed to the point where there are virtually no recruiting stocks left. These are truly cumulative effects, but we in the West had been largely unaware of them.

With these conclusions, my exploration of pollution effects that began more than 30 yr ago in the much-abused New York Bight — at the Sandy Hook Laboratory of the National Marine Fisheries Service — comes to an end. This revision of my earlier book on ocean pollution contains much of what I would like to say about the effects of coastal pollution on living marine resources and on humans. Of course there is far more that could be written — but not by me.

The research database that is the foundation for the text will be, I am sure, exploited from other perspectives by other writers who will also have the advantage of access to new information that is being published in overwhelming quantity. I have reached conclusions and drawn inferences that seem to be consistent with the available scientific literature — conclusions and inferences that may be supported or refuted in someone else's book a few years from now. Producing this revision of the original volume (Sindermann 1996) has been an exhilarating experience, extending as it has over a span of almost 10 yr (with frequent sterile periods). My hope continues to be that this new document will provide readers with a small window on some of the principal resource and related problems that have been created by coastal pollution.

REFERENCES

Huddle, N. and M. Reich. 1975. *Island of Dreams: Environmental Crisis in Japan.* Autumn Press, New York. 225 pp.

Longwell, A.C. 1976. Chromosome mutagenesis in developing mackerel eggs sampled from the New York Bight. U.S. Dept. Commerce. NOAA Tech. Memo. ERL-MESA-7, 61 pp.

Sindermann, C.J. 1996. *Ocean Pollution: Effects on Living Resources and Humans.* CRC Press, Boca Raton, FL. 275 pp.

Valentine, D.W. 1975. Skeletal anomalies in marine teleosts, pp. 695–718. In: W.E. Ribelin and G. Migaki (eds.), *The Pathology of Fishes.* Univ. Wisconsin Press, Madison.

Valentine, D.W. and K.W. Bridges. 1969. High incidence of deformities in the serranid fish *Paralabrax nebulifer* from southern California. *Copeia 1969*(3): 637–638.

Index